The Canadian Renewable Energy Guide

Second Edition

Edited by:
Alexandra R. McKirdy

SESCI
Solar Energy Society of Canada Inc.

Published by:

GENERAL STORE
PUBLISHING HOUSE

Box 28, 1694 Burnstown Road, Burnstown, Ontario K0J 1G0
1-800-465-6072 or Fax: (613) 432-7184
Email: orders@gsph.com

The material contained in this publication is based on current information available in regard to the subject matter covered and has been gathered with the utmost care for accuracy. However, due to the evolving nature of information in regard to the subject matter covered, this publication should be used for information purposes only. It is not intended to replace the services of competent professionals working in renewable energy and related fields. The publisher and editor assume no liability for any damage, injury, or expense that may be incurred or suffered as the result of use of this publication.

Canadian Catalogue in Publication Data

Canadian Renewable Energy Guide, Second Edition

Includes bibliographical references and index
ISBN 1-894263-07-3

1. Renewable energy sources – Canada.
2. Renewable energy sources – Canada – Directories.
I. McKirdy, Alexandra R. II. Solar Energy Society of Canada Inc.

TJ807.9.C3C35 1999 333.79'4'0971 C99-900462-X

Cover Design by
GLS Dezign

Design by
Derek McEwen Design and Typesetting

Printed and bound in Canada by
Tyrell Press Ltd.

First Printing June 1999

Sponsors

The preparation and publication of the *Canadian Renewable Energy Guide,* Second Edition was made possible by the financial assistance of the sponsors listed below.

Industry Canada

Canadian Gas Association

Environment Canada

Ministry of Energy, Science & Technology—Ontario

Natural Resources Canada

Solar Energy Society of Canada Inc.

Acknowledgements

The *Canadian Renewable Energy Guide*, Second Edition, would not have been possible without the vision, energy, effort and support of many people.

At the Solar Energy Society of Canada Inc., the past and current Boards' of Directors have continued to support and develop this and other important renewable energy projects. Specifically, Joe Thwaites, Terry White and Tim Egan have always given freely of their time for consultation throughout the project.

Many people took the time to review parts of the manuscript. They include SESCI Board Members Ezra Auerbach, Alanna Dean, Tim Egan, Steve Harrison, Jennifer Keyes, Andrew Pape, Stephen F. Pope, Joe Thwaites and Terry White. Other reviewers included Richard Kadulski, Sylvain Martel, Bob Mathews, Ian McKirdy, Sheila McKirdy, John Oliver, Ljubisav Stamenic and Elizabeth White.

The list of contributors for this Second Edition is long and I would like to apologise in advance for missing any additional contributors. I would like to thank Diane Bailey, Mike Cannon, Bill Eggertson, Drew McKibben, John Oliver, Ljubisav Stamenic, Joe Thwaites, Elizabeth White, Terry White, Kim Ah-You, Mike Bourns, Leon Burn, Sylvie Charette, Scott Davis, Guy Drouin, Jason Edworthy, Dave Egles, Alexandra Hampson, John Hollick, D. Gordon Howell, Frank Ilczyszyn, Christopher Ives, Richard Kadulski, Judy Kitto, Ray Lalonde, Tom Lawand, Gregory Leng, Sylvain Martel, Bob Mathews, Alexandre Monarque, Ian McKirdy, Joel Nodelman, Jorg Ostrowski, Michael Ross, Jimmy Royer, Roger Samson, Craig Shearer, Eric Usher and Brian Wilkinson.

I would also like to thank Tim Gordon, publisher at General Store Publishing House, Rosemary Kenopic, Susan Code, editor, Vicki Cameron, editor and Derek McEwen, graphic designer, at General Store Publishing House for their ongoing support and assistance.

On a personal note, I would like to thank Michael Wilgosh for his support throughout this project. I would further like to thank Sheila McKirdy for her unwavering support and encouragement during this project. Her opinions and assistance at critical times were invaluable in assuring the completion of the *Canadian Renewable Energy Guide, Second Edition.*

Alexandra McKirdy
Ottawa, Ontario
March, 1999

Table of Contents

Introduction

THE USE OF RENEWABLE ENERGY sources and the move towards sustainable energy are fundamental in redefining Canada's energy policy. Through the ongoing global climate change debate, it is becoming clear that many Canadians, are concerned about our growing energy demands. The recognition of environmental damage and pollution from primary and secondary industries will set the stage for unprecedented use of renewable energy, in combination with energy efficiency—displacing fossil fuels and nuclear energy resources over time. Using renewable energy preserves resources, while ensuring the security and diversity of our energy supply with minimal environmental impact.

This book shows that renewable energy can provide Canadians with a safe, secure and clean energy alternative. Realistic options are presented demonstrating the wealth of renewable energy applications already used successfully in Canada.

Renewables have been used as an energy source since the beginning of recorded time. Biomass, in the form of wood, has cooked our food and warmed our hearths for millennia. Micro-hydro has been used to power mills since the beginning of the Industrial Revolution, over two hundred years ago. Farmers in Europe and North America have long used wind to pump water for their homes, their crops and their livestock. The first basic solar water heater was built and sold in the 1890s. Photovoltaics have been commercially available since the 1950s, when they began to generate electricity manufactured for the United States space program. Renewable energy is not a new technology. It is a completely under-utilized energy source.

Fossil fuels are Canada's primary source of energy, with over 70% of our energy demands supplied by oil, coal and natural gas. In the early 1970s Canadians became increasingly aware of the problems associated with dependency upon fossil fuels. The 70s oil crisis resulted in increased oil prices, long lineups at gas pumps and a heightened awareness of the link between politics, domination of fuel supply and price. In the 1980s, the fossil fuel industry saw unprecedented growth and expansion including the development of new oil fields. More recently, a few multinational oil and gas companies, have publicly declared, and acted on, a growing commitment to renewable energy, through intensive marketing and new manufacturing.

Renewable energy has the potential to be one of the keys to the economic growth that has historically accompanied major energy source changes. But without the cooperation of local and provincial utilities, renewables cannot be effectively commercialised.

Other benefits of renewable energy technology include[1]:

- A sustainable energy supply
- Increased employment, mainly in small and medium-sized enterprises, fueled by industry growth

[1] EUREC Agency, The Future for Renewable Energy: Prospects and Directions, (James and James (Science Publishers) Ltd: London, 1996), 3.

- Decentralised production of energy which stimulates regional development
- Increased opportunities through technology & product transfer with developing countries
- The long lifetime of energy systems
- The modular character of the technologies allowing gradual transition to renewable energy sources, and,
- Reliable electricity supply in decentralised applications.

The potential of renewable energy in Canada and around the world is great, but currently only 6% of our nation's energy comes from renewable energy sources.[2] Natural Resources Canada predicts this will rise to 7% by 2020. The share of nuclear energy will drop from 11% to 6% over the same period, mainly due to the decommissioning of nuclear plants. Coal use will increase from 10% to 12% and natural gas use from 32% to 34%. Overall energy demand is expected to increase by over 23%, averaging 0.8% per year.[3]

This estimated 1 percentage point growth of renewable energy use is minimal and one could reasonably expect much more, especially in this time of utility restructuring, nuclear plant decommissioning and the changing nature of energy sources. In Europe, renewables are currently the fastest growing energy source. The barriers to renewable energy use include[4]:

- Lack of a publicly stated long range vision on energy policy
- Unwillingness by utilities to replace existing generating capacity
- Consumers' unfamiliarity with renewable technologies
- Lack of investment in research and development

The *Canadian Renewable Energy Guide*, Second Edition, has been published to address those barriers. It provides information on renewable energy applications, including, wind, microhydro, solar energy, earth energy and biomass. The detailed case studies of real systems were provided by Canadian installers and manufacturers of renewable energy products, it details examples of real systems.

In "Homes and the Canadian Experience," the Guide offers suggestions on incorporating passive solar design into a new home and adapting an existing home for renewables. It provides information on how to minimise a home's impact on the environment with the zero-energy house. The chapter "Power Systems" outlines how renewable energy sources, such as the sun, wind and water, are harnessed to generate electricity for domestic use. It also includes detailed information on the equipment used in renewable energy systems and how consumers can properly size a system.

The chapters "Commercial and Industrial Applications," "Remote, Northern and Rural Application" and "International Applications," outline specific renewable energy applications that have been implemented in settings across Canada and around the world. "Power Supply in North America" examines the role of the utilities in a changing power market. Examples include a utility-intertied system in Edmonton, Alberta, green power systems in Canada in the post-Kyoto period and the green energy initiatives of the Sacramento Municipal Utility District in California. Chapter 7, "Renewable Energy and Climate Change," discusses the effect of energy consumption on climate change.

Chapter 8, "Canadian Industry Overviews," specifically looks at the current status and the potential of the renewable energy industries. These overviews have been contributed by dealers currently active in their respective industries, specifically Active Solar Energy, Biomass, Earth Energy, Microhydro Energy, Photovoltaics and Wind Energy. The last two chapters provide detailed contact information for the Canadian consumer to further explore an action plan for incorporating renewable energy into their daily lives.

The potential of renewable energy in Canada is almost unlimited. Renewable energy resources

---

2 Natural Resources Canada, Energy Forecasting Division, Energy Policy Branch, Energy Sector, Canada's Energy Outlook 1996-2020. (Ministry of Supply and Services: Ottawa, 1997), 63.

3 Canada's Energy Outlook, 63.

4 Nancy Cole and P.J. Skerrett, Renewables are Ready: People Creating Renewable Energy Solutions. (Chelsea Green Publishing Company: White River Junction, 1995), 2.

can meet Canada's energy demand many times over as they can meet all of our current electricity, heat and fuel requirements. The ultimate goal of total energy sustainability can only be achieved if Canadians reassess and reduce their energy demands. Renewables are sustainable, environmentally acceptable, readily available and applicable in Canada's diverse climate, geography and demography.

Many renewable energy technologies are currently available as reliable and affordable products. This is a growing market opportunity for small and medium-sized companies in Canada. The economic competitiveness of the renewables industry in Canada would be improved by developing new technology and increasing current manufacturing capability to compete with existing large industries in the United States, Europe and Japan. The replacement of fossil fuels with renewable energy could lead to significant employment opportunities in both the manufacturing and high tech sectors.

Through the publication of the *Canadian Renewable Energy Guide*, Second Edition, it is hoped that Canadians become more aware of their renewable energy options, understand the true potential of renewable energy and are encouraged to use renewable energy in Canada.

Chapter I

Homes & the Canadian Experience

OUR HOMES shelter us from the elements, provide a sense of security, and symbolize our connections to family. They are also a link with the broader environment. The Greek word for house, *ecos*, is the root of the word ecology. As we strive to understand and honour our interconnections with the planetary environment, we begin to appreciate that our homes influence the world around us. Every home has the potential to affect the environment positively or negatively. If we are aware of the positive ecological potential for our housing and develop an understanding of what is needed from us to achieve that potential, we will be well on the road to developing truly healthy homes and communities.

A large number of factors, including occupant habits, landscaping, insulation, solar exposure, construction materials, finishes and furnishings, equipment and appliances, and municipal and utility company policies and practices, interact to determine the home's environmental impact. The biggest single item is its energy demand. The use of renewable energy is fundamental to environmentally appropriate housing. But renewable energy, even passive solar energy, is rarely successful if it is just "tacked on" to a conventional house.

The following pages review some of the current thinking on environmentally appropriate homes, with a focus on applications of renewable energy. The ideas presented are, for the most part, applicable to both new and retrofit projects. The case studies demonstrate how individuals across Canada have successfully put their ideas into practice.

Design — The Zero Energy House

The starting point for any new design or renovation of a home is the examination of the home's environmental contribution. Just as planting a tree has a positive effect on the environment, so, in theory, can a thoughtfully designed home. The essential design premise of a successful low-energy home is that little or no purchased non-renewable energy will be used for the house's operation. When this concept is applied from the outset and held in mind throughout design and construction, the result is often a home that uses 10% or less of the energy used by an equivalent conventional home, at little or no extra cost premium in some cases. This design philosophy works regardless of project

scale or budget and whether or not the house is grid-connected. As a comparison, the R-2000 approach to energy conservation results in energy use reduction of about 40% or 50% compared with conventional housing built to Building Code standards. However, the R-2000 technical standard should be a minimum target when designing with solar.

Design Priorities

The shape and orientation of the home affect space heating and cooling, artificial lighting, and mechanical ventilation requirements. Good solar exposure is essential for high passive solar contributions. Eliminating a basement reduces space heating loads and construction costs. Slab-on-grade designs and raised bungalows are alternatives to a full foundation. Minimising overheating can be accomplished by a combination of external shading, internal thermal mass to reduce temperature swings and natural ventilation to allow good cross-ventilation and vertical air flow. Artificial lighting requirements can be reduced by ensuring good daylighting. Natural ventilation, through operable windows, can usefully supplement forced-ventilation systems.

Organising the Space

By anticipating the natural indoor micro-climate, extra heating and cooling needs can be reduced. For example, because warm air rises, upper floors will always tend to be warmer than lower floors. If cool bedrooms and workrooms are desired, it makes sense to locate them on the lowest level. Sitting areas usually need to be warm places; consider a location upstairs or near south windows and a wood heater. Food storage uses a lot of energy in a conventional house, much of which can be saved by incorporating a root cellar and cold pantry in the design. For regions with cold dry winters, a ventilated closet warmed by the wood heater or solar-heated water can be used seasonally to dry clothes indoors.

The house design and function extend beyond its walls. Consider the micro-climate and food production potential of the outdoor living space. Plants also contribute to the indoor environment by improving air quality, providing beauty and a sense of tranquillity, and supplying food. Deep window-sills, bay windows, sunspaces and attached greenhouses provide places for indoor plants, herbs and vegetables.

Water Conservation

A zero-energy house has, by definition, reduced water needs. Six-litre toilets and low-flow showerheads can be considered minimal measures. A wide range of fixtures meeting this standard are now available on the market. Micro-flo and composting toilets, solar waste digesters, and rainwater cisterns may all be appropriate.

Summary

The more thought and subtlety (not complexity) that go into the design of a home, the better its performance and the simpler and less costly its mechanical and electrical systems will be.

What is . . . Active Solar Energy?

Active solar energy systems mechanically redistribute the sun's energy. The heart of the collector is the absorber plate, which converts sunlight into heat. The heat is transferred by air or liquid (water and/or antifreeze) to where it is used or stored. Practical applications for active solar heating include swimming pools, domestic and commercial hot water, and warming ventilation air.

Pool Heating

Active solar heating can significantly extend the Canadian swimming pool season,

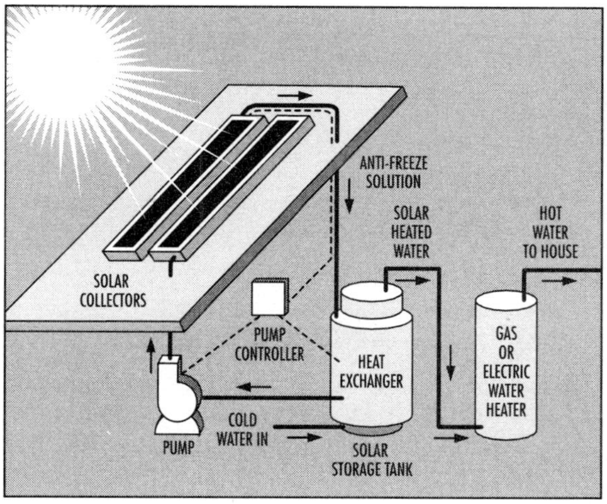

particularly if used with an insulating pool blanket. Pool blankets also save on chemicals, by reducing evaporation.

Air and Space Heating

Solar energy can warm a building and the building's ventilation air. The simplest form of solar heating is passive. It is completely determined by how the structural design and choice of materials work with sunshine and the elements. Mechanically assisted solar heating is more common, where pumps move warmed liquid from the collectors to radiators, such as in radiant floors.

Solar Water Heating

Solar water heaters can efficiently provide 80% of a normal domestic requirement for four months of the year, and 20% for the other eight months of the year. Of course that will vary with climate and demand. Typically, water heating accounts for about 20% of home energy use.

What is . . . Earth Energy?

Ground source or earth energy heat pumps offer a low life-cycle cost for heating and cooling (i.e. the total cost for installation and annual operating costs will be lower than the cost to install a conventional heating/cooling system and pay for fuel over the lifetime of the system). An earth energy system transfers heat between the soil and the house. It can heat the home in winter and cool it in summer. Most earth energy units can also be designed to heat water for domestic consumption.

Earth energy is replenished by sunshine. It is not to be confused with geothermal energy, which uses heat long stored deep in the earth's crust, or with air-source heat exchangers, which are of little or no use in a Canadian heating season.

The coefficient of performance (COP) is a measure of a heat pump's efficiency. It is determined by dividing the energy output of the heat pump by the electrical energy needed to run the heat pump, at a specific temperature. The higher the COP, the more efficient the heat pump. The number given is comparable to the steady-state efficiency of oil- and gas-fired furnaces. In an earth energy system, the energy that operates the system can move several times more energy from the earth to where it's required. The COP can range from a little more than one when the earth is frozen, to more than five in soil that is more than 10°C. Under Canada's *Energy Efficiency Act*, earth energy units sold must have a COP, of at least 2.8 (2,800 watts of solar heat delivered for every kilowatt of motive power consumed). Many units on the market are approaching a COP of 4.0, that is four units of space heating or cooling for every unit of electricity that is consumed. Some systems, such as the experimental Pilot Butte Conservation House in Saskatchewan, have been installed with solar collectors that pour their excess energy into the earth during the fall to increase the COP of the heat pump.

Earth energy technology is environmentally benign with no direct emissions or exhausts. By

requiring less electricity to heat equivalent space, there are significant reductions in the levels of CO_2, SO_2 and NO_x emissions.

The Earth Energy Society of Canada maintains a list of members (manufacturers and contractors) and also consumer information which is available on request. Contact them at:

Earth Energy Society of Canada
Suite 1050, 130 Slater Street
Ottawa, Ontario K1P 6E2
Tel: (613) 230-2332
Fax: (613) 237-1480
E-mail: info@earthenergy.ca
Web: www.earthenergy.org

What is . . . Passive Solar Energy?

Passive solar heating represents one of the most important strategies for the displacement of traditional energy sources in buildings. Passive solar techniques make use of the steady supply of solar energy through building designs that carefully balance energy requirements with the advantages of a building's site and window orientation. The term 'passive' expresses the fact that no equipment is used to provide a building with much of its heat and daytime light, other than the normal building elements, such as glass, insulation and heat absorbing floors and walls. All solar heat gains are brought in through windows and minimum use is made of pumps or fans to distribute heat or effect cooling.

All passive techniques use building elements, such as walls, windows, floors and roofs, to control heat generated by solar radiation. Solar heating designs collect and store thermal energy from direct solar gains. Control of overheating during summer is handled by making use of design and building elements to minimise the effects of solar radiation. Another solar concept is the use of daylighting design, which optimises the use of natural daylight and contributes greatly to energy efficiency. This is especially important in commercial and institutional environments that have fixed work stations. Arguments for using passive solar techniques include simplicity, price and the design elegance of fulfilling one's needs with materials at hand.

Passive Solar Heating

Passive solar heating of buildings occurs when the sunlight passes through a window, hits an object, is absorbed and converted to heat. The most efficient window orientation for heat gain is due south, but any orientation within 30 degrees of due south is acceptable. Once the heat has entered the building, various techniques come into play to keep and distribute it. Even in the Canadian winter, preventing overheating in the sun space presents a challenge.

Once the heat is in, a well-insulated and air tight building envelope helps prevent heat loss and allows the solar heat to serve more of the heating need. A crucial component of the energy efficient building envelope is the window system. While common double-glazed windows let heat escape, high performance windows with insulated frames, multiple glazing, low-e coatings, insulating glass spacers and inert gas fills can reduce heat loss by 50% to 75%. High efficiency windows and R-2000 levels of insulation and air tightness allow passive solar heating to cover a large proportion of heating needs in many locations. With the heat contained, often a simple ceiling fan, or a forced air furnace fan is all that is required for heat distribution. Using building envelope upgrades alone, up to 25% of a building's heating requirement can be gained with passive solar techniques.

In the absence of window upgrading, much heat loss can be controlled by the diligent use of drapes. Select thermal backed fabric and run the drapes floor to ceiling. If possible, install a valance in front of the drapes. When closed, the drapes prevent heat loss through the windows. Cold air is trapped behind the drapes and cannot leak into the room under them. The warm room air circulates in front of the drapes, and cannot flow through them, to the windows. In Canadian winters, this means closing the drapes at dusk, and occasionally keeping the drapes closed on stormy days.

A necessary technique to control overheating and extend warm conditions in the sun space once the sun has set is the use of heavy mass materials in the walls and on floors. Quarry tile or stone on floors in a mortar bed and one wythe of brick or double layers of gypsum board on walls will absorb solar radiation, smooth out the peaks of solar gain and slowly radiate heat back into the room after the sun has gone down. Some solar homes use a centrally located masonry wood heater to store heat. The bricks and stones surrounding the firebox absorb the solar gain and slowly radiate it into the room.

Passive Cooling

The need for air-conditioning in homes can be greatly reduced or even eliminated by using passive solar design which keeps solar radiation and heated air from reaching the building. There are two factors to deal with, internal and external heat gains.

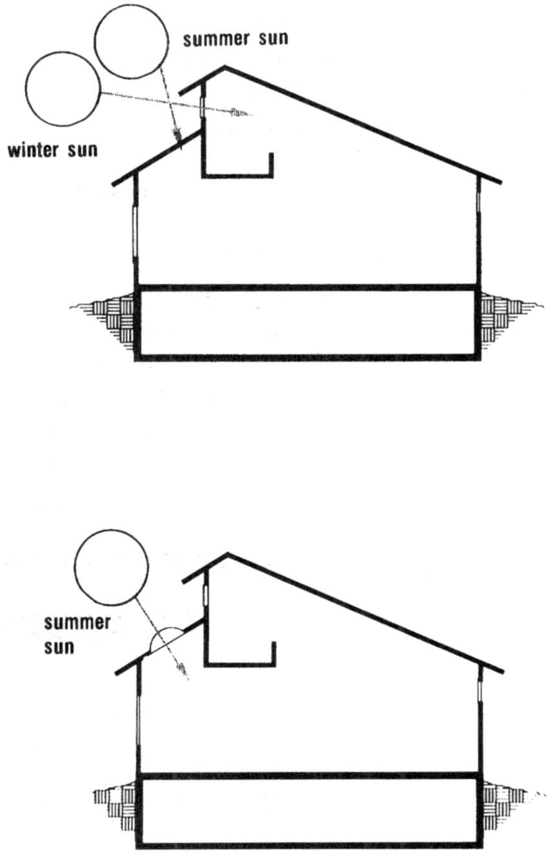

Celestory Windows Better for Winter Solar Gain

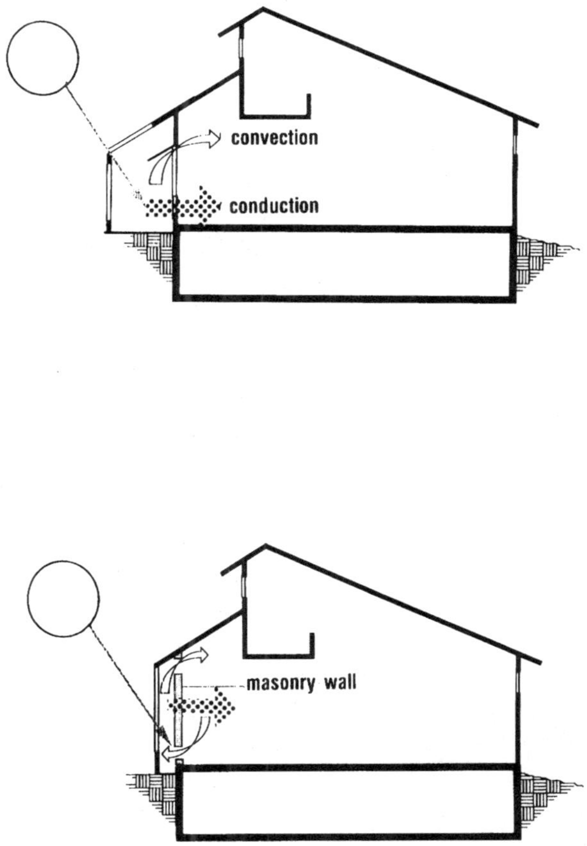

Sunspaces and Thermal Mass for Indirect Solar Gain

Source: *Passive Solar House Designs for Canada*, CMHC

Internal heat gains from appliances and occupants can be minimised and exhausted by natural ventilation. The heat sink, such as floors and walls help absorb heat from a room.

External solar radiation can be reduced by fixed or adjustable shading devices, including the eaves, providing shading by using vegetation or by using special glazing in windows. External shading devices can reduce solar gains by up to 90%, while still admitting a significant amount of indirect light. Controlling external heat gain can also be aided by maximising insulation, using adequate depth of eaves, reducing window size and using reflective materials in the external walls and roof.

In the building design stage, attention should be paid to cross-ventilation and the direction of prevailing winds which can be the source of cooling nighttime breezes.

Daylighting

In designing a building, it is important to consider the optimal use of daylight. The term 'daylighting' refers to making use of the overall light of the surrounding sky to illuminate building spaces, not just direct sunlight. In large commercial buildings, daylighting can significantly reduce energy consumption and provide a more comfortable working environment. Correct daylighting design will not only reduce costs related to electrical lighting, but will also reduce the need for air-conditioning in rooms heated by light bulbs or ballasts. A good daylighting system will consider the following elements: the orientation and space planning of the spaces to be lit; the location, form and dimensions of the opening through which the daylight will pass; the location of internal surfaces which may be able to reflect the daylight; and the location of movable or permanent objects that provide protection from excessive light or glare.

Design Software

Much of the art of passive solar design involves finding the right balance between glazing type, size and location, thermal storage, air circulation, and the efficiency of the building envelope. Once preliminary drawings are prepared, a detailed analysis should be done to fine-tune the design and confirm predicted performance. The following programs can be used:

HOT-2000: The official analysis tool of the R-2000 program. A list of certified R-2000 plan evaluators is available through provincial home builders' associations. The software is available for Windows 95 and Macintosh formats from:

Canadian Home Builders Association (CHBA)
150 Laurier Ave. W., Suite 200
Ottawa, Ontario K1P 5J4
Tel: (613) 230-3060
Fax: (613) 232-8214
E-mail: chba@chba.ca
Web: www.chba.ca

ENERPASS: a detailed and sophisticated analysis tool that gives hour-by-hour or seasonal total building performance simulations. It is suitable for analysing complex designs, particularly when unusual design strategies are employed. The software, and a consulting service for those designers not wishing to carry out their own analysis, are available from:

Enermodal Engineering
650 Riverbend Drive
Kitchener, Ontario N2K 3S2
Tel: (519) 884-6421
Fax: (519) 884-0103
E-mail: office@enermodal.com
Web: www.enermodal.com

Another useful tool is a spreadsheet program, 'The Comfort Design Checker Software', which comes with the book *Tap The Sun* published by Canada Mortgage and Housing Corporation and runs on Microsoft Windows Excel 5. This software, allows the owner to check the house design for excessive overheating in the winter.

How do you . . . Solar Retrofit your home?

Retrofits are solar applications for an existing home that take advantage of various applications of renewable, especially solar energy. The sun's energy can be used to provide space heating, domestic hot water, swimming pool heating and electricity for lights and household appliances.

How can solar energy be used in a retrofit situation?

To take the best advantage of the sun, it is important to identify how the energy is used, because unlike conventional fuel sources, solar energy is not brought to a site by wire or a pipe. Knowing how to use it will help determine the most effective manner of capturing the sun's energy. Is it to provide electricity for lights or small appliances, hot water for domestic use, or space heating?

The first step in a solar retrofit is to list how energy is used in the home so that the most appropriate applications can be selected. The next step is to decide how much solar energy is available. This is dependent on the time of year, the house's location and the location of the solar panels. In the winter, the days are shorter and the sun is lower on the horizon; in summer, the days are longer and the sun is higher. Therefore, the optimum collecting angle differs with each season. Further north, the changes in the sun's position are more pronounced, so the position of solar energy collectors will differ based on latitude and time of year for which the system is optimized.

Southeast to southwest is the prime solar orientation. Determining how much sun is available can be done by looking at the obstructions on the south side of the house. Trees, other buildings or natural features that cast shadows must be evaluated to determine if or when they will cause shading at various times of the year. Local geographical conditions that affect solar availability must also be taken into account. Rural or urban location will also have some bearing on determining the appropriate solar options to harness.

Deciding the best options for you

When considering a solar retrofit, the homeowner should take serious measures towards energy conservation. A consumer should make all the energy conservation upgrades that can be made to a house before investing in new solar equipment or hardware. For space heating applications, draft-proofing the house and upgrading insulation are important steps in heat and energy conservation.

If the proposed work coincides with renovations, repairs or general maintenance, the consumer can undertake significant upgrades, including the upgrade of wall and ceiling insulation, extensive air sealing, and the installation of new windows. If windows are being replaced or added, the new units should be high performance windows with low-e glass, gas fill and insulating spacers.

For further information on active solar energy, passive solar energy or solar retrofitting contact:

Canadian Solar Industries Association
2415 Holly Lane-Suite 250
Ottawa, Ontario K1V 7P2 Canada
Tel: (613) 736-9077
Fax: (613) 736-8938
E-mail: cansia@magmacom.com
Web: www.cansia.ca

or the

Solar Energy Society of Canada Inc.
116 Lisgar Street-Suite 702
Ottawa, Ontario K2P 0C2 Canada
Tel: (613) 234-4151
Fax: (613) 234-2988
E-mail: sesci@cyberus.ca
Web: www.solarenergysociety.ca

CASE STUDY:

Green Housing and Sustainable Living
Gabriola Island, British Columbia

LOCATED in the Gulf Islands near Vancouver, John Hague's home was designed to minimize environmental impact by incorporating a host of material- and energy-saving techniques:

- rehabilitatation of valuable land by correcting man-made surface water problems
- environmentally sensitive architecture—right sized for needs
- use of shallow ground water and 2500 gallon cistern for most water needs
- use of low flow water faucets and toilet
- use of a solar water heater with a glazed parabolic batch solar collector
- orientation of glazing to maximise passive solar heat gains
- use of high performance low-e, gas filled vinyl-framed windows
- use of an air-tight wood stove with interior chimney and central paddle fan
- installation of durable long life vinyl siding, and aluminum soffits and facia that minimize the need for painting
- installation of long-life fibreglass shingles

John is managing a woodlot sustainably to heat his home—the wood comes mostly from fallen and standing alders. Selective logging has helped to minimize fire and windfall hazards, while tree planting replenishes the resource and creates a wind break around the house. Special "slow grow-no mow" grass reduces lawn maintenance and John's electric power mulching lawn mower is quiet and emission-free. Kitchen compost is helping to regenerate the topsoil.

The Big Picture

A fundamental change in the public attitude towards the environment will increase the demand for green housing. Green houses are cheaper to operate and maintain, and are more comfortable and healthier to live in — the extra costs are usually very modest, and sometimes zero.

Contact
John Hague, BA, CA
88227 Delcrest Drive
Delta British Columbia V4C 3Z5
E-mail: delwood@uniserve.com

Architect
Richard Kadulski, MAIBC
204-1037 West Broadway
Vancouver, British Columbia V6H 1E3
E-mail: kadulski@direct.ca

High Performance Windows

High performance windows are an essential component of a zero-energy home. They allow passive solar gain with minimal heat loss in winter and reduce summer heat gain. Other benefits include the reduction of drafts and radiant heat loss, elimination of window condensation, and reduction of sound transmission from outside.

A high-performance window will have four features that differentiate it from a conventional window:

1. Low-E Coatings. A low emissivity (low-E) coating consists of a very thin metallic layer applied to glass or polyester film. This coating reduces radiant heat loss through the glass. A high performance window may have one or more low-E coatings. Each coating reduces the transmissivity of the window, that is the amount of solar gain admitted to the house and the amount of UV light transmitted. Therefore, the number of coatings selected for a given situation will depend on the area and orientation of the window. For example, a north-facing window in a cold climate needs the most insulative glazing possible with two or more low-E coatings, whereas a house with a modest amount of south-facing glass in a moderate climate, probably only needs one low-E coating.
2. Gas Fill. A large molecule inert gas is used to fill the space between the glazing in a sealed unit. The gas reduces conduction and convection losses through the unit. Argon is usually used for this purpose in standard (1/2 inch) spaces, although the more expensive krypton gas works better in narrow (1/4 inch) spaces.
3. Low Conduction Spacers. The sheets of glass in a conventional sealed unit are held apart at the perimeter by an aluminum spacer. As aluminum is an excellent conductor of heat, frost and condensation occur mostly around the edge of a sealed unit. The heat loss through an aluminum spacer considerably reduces the R-value of the sealed unit. The more insulative the glass, the more pronounced the effect, and the greater the thermal stress on the glass. For these reasons, high-performance windows use insulative, non-metallic spacers that can reduce heat loss through the unit by as much as 20%.
4. Insulative Frames. Window frames and sash materials account for significant conductive heat loss since they constitute up to one-third of the overall window area. Wood windows are readily available and are reasonably insulative, but the best thermal material on the market today is a fibreglass frame with non-CFC foam insulated cavities.

Selecting High-Performance Windows

A window with the foregoing features will provide superior thermal performance. The window's energy rating (ER) provides a measure of that performance. This rating includes factors for solar heat gain, heat loss by transmission through frames, spacer, and glass; and heat loss by air leakage. The ER number lets you compare different products. A positive number (+) indicates that on average, the window could add more heat to the home (through solar gains) than it loses during the annual heating season. A negative number (-) indicates the window will lose more heat than it gains. Most window units have negative numbers. The best high performance windows have positive ratings.

The ER numbers do not consider how the unit is installed. It is important to air-seal around the window when it is installed, to avoid losing the benefits of the low infiltration rate. Spray in-place expanding foam insulation is commonly used when installing windows. The foam insulation fills and seals the gap. In new construction, care must be taken to seal the window to the vapour or air barrier.

Retrofit situations often lack an air or vapour barrier as such, in which case the window should be sealed to the interior finish, i.e. drywall or plaster.

The ER value was developed for the heating dominated Canadian climate, and therefore does not calculate the effect of the window on summer cooling. Highly insulative windows can reduce cooling loads as well as heating loads. Overheating can also be controlled by house design. If there is a concern about overheating, some operable windows, including high and low windows for stack-effect night cooling, should be incorporated. The formula for calculating the ER ratings is defined in CSA standard A440.2. The formula for calculating the ER rating is different between window styles, so for equivalent comparisons, you must compare fixed-to-fixed, casement-to-casement, etc. Skylights have different solar heat gain characteristics so they are not yet covered by ER ratings.

The ER calculation incorporates an average solar gain. For passive solar designs, an ERS (energy rating specific) calculation, according to CSA Standard A440.2 Energy Performance of Windows and Other Fenestration Systems, can be used to evaluate the solar contributions of the windows to the building energy load. The software programs Frame and Vision can also be used to determine the specific characteristics of custom windows. A window located where it receives no direct sunlight will not provide much passive solar heat but if it has a positive ER number, it will lose less heat than it gains, still making it a good environmental choice.

may provide thermal mass and support exterior shading devices. It should not incorporate heat-losing conductive thermal bridges, nor should it use excessive amounts of material, toxic materials, or materials with high embodied energy. There are no universally perfect materials or systems, but there are many options that may be appropriate in various circumstances.

Typical ER Ratings

Window Category	*Type of Spacer*	*Type of Glazing*	*Fixed Window*	*Operable Window*
Common (poor energy performance)	Aluminum	Double	-35	-50
Common (acceptable energy performance)	Aluminum	Double	-15	-30
Minimum for High Performance		+2	-11	
Moderate-cost High-Performance	Insulated	Double, low-E, argon	+5	-8
Best High-Performance	Insulated	Triple, two low-E coatings, argon	+15	+8

Construction Materials & Techniques

In planning, designing and building a zero-energy house, the selection of materials should consider performance criteria and circumstances of production, distribution and maintenance. The builder/owner of the house must determine the minimum amount of energy required to build and operate the house, including the amount of energy required to manufacture the materials used in the house. It is also important that the zero-energy house have a minimum impact upon the environment. A builder/owner of a zero-energy home can minimise their energy demand and their impact on the environment by purchasing their materials locally, using recycled materials and utilising different forms of renewable energy in their home.

Structural Systems

In an environmentally appropriate zero-energy house, the structural system performs its normal function while supporting an increased level of insulation and an improved air barrier. It

Foundations

Concrete is probably the best structural material to use below grade. Several manufacturers have developed insulated forming systems using non-CFC polystyrene that provide insulation and reduce the amount of concrete required. A disadvantage of these systems for passive solar houses is that the insulation on the interior surface stops the use of the concrete as thermal mass. Concrete, insulated on the exterior, avoids this difficulty but may be expensive. Concrete floor slabs, insulated underneath or with proper perimeter insulation, make for an inexpensive, good-looking finish, which can provide thermal mass. Concrete can also be a good medium for a radiant heating system, which is compatible with solar or wood heat sources. Earthcrete (also known as 'soil cement' in Western Canada), a mixture of cement and sandy subsoil similar to adobe, is environmentally preferred to concrete and can make a good floor slab.

Above-Grade Sructure

A large number of wood frame techniques seek to build as thick a wall (or roof) section as possible using the least amount of wood. Site-built houses generally use more wood than manufactured houses. Canada is the world leader in energy-efficient wood frame construction, and an expanded wood frame is a good choice. Prefabricated or site-built 'eco-studs' use less lumber than standard framing, achieve wall thickness of 6 to 12 inches and allow high levels of insulation. Some builders of low-energy homes are using this type of system in preference to 2" x 6" framing.

Structural Insulated Panels (SIP) or stressed-skin panels are a fast way to build a well-insulated and airtight building. There are various types of SIP; a common type is a panel with oriented strand-board glued to both sides of a sheet of expanded polystyrene. The SIP panels are joined with plywood splines. The panels are usually loadbearing, but can also be used as infill in a post and beam frame. Drawings need to be accurate if SIP are to be used, and plumbing and wiring needs to be planned ahead and carefully installed to avoid damage to the insulation. While SIP do not permit last minute design changes on site, they are still growing in popularity because of how quickly they can be installed and their good thermal performance.

Stack-Wall Construction

Stack-wall construction consists of stacked cordwood, mortared on the inner and outer face with the centre spaces filled with insulation, usually vermiculite. Low-cost and labour intensive, stack-wall construction may be appropriate for owner-builders. Since wood has an R-value of about one per inch and a stack-wall is rarely thicker than twenty inches, it is necessary to add additional insulation, plus an air barrier, to one side of the wall to achieve high enough R-values, R-32 to R-40, depending on climate, to satisfy the requirements of a low energy house.

Straw Bale Construction

In the last decade, a small niche of builders have become interested in strawbale construction and its environmental and energy conservation benefits. Hands-on workshops, electronic bulletin boards, exhibits, conferences and testing have stimulated further applications.

Straw bale construction, like many other indigenous or alternative construction materials and systems, must be considered carefully. They may only be suited for some climatic conditions and applications. To avoid problems during construction, construction details and quality control must be suitable for the application. For example, straw bale construction is not an appropriate material for use in a mild, wet climate where continuous wetting may create ideal conditions for biological deterioration.

There are two distinct types of strawbale construction: "bearing walls" where bales act as both the structural load-bearing component and as insulation simultaneously, and "infill" systems in which bales act only as insulation with other elements (usually wood) acting as structure. Wall thickness, bale density, bearing capacities, types of straw, finishing materials, foundation systems and application methods vary according to local conditions. In all cases, straw bales are used as insulation.

Strawbales (not hay bales) can be used in many different settings including new and retrofit housing, light industrial buildings, city and rural buildings, developing and developed countries in warm, cold, and dry climates, and for a wide variety of buildings. There are few urban examples of strawbale construction due to the limitations of lot size and relatively higher taxes per finished area. But they are still ideal for rural sites in the prairie regions of North America where there is an inexpensive and readily available supply of straw from local farmers.

The advantages of strawbale construction include: greater sustainability, environmental soundness, short rejuvenation period, good insulation value, adequate structural capabilities, good local availability and good sound absorption. Some disadvantages of strawbale construction include: unfamiliarity of code by local and municipal officials, unwilling insurance and mortgage companies, inconsistent quality control, uneven wall surfaces, higher property taxes per unit of habitable space and the imperatives of moisture, fire and vermin protection.

Rammed Earth Construction

In rammed-earth construction, local earth (70% gravel/sand, 30% clay) is manually or pneumatically tamped between standard concrete forms until load-bearing strength is achieved. Puddled mud, adobe, cob, and *leichtlehmbau* (German for light clay construction) are related earth construction methods. Most walls are between 12 inches and 24 inches thick and are suitable for passive solar heat sinks, structural walls, and fire and acoustic barriers. The compressive strength is lower than for concrete, so the walls are thicker. In Canada, exterior rammed-earth walls will require insulation, which can be installed by placing rigid insulation in the middle of the forms before ramming. Rammed-earth can be stabilised by adding cement to the mix and reinforced using steel rebar. Rammed earth construction uses more labour than other construction types, and like straw bale construction, the design must compensate for the thicker walls.

Insulation Materials

The type of insulation used depends partly on where it will be installed and the job it has to do. For example, blown-in or foamed insulation fills gaps better than batts. The thermal mass of the structure (i.e. how long it takes to heat up and cool down) may affect whether one puts the insulation inside, outside, or within the structure. Cool surfaces and moist air combine to produce condensation. Insulation performs poorly when damp or wet so its design and installation must avoid moisture penetration of the insulation. Some insulation types can also function as an air or vapour barrier.

High recycled content of an insulation material reduces its negative environmental impact, as does the distance between manufacturer and installation. Most insulation materials contain some toxic particles or gases that may affect producers, installers, and building users, especially during installation.

Dry-blown Cellulose Insulation

R 3.6/inch (RSI .025/mm). Dry-blown cellulose is used for attic insulation in new and existing buildings and for retrofitting empty wall cavities. This is a recycled material consisting of finely-shredded newspapers and other cellulose materials with minimal energy used in production. The insulation is treated with boric acid as a fire retardent and borax as a biocide. Cellulose insulation is sold by most building suppliers, who often have blowing machines available for rent.

Wall and Ceiling Insulation

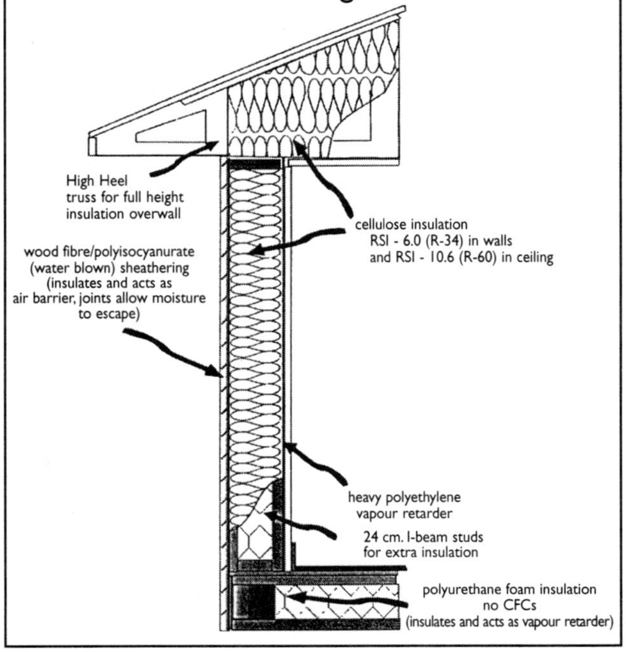

Source: Waterloo Regional Green Home

Wet-Blown Cellulose Insulation

R 3.6/inch (RSI .025/mm). Wet-blown cellulose, used for wall cavities in new buildings, is identical to dry-blown cellulose but is mixed with a small amount of water, and sometimes a latex binder, to make it stick in place. The cellulose is allowed to dry for a few days before the wall is closed in. Wet-blown cellulose out-performs batt insulation in its ability to fill all nooks and crannies and has a higher percentage of recycled material. All cellulose products create dust during installation. Wet blown cellulose insulation must be installed by qualified installers.

Foamed Isocyanurate Insulation

R 5/inch (RSI .034/mm). This is a foamed-in-place insulation that can function as insulation, air barrier, and as a vapour-diffusion retardant. It uses carbon dioxide rather than CFCs or HCFCs as a blowing agent. And, because it is shipped as a concentrated liquid to be foamed at the site, it saves considerable transportation energy. While usually used in new construction, it can also be used in existing structures. It must be installed by qualified installers.

Fiberglass Batt Insulation

R 3.4/inch (RSI .024/mm). Batt insulation is suitable for regularly spaced framing. It must be installed with no gaps at framing members, otherwise convection currents will short-circuit the insulation. Manufacturers are beginning to use recycled glass (one Ottawa manufacturer is now using at least 60% of recycled post-consumer glass) requiring less energy and reducing pollution in the manufacturing process.

Mineral Wool Batt Insulation

R 3.2/inch (RSI .022/mm). Mineral wool batts are similar to fibreglass batts and the same comments apply regarding installation. Mineral wool batts typically include about 50% recycled content, including slag materials, and have no flammability concerns. The individual fibers are longer than fibreglass and are coated with oil.

Polystyrene, Polyisocyanurate and Polyurethane Foam Boards

These materials range in R value from R 3.8/inch (RSI .026/mm) for Type 2 expanded polystyrene to R 6/inch (RSI .041/mm) for polyisocyanurate. From an environmental standpoint, all these products have toxic manufacturing processes. Expanded polystyrene uses CO2 as its foaming agent rather than ozone depleting CFC or HCFC and is therefore environmentally preferred at present. Expanded polystyrene insulation (except Type 1) may be used below grade, which makes it useful for insulating the outside of foundations.

Compressed Straw Board

R 2.0/inch (RSI .014/mm). An environmentally friendly product, but it is less insulative than the foam boards. At this time, there is a limited production of straw boards in North America.

Heating and Cooling the Zero-Energy House

In a zero-energy house, there is little need for supplementary space heating and cooling. However, most houses in Canada will need some space heating and cooling, but because of the well insulated building envelope, a house approaching 'zero-energy' will need a much smaller supplementary system. Hot water loads will often exceed space heating loads and electrical energy costs for lighting and appliances can exceed heating fuel purchases. High-efficiency furnaces and boilers are useful replacements for inefficient equipment in conventional homes, but must be sized appropriately. However, there are many space heating options to choose from. High-efficiency fireplaces and wood heaters, solar thermal systems, and a multitude of combination systems can provide both domestic hot water and space heating. Various types of heat pumps and radiant distribution systems make low temperature sources, such as the hot water tank or solar collectors, feasible. The following factors need to be considered:

How much heating is needed? It is quite possible to heat a small, well-insulated low-energy house from a energy efficient hot water tank, which should result in capital and operating cost savings. The HOT-2000 computer software is useful to determine annual heating loads and equipment sizing. Equipment that is oversized for the application is always a poor choice.

What type of distribution system is preferred? In addition to supplying warm air, ductwork can be used to distribute cool air in the summer, and fresh ventilation air year round. When ductwork is installed it is often leaky and poorly designed so that much of the heat from the furnace may not get to where it is needed. Hydronic radiators cost more to install than ductwork and can be used only for space heating, but they provide clean, dust-free heat. Ventilation air and air-conditioning can sometimes be retrofitted to a house with radiators by installing a simple, central ceiling plenum. For new construction, a radiant floor heating system may be appropriate, with a separate dedicated ventilation system.

What special needs or circumstances must be met? Often the choice of equipment depends on individual circumstances. For example, if a family member has chemical sensitivities, oil, natural gas and wood will likely be excluded as fuels, and heat distribution will probably be low-temperature radiant.

Heating with Renewable Energy Sources

Passive solar energy, active solar energy and sustainably harvested wood are the three currently available renewable energy heating sources.

Domestic Hot Water Systems

Domestic hot water (DHW), used for such things as laundry, showers, and dishes, can consume up to 20% of the home's energy. Next to space heating, it is the home's largest energy requirement. Most hot water heaters currently use natural gas or electricity. Heating water with solar is an excellent way to reduce pollution. A solar domestic hot water (SDHW) system can be a straightforward retrofit to most homes, and can typically provide over 50% of the annual hot water needs. With a projected life of twenty to forty years, most SDHW systems pay for themselves in 15 to 30 years, while they reduce dependence on non-renewable resources.

The Typical Canadian SDHW System

Solar energy, leaving the sun in the form of radiation, is captured by the black absorber surface of the solar collector which converts it to heat. The solar collectors are the heart of the system, and are usually placed on the south-facing roof of the building. There are several different collector technologies that can be used with a SDHW system. A pump circulates a heat transfer fluid through the pipes to pick up heat from the collector, and transports it to the heat exchanger. The heat exchanger transfers the heat from the pipes to the water in the solar storage tank, which stores the heat. When a hot water tap is opened, solar heated water flows into the conventional tank, then on to the tap. The conventional water heater ensures the household doesn't run out of hot water when there is a high hot water demand, or when the weather is poor for several days.

Freeze Protection

For year-round systems, freeze protection is required to avoid freezing the pipes and the collectors on a cold night. In some systems, antifreeze is pumped through the collector, and into a heat exchanger to transfer the heat to the hot water tank for storage. Propylene glycol based heat transfer fluids are typically used in Canada. The heat transfer fluid, just like coolant in a car, needs to be changed regularly, typically every five years.

An approach that requires no anti-freeze is a 'drainback' system. Plain water is used as the heat transfer fluid. Whenever solar energy is available, the solar pump circulates the water through the collectors and back to the solar tank. When the pump is shut off, the water drains back to the tank through the solar pipes that are all sloped from the collectors down to the tank. The advantage of this system is the reliability of gravity as freeze protection. The correct installation of this design is critical, however, as the piping must slope correctly, and the panels must be mounted higher than the tank.

Controller

Solar water heaters require a controller to turn them on whenever solar energy is available, and off at night, or when the storage tank reaches its maximum temperature. The most common controller is a differential temperature controller, which compares the temperature of the solar collector to the temperature of the storage tank. Whenever the solar collector is warmer than the tank, the controller turns the pump on. When the collectors are colder than the tank, such as when the sun goes behind a cloud, or at night, the controller shuts off the pump. The controller also prevents overheating by turning off the pump whenever the tank reaches the maximum temperature, typically set to 80°C.

PV-Pump

An interesting approach used in many Canadian systems is the use of a photovoltaic (PV) panel to directly power the solar circulation pump. By operating the pump only when the sun is shining, the PV panel acts as both the power source and the controller for the system. This is an elegant approach which is also very reliable; the system will continue to operate regardless of power failures on the utility grid.

Seasonal Systems

Solar DHW systems used in summer applications, such as cottages or campgrounds, may not need freeze protection. In these systems, tap water can be circulated directly from the hot water tank through the collectors. Direct systems have a couple of advantages. Because they heat the water directly, they are more efficient than systems that use a heat exchanger. Furthermore, they are less expensive as the cost of the heat exchanger is avoided.

Types of Collectors

Unglazed Flat Plate

Unglazed flat plate collectors are the most common solar collectors, typically used for heating backyard swimming pools. They are made of UV stabilized polymers or rubber and are installed without glazing or insulation.

Glazed Flat Plate

A glazed flat-plate collector consists of tubes attached to a black metal absorber plate. The whole collector is contained within an insulated box to prevent heat loss and covered over with a sheet of glass.

Evacuated Tube

With an evacuated tube collector, the absorber is contained within individual sealed vacuum tubes. The vacuum, as with thermos bottles, is very effective at preventing heat loss. Consequently, these collectors are most commonly used in cold climates.

Other Strategies

Thermosyphon

Thermosyphon systems circulate the collection fluid through the solar collectors to the tank without the use of a pump. Instead, they use natural circulation to drive the process. The tank must be mounted higher than the collectors. When the sun heats the collectors, the heated water rises naturally and displaces cooler water in the tank. The cold water in the bottom of the tank sinks down and returns to the bottom of the collector, and the cycle continues as long as there is solar energy available. These systems require a check valve on the return line to prevent reverse thermosyphoning at night. Some thermosyphon systems extend their operating season by operating with heat transfer fluid and heat exchanger.

The primary appeal of thermosyphon systems is their simplicity. Without the use of pumps or controllers, no electricity is required. That is why these are especially useful for cottages and off-grid locations, where electricity may not be available.

Batch Heater

A batch heater, or integral collector-storage (ICS) system, is another design strategy that requires no electricity. It is a simple system that combines the collector function and the storage function in one unit. The system consists of a black painted tank, typically in a glazed and insulated box. The water in the tank is directly heated by the sun. While these are considered seasonal systems in Canada, the relatively high mass of water in the tank makes these units resistant to freezing, enabling longer operating seasons than other direct systems, such as thermosyphon. Depending on location and piping layout, these should perform well in conditions with cool nighttime temperatures.

Biomass-Burning Appliances

Many Canadians burn wood to heat their homes. Four hundred thousand Canadian dwellings are heated only by wood with another 950,000 dwellings using it as a supplementary heat source. Wood is a renewable energy source, but it must be used efficiently. Burning wood efficiently and ensuring it is cut down sustainably is a good environmental choice for home heating.

Wood burning technology in homes has changed in recent years. Advances in wood burning technology include new internal designs that burn wood more cleanly and at higher efficiencies, technology that keeps the door glass clear combining efficient heating with flame viewing, pellet stoves that use special fuels and are capable of long periods of unattended heating, and new safety and industry training standards.

Think safety and efficiency first and then consider issues of cost, convenience, viewing pleasure and use as an emergency heating source. By heating with wood, you are contributing to the local economy, reducing your output of greenhouse gases.

Safety

Safety should be the primary concern when deciding to heat with wood. Burning wood carelessly for home space heating has caused many house fires, proportionally more than those caused by either oil or gas heating. The Canadian Standards Association (CSA) has developed an installation code (CSA B365) that combined with safety tests for all wood burning components has set the stage for consistent and safe installation. Biomass energy system installers should also have completed a Wood Energy Technical Training (WETT) course to ensure their competency. Safety can not be emphasized enough. Fires caused by buildup of creosote in the chimney through improper installation, operation or maintenance are common and kill many Canadians every year.

Consumers have a range of options in how they can burn wood for home heating. There are now different ways to burn wood beyond a traditional stove connected to a metal chimney. Much of this technology has developed in conjunction with the improvements in house building techniques. As stoves and homes become more efficient, less energy is required to reach and maintain a comfortable temperature. An average new house can comfortably be heated by wood all year.

Standards

The efficiency of the wood burning appliance is an important consideration, requiring fewer trips to the wood pile and resulting in less pollutants being released into the environment. Since 1990, all wood stoves manufactured in the United States have been required to meet *Environmental Protection Act* (EPA) emission standards, which set a limit of 4.1 grammes of particulate matter per hour of operation. In Canada, the CSA also has set emission standards; however most manufacturers literature will generally mention only that it meets EPA standards.

Wood Heating Performance Considerations

The performance of any wood burning fireplace is most greatly affected by the type of wood burned. Green, freshly cut wood has a moisture content of 35% to 50%. The more water the lower the combustion temperature will be. This reduces efficiency of combustion which releases more pollutants into the atmosphere in the form of unburned gases and solid particles. Blue smoke from a chimney is a sign of bad combustion and could be due to poorly dried or seasoned wood. Firewood should have seasoned in a covered pile after being cut and split in late winter or early spring. No stove can overcome the problems caused by trying to burn green wood as they have been designed for peak performance when using properly seasoned wood.

Performance Characteristics

There are important performance differences between stoves that deliver heat to the room by radiation, and those that deliver heat by air flow or convection. Radiant stoves send heat out in all directions. The surface of objects, such as walls, floors, ceilings, furniture and people, are warmed directly and these in turn will warm air. Convection stoves heat air that flows between the stove body and a metal casing. The air enters openings at the bottom of the casing and flows into the room through a grille at the top of the stove.

While the heating efficiency of radiant and convection stoves is about the same, there are different advantages to using one or the other. For example, a radiant stove is effective in an open area where the radiant warmth is not blocked by walls. The heat from a radiant stove is somewhat more difficult to distribute to other areas of the home. A convection stove, on the other hand, produces hot air which can be more easily distributed to other areas. A fan uses forced convection to transfer the stove heat to an air stream that can then be distributed by ducts. A person can sit quite close to a convection stove without feeling uncomfortably warm as is the case sitting close to a radiant stove. A convection stove is a good choice if the stove is located in a basement family room or a room that is not fully open to the main part of the house. Many modern stoves have a blend of both radiant and convection characteristics. That is, the back and sides have shields behind which air flows, whereas the front and top have radiant surfaces. In operation, these stoves produce both radiant and convective heat and will be suitable for most installations.

Canadian manufacturers have established an international reputation as designers of some of the most effective advanced combustion appliances in the world. Improvements in combustion are achieved in modern stoves either by ensuring that the wood gases are not sent up the flue unburnt or by requiring a special fuel that burns more completely. They have a number of characteristics to ensure that particulate matter which appears as smoke and gases are completely burnt before going up the chimney. These stoves have extra firebox insulation to keep temperatures high, devices to reflect heat back into the firebox creating increased gas turbulence for better combustion, and devices giving the gases a longer, hotter route so they will burn before being cooled. They also have heated secondary air supplies that usually feed air above the fire through small ducts. When an advanced combustion stove is burning, you will often see little jets of flame coming from these small air inlet ducts. Because of the high temperatures, these combustion devices or baffles will need replacing at regular intervals.

Catalytic combustion ensures that unburned gases, which would otherwise escape up the chimney, are burned to produce heat, carbon dioxide and steam. The catalyst in a catalytic combustion appliance is a coated ceramic honeycomb through which the exhaust gas is routed. The catalytic coating lowers the ignition temperature required for combustion of the gases. Because the catalyst restricts gas flow through the appliance, these units should always include a by-pass damper into the flue. The damper is opened until a hot fire is established, then it is closed, forcing the gases through the catalytic combustion. Catalytic honeycombs need to be replaced at regular intervals of one to six years, depending on usage.

Current wood stoves allow for the combustion air supply to be taken directly from the outside via insulated ducts. This adds efficiency as it avoids sending already heated room air up the chimney to the outside and avoids creating drafts near the stove. Care must be taken when hooking up the inlet duct that prevailing or occasional winds will not reverse the stove draft, sending flames backwards into the ducting and potentially causing a fire.

Different Stove Types

Wood Stoves

A wood stove is the most common, flexible and affordable space heating option. A wood stove can be located almost anywhere there is enough space and where a chimney can be properly routed. Freestanding wood stoves come in a wide variety of styles incorporating some or all of the performance characteristics detailed earlier. Often home owners select a stove that is too hot for the space to be heated. Many stove owners compensate by underfiring the stove, which causes poor performance, increased emissions and potential chimney fires caused by creosote buildup.

Cookstoves

While wood-burning cookstoves are not designed to heat an entire home, many rural Canadians have found that they are great for cooking food and can provide supplemental space heating. Cookstoves cost more than many wood heaters, but some models can warm several rooms of a house in addition to cooking, baking, and generating hot water. The cookstove is useful in areas susceptible to power brownouts or power failures. While cookstove manufacturers have not yet been able to build advanced, clean-burning technology into their ranges, some models with gasketed firebox doors are noticeably more efficient than the original turn-of-the-century versions.

Masonry Heaters

A masonry heater's main feature is its ability to store a large amount of heat. These units are built with tonnes of bricks or stone to store and later release the heat of the burned wood. Masonry heaters are different from the conventional open English fireplaces. The core of the masonry heater is usually built from interlocking components that form the firebox and heat transfer passages. This core is surrounded by brick or stone veneer finish. These fireplaces are sometimes called Finnish or Russian stoves.

With a masonry heater, the owner builds one or two fires per day and allows the fire to go out. The heat stored in the masonry continues to radiate warmth for many hours. If a very small amount of heat is needed to take off the chill, such as in the spring or fall, the owner simply burns a smaller fuel charge. As the wood is burned hot, it is a clean fuel and emissions are similar to the best wood stoves. A masonry heater is ideally installed in a central location in a new home, although retrofits are possible. Kits are available, but site-built heaters should be done by masons experienced in their design and construction, and they often require additional foundation support.

Pellet Stoves

Pellet-burning appliances include a hopper to hold at least 20 kilograms of fuel and a screw auger to move the pellets from the hopper into the combustion chamber at a controlled rate. In the combustion chamber, the pellets and forced air from a electric fan are mixed in the right proportions to ensure a very high temperature of combustion. With the high temperature, and low moisture content of the pellet, low emissions are possible. Other advantages to pellet stoves include convenient handling of fuel and leaving the stove for up to 24 hours on one load of fuel. In many cases the pellet stove may be vented through the wall without using a conventional chimney. Its disadvantages include more costly and complicated stoves, greater costs for pellets than wood, electricity required to operate, noisy fans and a flame that is not as natural looking as in a wood stove.

Pellet technology improves combustion efficiency by controlling the fuel supply. Pellets are dried ground wood or other biomass waste, such as corn or straw, compressed into small cylinders about the size of a paper clip. The pressure and heat created during their production binds the pellets together and lowers the water content to 5%. The low water content of pellets and electric control of combustion air supply allow pellets to be burned at very high temperatures for complete combustion.

Conventional Fireplaces

Traditional fireplaces are designed for fire viewing and as such are very inefficient. They generally send more energy up the chimney than they supply to the house. They should not be considered for new homes. However, there are three options to improve the efficiency of an existing fireplace. The Model National Energy Code for Houses requires that all fireplaces be fit with tightly closing doors. First, a glass door will reduce the air leakage when the fireplace is not in use, although this will not improve combustion. Second, a firebox liner or tubular grate could be installed. However, this produces only minor combustion improvements and could be dangerous if used continuously. The third and most efficient option is to install a fireplace insert.

Fireplace Inserts

Fireplace inserts are wood stoves designed to fit within masonry fireplaces. The stove firebox is surrounded by an outer convection shell. Heated air has a direct path to the room where the heat is required, rather than being trapped in the fireplace. Present day inserts require a stainless steel liner from the wood stove to the top of the chimney for safety and efficiency. A properly installed fireplace insert is now almost as efficient as a free standing wood stove and will meet emission standards.

Central Heating

Central heating with wood-fired furnaces and boilers is no longer very common. As our houses have become more energy efficient, they are easier to heat with stoves and fireplaces which also offer the benefits of fire watching. Furnace and boiler combustion technology has not kept pace with that of wood stoves. However, central heating may still be an option if the house is old, large or not very energy efficient, there is no place to install a fireplace or wood stove, or large amounts of low-cost wood are available. Wood furnaces and boilers can be installed to work either together or as an add-on unit with other energy systems using oil, natural gas or electricity. Combined wood/oil or wood/electric furnaces come in a single packaged unit. Add-on furnaces and boilers would be installed beside existing furnaces and boilers. All such units must be safety tested and certified.

Solar Air Heating

A unique Canadian product, the Conserval Solarwall® solar air heater, was developed for commercial application by Conserval Engineering Inc. in the early 1990s, and is now sold across North America, Japan and Europe. It consists of a dark-coloured perforated aluminum sheet mounted on a south-facing wall which allows the sun to pre-heat air used for ventilation. Air picks up solar heat from the dark aluminum and is drawn through many tiny holes in the perforated sheet by a fan inside the building. A Solarwall® saves up to 30% of the heating needs of the Ford assembly plant in Oakville, Ontario. The Solarwall has been used on apartment buildings, but may have limited applications in single family homes.

Pool Heating

Solar heaters are a cost-effective and the environmentally preferred method for heating swimming pools. Inexpensive unglazed collectors are suitable for seasonal pools. A solar pool heating system extends the swimming season and increases the return on the investment in the pool. When used with an insulating thermal blanket, a properly sized solar system eliminates the need for, and the consequent operating costs, of a conventional heater. Year-round pools may require glazed collectors. Because non-renewable fuels, such as natural gas, propane, oil and electricity, are both expensive and environmentally damaging, solar pool heating is a very attractive option.

Heating Swimming Pools with Solar Energy: Questions and Answers

With the rising costs of energy, more swimming pools are being heated with solar energy. The following information provides answers to the most frequently asked questions concerning solar pool heating in Canada. Much of this information is "rule of thumb"; each situation when analyzed in detail may differ slightly.

Q: How does a solar pool heater work?

Solar pool heaters use simple effective solar collectors to heat the pool. They require no separate storage tank, since the pool itself will store the energy. In most cases the existing filtration pump is used to circulate the pool water through the solar panels. In some situations, a separate pump may be required to help pump the pool water up to the collectors.

When there is adequate sunshine available, the filtered pool water is circulated through the solar collectors, where it is heated by the sun and returned to the pool. Mechanically, the water travels from the pool through the pump, from the pump to the filter, then through the solar collectors. From the collectors, the water flows to the back-up heater (if there is one), to the automatic chlorinator (if there is one), and then back to the pool. When the sun is not available, the solar collectors are by-passed.

Automatic controllers are often used in conjunction with a motorized valve to direct the pool water to the collectors whenever solar heat is available. The system can be controlled manually with the use of a diverter valve. Normally, systems are designed to drain down for ease of winterizing in the fall.

Q: What types of solar collectors are used?

The primary collector types used in Canada are black plastic panels and rubber mats. These collectors have headers on either end — either 1 1/2" or 2" diameter — linked by a series of smaller water passages. Collectors are available in a variety of sizes, with 4'x 8', 4'x10' and 4'x12' being the most common. Glazed collectors are commonly used for domestic hot water systems and are not usually used for pool heating.

Q: How do the different types of collectors perform?

The performance of a solar collector depends on its materials and configuration. Plastic panels generally capture more energy per square foot than rubber mats. The thermal performance of all collectors is tested by the Florida Solar Energy Centre. A report comparing the performance of all manufacturers is updated and published annually.

Q: How much does a solar pool heater cost?

The cost of a residential solar pool heater varies dramatically depending on the size of the pool. Most systems average between $3,000 and $4,000. A solar system for small above ground pools may be had for less than $1,000, while systems for large pools can cost over $5,000.

Q: How long will solar pool heaters last?

Solar collectors incorporate ultraviolet inhibitors that protect the material from breaking down in the sun. Most collectors should last at least 10 to 15 years, with some lasting more than 20 years.

Q: What is the proper orientation and tilt of the collector array?

Ideally, solar collectors should face due south, though an orientation of up to 45° to the east or west of due south will not affect performance dramatically. The optimum collector tilt for summer heating of pools is between 15° to 45°. Collectors are mounted directly on the slope of a roof. Any reduction in optimising orientation and tilt must be accommodated for by an increase in collector area.

Q: Can I install a solar pool heater myself?

Only a person who is comfortable with plumbing and electrical wiring should consider installing a solar pool heater. The first time around, it could take two people at least two days to install a system that an experienced crew could do in about six hours. Hard work on a hot roof can save the owner/installer several hundred dollars over the cost of a contractor installed system.

If you plan to install a system yourself, the following suggestions may save the owner/installer time and money:

- Prepare a schematic drawing of the system
- Make sure the collectors will fill and drain properly
- Determine there is enough room on the roof for your system
- Pay careful attention to the flow requirements of the collectors, the system pressure drop, and the capacity of the pump.

Q: How many collectors do I need for my pool?

A pool loses virtually all of its heat from its surface through evaporation. Because the heat loss is a factor of surface area, the solar heater can be based on pool area. Most pools need a solar system that is 50% to 75% of the pool area. The exact ratio will depend on the roof orientation and tilt, amount of shading on the pool surface, and preferred temperature of the pool.

For example, a 16′ x 32′ in-ground pool with an area of 512 square feet would typically need between 256 and 384 square feet of solar collectors to adequately heat it. This could be obtained by using six to eight 4′ x 12′ collectors.

Q: How much maintenance does a solar heater need?

A properly installed solar heater requires virtually no maintenance aside from fall winterizing and spring start-up. Regular cleaning of the filter (which is required for normal pool maintenance) will ensure adequate flow through the collectors. Check the system for proper operation annually, particularly if it uses an automatic controller. A properly operating solar heater feels uniformly cool to the touch and controls the temperature rise of the pool water slowly and steadily, about 1°C to 3°C.

Q: Do I need a pool cover?

Using a pool cover is highly recommended, regardless of how the pool is heated. The greatest heat loss is due to evaporation from the surface. Heating a pool without a cover is like heating a house without a roof on it. The other benefits of using a cover include reduced chemical costs and cleaning maintenance.

Quick Solar Pool Heating Facts

- There are approximately 600,000 pools in Canada.
- A pool can use as much energy to heat in summer as the house does in winter.
- Pool heating is the most cost effective solar application for the home.
- Over 250,000 pools in North America use solar heating.
- A solar pool heater can displace 5 to 10 tonnes of greenhouse gases annually.

Energy Efficient Appliances

The large energy-using appliances (refrigerators, freezers, clothes washers and dryers, and dishwashers) and other major household electrical appliances sold in Canada must be EnerGuide labeled. The EnerGuide label compares the annual energy consumption of the appliance to others in its category. The EnerGuide directory is published annually and provides ratings for most large appliances sold in Canada. Some of the super-efficient products imported for off-grid homes do not appear in the EnerGuide directory. Free copies are available from:

Energy Publications
c/o Canada Communications Group
Publishing Centre
Ottawa, Ontario K1A 0S9
Tel: (819) 956-4800;
Fax: (819) 956-1498.

Refrigerators and Freezers

Refrigerators use more energy than other appliances, however, manufacturers are improving energy performance. For example, a typical refrigerator sold in January 1994 used 720 kWh/year compared to 1560 kWh/year for the same sized model sold in January 1990. This is a reduction in energy use of over 50% in four years. The annual savings in operating costs at $0.08 per kilowatt-hour are about $67. The most energy-efficient refrigerators, suitable for off-grid applications, use less than 400 kilowatt-hours per year. Propane refrigerators are an alternative for off-grid homes.

The energy performance of a refrigerator can be improved by adding insulation to the outside, removing any frost build-up, setting the temperature no lower than 3°C, locating the appliance in a cool space not beside ovens and stoves, vacuuming the cooling coils at least once a year, and replacing faulty refrigerator door seals.

Root cellars and cold pantries are environmentally preferred, inexpensive alternatives to running two refrigerators or buying an oversized one. Root cellars and cold pantries are insulated storage spaces, vented to the outside, and equipped with storage bins, shelves or racks. They are suitable for storing fruit, vegetables, preserves and wine. An old chest freezer (compressor removed) can be converted into a small root cellar by adding a flexible duct with damper and venting it outside. A north-vented cold pantry should be considered for new homes or kitchen renovations.

Chest freezers are by far the best choice for energy-efficiency among freezers. Large freezers use more energy than small ones. EnerGuide suggests allowing 130 litres (4.5 cu ft) per person when selecting size. However, families who buy a side of beef or who grow their own vegetables may need more space. As of January 1996, all new refrigerators and freezers use HFCs as the refrigerant. These are less damaging to the ozone layer than the CFCs used in older appliances.

Clothes Washers

North American vertical axis clothes washers are not particularly efficient in their use of energy or water, although models offering a choice of temperature settings and water levels can be operated relatively efficiently by selecting cold water and low water levels for small loads. Sudsaver models save water, energy and detergent. European front-loading horizontal axis washers have a reputation for superior washing performance and for water and energy conservation. Better spin drying reduces energy if a clothes dryer is used. Disadvantages of European front loading machines, include the higher purchase price, somewhat smaller capacity, and slower wash time.

A 1992 study by Ortech for Ontario Hydro compared the normal cycle of a mid-range North American washer with a mid-range European washer and found that the European washer provided 75% savings in energy, 40% savings in water, 66% savings in detergent, and superior cleaning power. The energy, detergent, and water-saving factors combine to make the total cost of a European washer (purchase price plus operating costs) comparable to a standard North American appliance. U.S. manufacturers are now making European-style horizontal axis washers, and they are available in Canada.

Fluorescent Lighting

Fluorescent lights provide about 3.75 times as much light per watt as incandescent lights and last at least 10 times longer. Improved technology in the form of electronic ballasts and compact fluorescent lamps that approximate incandescent light quality have made fluorescent lighting more suitable for residential use.

Electronic ballasts are now widely available and are quieter and more energy-efficient than older electromagnetic ballasts. Electronic ballasts eliminate the flicker which occurs 120 times per second with electromagnetic ballasts. About one third of the population appears to be sensitive to this flicker, and the associated electronic static emitted from the electromagnetic ballasts has been linked to learning problems, discomfort and irritability.

Fluorescent lamps have a higher initial cost than incandescent lamps, but these costs are more than offset by lower operating costs. With the increased use and availability of compact florescent lamps, the price increment compared to conventional incandescent lamps is constantly decreasing.

In any room where artificial light is used for extended periods — typically kitchens, living rooms, family rooms, home offices and workshops — fluorescent lights are an appropriate choice. Low output compact fluorescents provide more light than night lights of similar wattages, for greater safety in bathrooms and hallways. Tube lighting can be used effectively in kitchens and living areas, workshops and garages. The new, slimmer T8 fluorescent tubes use less energy (32 watts nominal versus 40 watts nominal for a 4 ft tube). The light quality of the T8 lamps is generally closer to incandescent lights than the T12 tubes. Electronic ballasts should be specified. Compact fluorescents are suitable for smaller rooms, task lighting and anywhere it is inconvenient to change a bulb. Some models can be used in enclosed outdoor fixtures, provided they are enclosed.

Toilets

Six-litre toilets are becoming widely available as many municipal and provincial building codes begin to require their use. The standard toilet on the market today is the so-called "water-saver" or 13-

litre type; however, 20-litre toilets are still available. The tank capacity should be clearly marked on the toilet and should be checked before purchase. The new toilets have a label stating the water consumption that is fired into the unit during production.

There are three categories of truly water-conserving toilets available: the 6-litre or 'ultra lo-flow,' the 'micro-flow' which use less than a litre, and composting units, which are waterless or micro-flo composting toilets. There are two types of 6-litre toilets; the gravity type and the pressure-assisted. The February 1995 issue of *Consumer Reports* found that the pressure-assisted generally performed better than the gravity types, but were more expensive and had a louder flush. Micro-flow toilets include the cottage or valve type available at most building suppliers.

Many Canadian communities are currently facing capacity problems with their water and sewage treatment systems. Upgrading or replacing existing treatment plants is a very expensive proposition. Replacing household toilets with 6-litre models may achieve the same results at less cost. In addition, the resulting water conservation will reduce electrical energy consumption for pumping and chemical use for water and sewage treatment. Also, municipalities which use activated sludge as a sewage-treatment process would operate more efficiently with less water flowing into the treatment plant.

Composting toilets are appropriate in many cottage or rural situations and, contrary to popular opinion, do not smell if properly installed and managed. They require no hook up to a sanitary drain or septic tank, although units with a micro-flow toilet usually require a cesspit. Also, a grey water disposal system will be required to handle waste water from sinks and showers. Some units require power to the composter for waste processing, while all require low-wattage vent fans to maintain the unit under negative pressure to prevent odour problems. A composting toilet is less expensive to install than a conventional septic tank and leach field, and is environmentally preferred because it reduces or eliminates risk of fecal and nitrate contamination of lakes and wells. The composting process does, however, produce greenhouse gases. The small quantity of compost produced is disposed of periodically on flower beds.

Because many health departments or municipal bylaws still require installation of a standard septic system—even with a composting toilet—it is wise to check with the relevant authorities when planning for residential sewage disposal.

CASE STUDY:

Alberta Sustainable House
Calgary, Alberta

THIS THREE-BEDROOM, 169-square metre (1,820 square foot) sustainable, autonomous and energy-credit urban home and office has no boiler, furnace or natural gas line and demonstrates the feasibility of zero-energy housing in very cold climates. Design considerations were occupant health, environmental foresight, renewable energy, resource conservation, recycling, low embodied energy, self-sufficiency and appropriate technology. The technologies include R-17 glazing, a multi-purpose masonry heater-fireplace (baking, space and hot water heating, future electrical generation), a solar greenhouse to produce vegetables, fish, oxygen, humidity and clean-air plants, non-additive concrete, shallow footings (rubble trench); "eco-studs", non-toxic drywall mud, waterless and ultra-low flush toilets, a solar collector, grey-water heat recovery and grey-water treatment, radiation shielded full-spectrum lighting, two "Sun Pipes", and outdoor solar cookers. The home has been occupied since April 1994. Work on hydrogen fuel cells, Stirling co-generators, ground-coupled high-performance heat pumps, Tesla blade-less steam turbine, in-house sewage treatment and other technologies for energy independence and a conserver lifestyle are being pursued as time and funds allow.

Energy target: 1 Wh/DD/m3 for all non-grid back-up energy. Actual electrical consumption over 16 months has been 5 kWh/day, or about 10 kilowatt hours per square metre per year.

Reduced Greenhouse Gas Emissions: It is calculated that about 17,100 kilograms of emissions are avoided.

Energy Conservation Features

- Insulation:
  Walls: R-50 "Can-Cell" cellulose
  Ceilings: R-74 cellulose
  Windows: R-3 to R-17 (centre of glass) "Wilmar" windows R-3.5 quilted window shades
  Foundations: R-10 to R-20 "Terra Foam" expanded polystyrene
  Trench: R-30 "Granulite"

- Air Tightness: The "Air-Tight Drywall Approach" to air-sealing was used.
- Air Lock: Two-door mud rooms at front and rear entries prevent heat loss and maintain comfort when people are coming and going
- Food Storage: The "Vestfrost" refrigerator uses 320 kilowatt-hours per year. A ground-level root cellar stores fruit and vegetables.

Renewable Energy Features

- **Passive Solar.** Direct gain, passive solar space heating through east and south windows provides about 60% of all space heating requirements. Breadbox solar water preheater (reused pipes/tanks with two kinds of selective surface).
- **Active Solar.** Vertically mounted through-the-wall hot air collector on the south wall which can act in thermosiphon mode if needed. Seasonally adjustable thermosiphoning evacuated tube, "Thermomax" solar hot water collector for domestic hot water and space heating.
- **Solar Preheat.** The entire south elevation acts as a passive solar preheater for the HRV air intake.
- **Biomass.** A wood-burning "Tempcast" masonry heater provides domestic hot water, future DC electricity (small steam turbine), baking/cooking, space heating (radiant floor and wall), passive solar direct gain heat storage in winter and thermal flywheel for daytime cooling in summer.
- **Water Conservation.** Water consumption has been an average of 2,000 per month.
- **Composting Toilet.** The "Phoenix" waterless toilet saves about 200,000 litres of water per year for a family of four. Other water conservation features include a 0.6-litre micro-flow toilet, low-flow shower heads and faucet aerators.
- **Rainwater Collection.** The rainwater collection system includes the roof, eaves trough, down spout, a strainer, underground piping, a 3,400-litre (750 gallon) settlement tank and a 14,550-litre (3,200 gallon) storage cistern.

- **Other Features**
  - Photovoltaics. A total peak power of 1.7 kW will be installed. A PV/wind system was used to power tools during the construction phase.
  - Straw bale and rammed-earth construction in a separate building in the backyard.
  - Geothermal. Ground-coupled piping has been installed under the backyard cistern 3 metres (10 feet) below grade (average temperature 5°C) for a future high-efficiency heat-pump (estimated COP 5.5).
  - Stirling heat engine. A small demonstration is currently working off ice cubes or passive solar heat. Work is under way on low-tech Stirling co-generators built entirely from salvaged materials.
  - Solar hydrogen fuel cell. The small demonstration unit will be upgraded to an integrated working co-generator of electricity and hot water.

Contact: Jorg or Helen Ostrowski or Orian Low
A.S.H. Inc.,
9211 Scurfield Dr. N.W.,
Calgary, Alberta T3L 1V9.
Tel: (403) 239-1882
Fax: (403) 547-2671
E-mail: jdo@ucalgary.ca

CASE STUDY:

The Great Canadian Reno-Demo Project

North Vancouver, British Columbia

THIS RENOVATION on a turn-of-the-century house in North Vancouver demonstrates recycling of housing stock, energy efficiency and environmentally healthy construction. The demonstration includes the conversion of a grade-level basement into a suite to increase the availability of affordable housing and accommodate an increasing population while maintaining neighbourhood character.

Energy Performance

Annual space heating energy consumption reduced by 60%.

Renewable Energy

- Passive solar design: Most windows in living area face south; exterior overhang provides summer shading, insulated concrete floor slab and ceramic tiles provide thermal mass. Passive solar contribution is about 25%.
- Solar water heater for preheating domestic hot water. Incoming city water passes through the roof-mounted batch type collector with a 158-litre storage tank.

Energy Conservation

- Foundations: Lite Form polystyrene insulation form work provided R-16.
- Existing exterior walls were 2″ x 4″ which had little insulation. R-12 batt insulation plus 1.5″ DOW Cladmate exterior polystyrene sheathing was added.
- The addition walls are R-28 stressed skin panels by ISO Building Systems for resource efficiency and good thermal performance without thermal bridging.
- High-performance windows by All Weather Windows were installed with an operable windows ER value of -8.7 and an average overall R value of three.
- A high-efficiency direct vent natural gas-fired water heater replaced the old furnace and hot water tank and provides both hot water and space heating for the whole house. A fan coil blower unit uses heat from the hot water tank to provide warm air to the upper part of the house. In-floor radiant heating in the basement suite uses hot water directly from the tank.
- Heat Recovery Ventilator provides continuous fresh air.
- Compact fluorescent and low-voltage halogen light fixtures were installed.

Other environmental features

- Low-emission materials and finishes were used to reduce pollutant load inside.
- Excavated materials were re-used on site.
- Brick garden walls used bricks from the original chimney.
- Prototype roof membrane on deck were made of vegetable oil-based, environmentally benign material.

Contact: **Richard Kadulski, Architect**
204-1037 West Broadway,
Vancouver, British Columbia V6H 1E3
Canada
Tel: (604) 689-1841,
Fax: (604) 689-1841,
E-mail: kadulski@direct.ca

CMHC's Healthy House In Toronto: No Connections To Municipal Water Or Sewers

Contributed by Canada Mortgage and Housing Corporation, Ottawa, Ontario

PICTURE THIS: a three-bedroom family home built on a small downtown lot in Metro Toronto, independent of city water, sewer, and natural gas. Sound like a hardship? Think again. Because these are just a few of the many features that make this home a clean, comfortable, environmentally responsible haven in the heart of the city.

In 1991, architect Martin Liefhebber's design for a self-sufficient house was one of two winners of Canada Mortgage and Housing Corporation's (CMHC) Healthy Housing Design Competition. The winning design met CMHC's criteria that new housing technologies and designs be workable, commercially available, and affordable for the occupants and the community while reducing impact on the planet.

Although many Canadian homes have energy and water conservation features, Liefhebber's design integrates the additional healthy housing principles of good indoor air quality, affordability and environmental responsibility. The four-storey, semi-detached house is located in an established neighbourhood in downtown Toronto. Because the house requires no water, sewer or gas connections, it was possible to build on a lot that normally would be too expensive to develop.

Some of the features of CMHC's Healthy House in Toronto include:

- Water consumption in this house is reduced to one-tenth of that in a typical household; 720 litres of water are used in the house each day, 600 litres of which are recycled.
- Rain and snow are the only sources of water for the house.
- The dishwasher and washing machine use 50% less water than standard appliances.
- The toilets use six litres of water per flush.
- Energy efficiency is achieved through the appliances selected, airtight wall and roof construction, thermally efficient windows, high levels of insulation in the building envelope, and the optimum use of solar energy and thermal

mass to maintain a stable temperature. The corrugated metal decking in the suspended floor/ceiling system holds 45 metric tonnes (50 tons) of concrete, which provides adequate thermal mass for the passive solar system.

- The thermal mass of the house absorbs heat during the day and releases it during the cooler hours of the night, thereby reducing daily swings in indoor temperature.
- Radiant hot water floor heating is used when there is not enough sunlight to provide passive heating.
- The house is designed for the installation of a back-up co-generator to be used when there is insufficient sunlight for more than four days. The expected cost to operate such a co-generator is $80 a year.
- The refrigerator uses 40% less energy than a typical energy efficient refrigerator.
- The heat recovery ventilator exhausts stale air to the outside to maintain good indoor air quality.
- Materials with reduced or no emissions, such as veneer plaster, were used to reduce air pollution at the source.
- Many recycled materials were used, including recycled glass in the flooring, slag from steel production in the mineral wool insulation, waste wood fibres in the concrete wall forms, and cellulose in the fibre-cement siding.

"This house may appear revolutionary, but it is actually based on ideas that are not new or high-tech," says Chris Ives, project manager for CMHC. "The result is not a demonstration house of tomorrow, but rather a demonstration of housing choices do-able and affordable today. These housing choices can apply to both new and renovated homes, and even to rental accommodation. Despite the home's high-tech appearance, most of the products and systems are actually simple and straightforward. Off-grid houses do not necessarily require hours of labour for upkeep. In fact, everything in the house is easy to maintain and available in today's marketplace. All of the components in the house have proven track records, but some have just never been used together in one house before as an integrated system."

The inner city lot that was chosen, however, posed a challenge for various reasons. The 161 m2 house sits on a 6.86 x 21.84 m lot with a 6.1 m cliff in the backyard. The vertical design of the house meant that it would be more visible than the garages typically built on the back lane where the lot was situated. Fortunately for the project, neighbours in the Riverdale area where the house is situated — many of whom are environmentalists — were very supportive of the project. In fact, they actually submitted a petition to the city supporting it.

When the project began, the Ontario Ministry of Environment and Energy, the City of Toronto Public Health and Ontario Hydro were concerned about a house that would treat all of its own sewage on site and would use photovoltaic systems for electricity. Generally, building code requirements dictate that a house in an urban centre such as Toronto must be hooked up to municipal services. Within the approval agencies, champions for the project came forward. These visionary people saw that this house could demonstrate how to minimize the environmental impact of housing and yet make homes more affordable to build and operate. It is expected that the total operating costs for one year for this house in Toronto will be between $200 and $300. These costs are primarily for back-up heating, and maintenance of the biological filters and ozonation equipment in the water treatment system.

The Water Treatment Systems

Canadians use a lot of water. In fact, water has the potential to be the most expensive utility we have. The average house is built anticipating consumption of 1600 litres per day. This is a heavy

Rainwater is stored in a 20,000L (4,400 gal.) underground cistern in the backyard. This is an adequate supply for six months based on Toronto's average annual precipitation.

volume for chemical treatment. The challenge in the approval process for this house was the waste water treatment system and the use of reclaimed water for bathing and showers. The design team looked for a system that would be easy to maintain, provide safeguards should any problems develop, and fit in a small space.

Finding an effective waste water treatment system that would meet environmental and health concerns was not the only problem — there are several excellent systems on the market. The problem was space. Only 11 m2 was available for disposal of waste water. Because the house would depend on rainfall and snow for its water supply, it was possible there would not be enough fresh water available. A conventional Class 6 septic system with an absorption trench leaching system on soil and using conservation measures (low flow fixtures and appliances) required 76.8 m2 plus distribution area for disposal. Adding a sand filter leaching system reduced the requirement to 16.9 m2. The final solution included all of the preceding measures plus a water reclaiming system, which eliminated the disposal problem and meant the house could be totally independent from municipal water and sewer services. By recycling 600 litres of the 720 litres of water available for consumption each day, disposal water is equal to treated potable water of only 120 litres per day and the disposal area required is 1.7 m2.

The system selected was developed by Dr. Craig Jowett at the University of Waterloo and is called the Waterloo Biofilter system. It is an efficient waste water treatment system able to digest biological loading effectively in a small basement space — 7 m2 in this case. No chemicals are involved in this system. Primary treatment in a septic system ferments and biodegrades large organic molecules into smaller dissolved organics which are more readily treated in the aerobic Waterloo Biofilter. A recirculation tank downstream from the septic system achieves denitrification in an aerobic environment. The secondary treatment in the Biofilter is an aerobic biological process which removes organic matter and effects denitrification. The Biofilter effluent passes through the recirculation tank several times.

Advanced treatment and disinfection of the waste water uses a staged approach typical of drinking water treatment technology (note that the potable water treatment and the waste water treatment streams are completely separate). The multi-stage filtration unit does not rely on the use of chemicals. A roughing filter of gravel layers, graded from coarse at the bottom to fine at the top, removes a large percentage of suspended solids by filtration throughout the depth of the bed. Then the effluent passes through a slow sand filter. This consists of a fine sand bed that develops a thin, biologically active layer at the sand surface, effective in absorbing particulates, bacteria and cysts. A significant percentage of dissolved organic matter is also consumed. Finally, the water passes through an activated carbon contactor to absorb colour, taste and odour. The unique aspect of the filter is that all three elements can be incorporated into a single tank.

The slow sand filter is cleaned by automatic back flushing about once a month and the carbon needs to be replaced about once a year. If the sand filter is not cleaned when required, it will simply become plugged-up and gradually produce less and less effluent. Water quality is not affected. If the carbon is not replaced, only the aesthetic quality will be affected. This serves as a reminder to the occupants (when using reclaimed water) that attention is required.

As a final barrier to pathogens, the effluent from the filter is passed twice through an ozone disinfection unit. This ozonation unit is equipped with a sensor that will automatically shut down the supply if there is insufficient disinfection due to a malfunction or fouled equipment. Treated water is discharged to a 1200 litre storage tank and is then pumped into the reclaimed water system.

Rolf Paloheimo of Creative Communities Research Inc., the builder and owner of the house, says that it was vital that the system produce safe and aesthetic water reliably. "While waste water technology is dedicated to removing waste from water, drinking water technology is dedicated to ensuring that water is safe for human consumption. For me, the purpose of the polishing of the water in the reclamation stage is to ensure that the finished water is always acceptable."

Not only is Paloheimo's company selling the waste water treatment system developed for CMHC's Healthy House in Toronto, but he and his family have been living in the house since November 1996. He therefore has a personal interest as well as a business interest in the quality of water being produced. "As a result of this project, I, as a builder, have become aware of the need for alternatives to existing on-site waste water systems and also to large scale collection, treatment and disposal as currently practiced by our cities and towns. I have great confidence in this system's ability to deliver safe and aesthetic water."

The waste water treatment system was a prototype. Mr. Paloheimo anticipates that residential application of the system would cost about $10,000, or even less if demand warranted quantity production of the system. Ten such systems have recently been installed in Yellowknife, Northwest Territories, for houses normally serviced by truck haulage of water and sewage.

The source of potable water is from rainfall and snow. The water purification system mimics the natural path that rain follows when it passes through the ground to a spring. Rainwater from the roof of the house is collected in a 20,000 litre concrete cistern. Its size is based on Toronto's average annual precipitation and, when full, should supply the household's potable needs for five months if there is no rainfall. In the unlikely event of a shortfall, levels can be supplemented from outside sources. The cistern contains a layer of limestone to neutralize acid rain, a problem prevalent in the Toronto area. The water passes through a sand filter, a charcoal filter to remove odour, organics and low levels of hydrogen sulfide, and finally ozone treatment.

The potable and reclaimed water systems are independent of each other to ensure purified rainwater is being consumed while reclaimed water is only used for secondary purposes. The potable water supplies all the sinks and the dishwasher. The reclaimed water supplies the tubs, showers, toilets and washing machine. Water is reclaimed and recycled three to five times. The purified surplus water — 120 litres — is discarded daily and used to water the front garden.

The Electrical System

Generating electricity using photovoltaic (PV) panels is not new. Solar energy has become a more appropriate and acceptable option for electricity because of improvements in materials and equipment technology and the greatly reduced requirements of energy efficient housing. To determine how many panels are required, the amount of power needed for the equipment and appliances in a home must first be calculated. The more power needed, the more PV panels and batteries are required, and the greater the expense. CMHC's Healthy House in Toronto has eight PV panels mounted on a metal structure attached to the roof. The panels act as an awning for the south facing windows on the fourth floor of the house to ensure that the rooms are not too bright or too hot. The PV panels generate up to 2.3 kW of power per hour at peak performance on sunny days. The electricity is stored in batteries in the basement utility room. A 4 kW-inverter changes direct current from the 48 volt battery into household alternating current at 120 volts.

It is expected that on the worst grey day of winter, the house will consume 9 to 12 kWh of electric power over a 24-hour period. On a typical day with five to eight hours of sunlight, it is possible to generate a total of 10 to 15 kW hours. If there is no sun at all, the batteries can provide about four days of power before they need recharging. Toronto has many grey days between November and April. However, there is usually enough sunlight to provide sufficient power to meet 75% of the household needs. Overall, this house will use about 3,600 kW hours in one year, 900 of which will be provided from back-up sources, such as an auxiliary co-generator that burns a standard fuel like gasohol.

The floors in the house, which absorb the sun's rays, are a key component of the heating system. The thermal mass provided by 45 tonnes of concrete poured into corrugated metal decking in the suspended floor/ceiling system is essential to the passive solar system performance. When solar heating is inadequate, radiant floor and ceiling heating provide back-up space heat. Loops of metal-reinforced plastic pipe are evenly embedded in the concrete floors and water is circulated through these pipes. The water in the pipes is a closed system, independent of the house potable water system. The concrete floor is poured in a light corrugated steel pan with reinforcing bars. The steel pan surface becomes the ceiling of the room below. The heat radiates into the rooms from the ceiling and the floor, providing even temperature without drafts and without raising dust.

Reducing the electrical demands was a prime consideration in choosing equipment such as bathroom fans, refrigerator, freezer, stove, washing machine, dishwasher, office equipment, circulatory pumps, fans and even the light bulbs. However, there is one appliance noticeably missing. The clothes dryer found in a traditional house has been replaced with a drying closet. The heat recovery ventilator supplies warm air to dry the clothes hanging in the drying closet and exhausts the moist air to the outside. Drying time is about two hours, which is longer than the usual forty-five minutes in a traditional dryer, but the cost is almost zero and the clothes last longer with such gentle treatment. The washing machine uses 50% less water and half the electricity of a standard machine, and also extracts more water in the spin cycle which helps reduce drying time.

An energy efficient refrigerator would still consume one-quarter of the PV panel output on a sunny winter day. The compressors and condensers of the custom-built refrigerator in this house are positioned outside the envelope of the house to improve efficiency in winter, reduce heat gain in summer and help decrease noise levels inside the house. Along with increased insulation, these modifications achieve a 40% reduction in energy requirements for this appliance.

The floor absorbs excess heat during the day and releases heat during the night when it is needed. A 2,200-litre thermal store retains the surplus heat energy. Thermostatic controls trigger the radiant floor heating system to supply heat to, or withdraw from, the thermal store during the natural daily cycle of the house. If solar heat is not sufficient, thermostat controls trigger circulation of warm water from the thermal store through the floors and ceiling.

Many features of the house provide more than one benefit. For instance, the large, south-facing windows provide passive solar heating and enough light for daily activities without the need for additional lighting. Solar panels provide the energy needed to heat water for domestic use, and heat the water in the closed-loop radiant floor heating system. The solar panels are hung on the third floor of the house and act as a shade for the windows. The storage tank for the solar panels is located over the fourth-floor bathroom, so the system can use gravity rather than electric pumps to return cold water to the panels. The position of the PV panels and the hot water solar panels allows air to flow past them to prevent overheating of the panels.

Indoor Air Quality

Researchers are finding that more people are becoming increasingly sensitive to their indoor environment. With at least 25% of the population bothered by asthma, allergies or chemical sensitivity of some kind, researchers believe that there is a link between a home's air quality and the health of the occupants. In any house, the lifestyle of the occupants and the chemicals brought into the home have an impact on indoor air quality. Even normal household cleaning products can add undesirable chemicals to the indoor environment.

"The most important Healthy Housing guideline to follow concerns occupant health by providing a clean indoor environment", says Chris Ives. "This need not be difficult but it certainly helps to have a keen sense of smell to do a "'sniff test'" on every item that is brought into the house. If a building material does not smell good to you, it probably is not good for your health. Much attention has been given to ventilation to provide good indoor air quality. CMHC's research has shown that source control must go hand-in-hand with continuous, controlled ventilation."

Through careful selection of building materials, biological and chemical contaminants have been reduced or eliminated in this house. For example, a thin finish coat of veneer plaster, which has no additives, was applied over "blue board" on all of the walls of the house. Veneer plaster is faster to install than traditional plaster, and provides a harder, abuse-resistant surface which does not need to be painted. Solid surface countertops in the bathroom and glass impregnated concrete countertops in the kitchen are benign. Water-based floor and trim finishes have a low content of volatile organic compounds (VOC). The radiant floor heating provides a comfortable, dust-free heating system.

Ventilation was also carefully considered. Air is continuously filtered to remove pollutants. The balanced ventilation system runs continuously. One fan exhausts stale air while a second one draws in outside air. The "ramped" control system, which progressively increases the fan speed, also permits lower energy consumption. A heat exchanger recovers energy from the exhaust air and preheats the incoming air.

Contact: **Christopher Ives**
Canada Mortgage and Housing Corporation
700 Montreal Road
Ottawa, Ontario
Tel: (613) 748-2312
Fax: (613) 748-2402
E-mail: cives@cmhc-schl.com

CMHC'S HEALTHY HOUSE IN TORONTO

CMHC SCHL
Helping to house Canadians

DRINKABLE WATER AND WASTE WATER MANAGEMENT

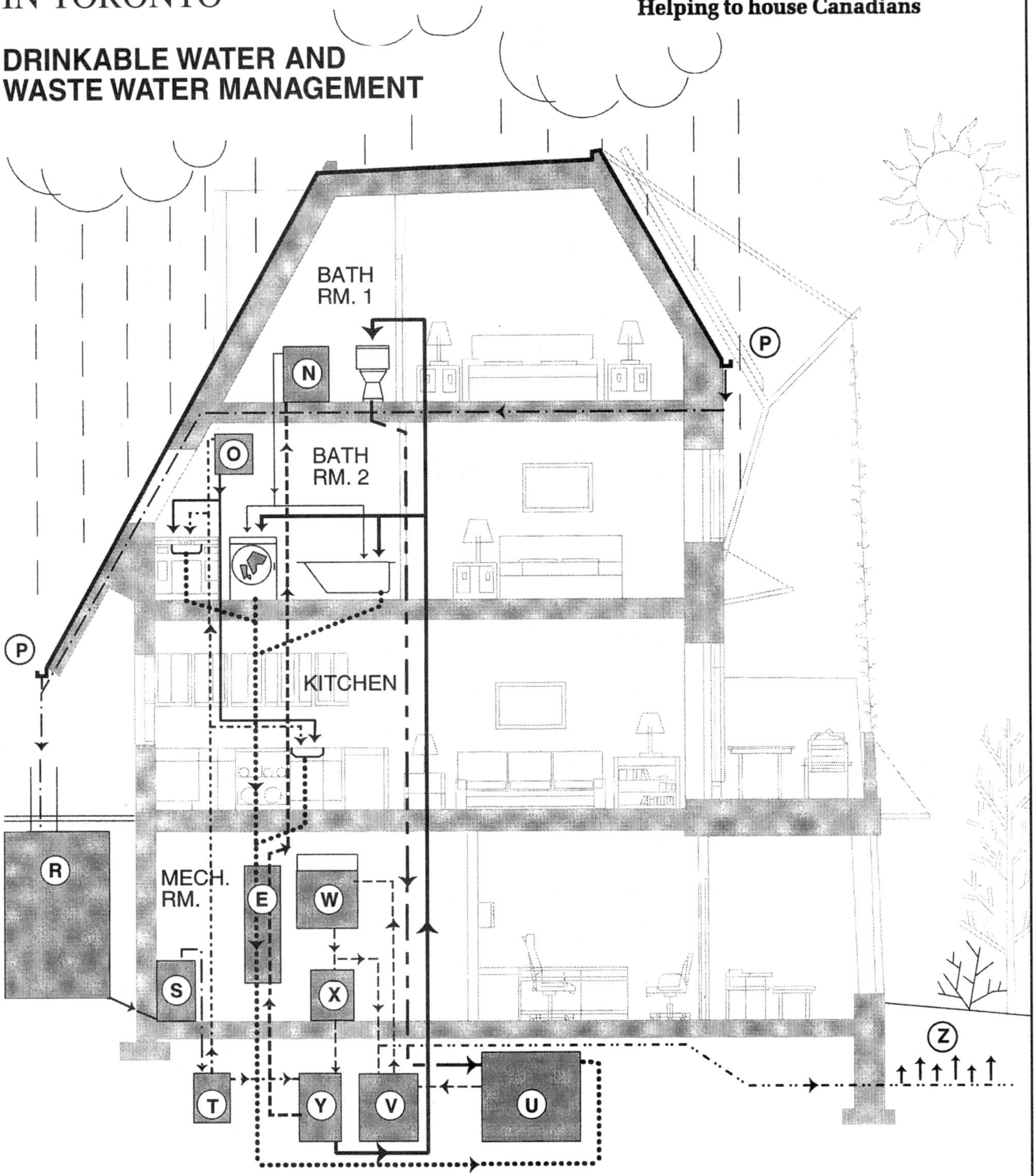

- (E) Grey water heat exchanger
- (N) Reclaimed-hot-water tank
- (O) Drinkable-hot-water tank
- (P) Eaves troughs
- (R) Rainwater cistern
- (S) Combination filter
- (T) Drinkable-cold-water tank
- (U) Septic tank
- (V) Recirculation tank
- (W) Waterloo Biofilter ™
- (X) Twin combination filters
- (Y) Reclaimed-cold-water tank
- (Z) Garden irrigation

CMHC'S HEALTHY HOUSE IN TORONTO

DRINKABLE WATER AND WASTE WATER MANAGEMENT

DRINKABLE WATER SYSTEM

(P) **Eaves troughs**
Collect roof rainwater, which passes through filter screens and then to cistern.

(R) **Rainwater cistern**
20,000 litres (normally sufficient for 6 months consumption).

(S) **Combination filter**
The rainwater passes through a combination roughing, slow sand, and carbon filter, and then through an ultra violet light disinfection unit before being stored for drinking.

(T) **Drinkable-cold-water tank** [600 L]
Supplies kitchen and bathroom sinks; overflow to reclaimed-cold-water tank.

(O) **Drinkable-hot-water tank** [140 L]
Supplies kitchen and bathroom sinks.

WASTE WATER SYSTEM

(E) **Grey water heat exchanger**

SYSTEM TO RECLAIM WATER

(U) **Septic tank** [3,600 L] (under driveway)
Anaerobic bacteria transforms waste water for treatment by the Waterloo Biofilter.™

(V) **Recirculation tank** [2,000 L] (under floor)
Provides de-nitrification in an aerobic environment.

(W) **Waterloo Biofilter ™**
Aerobic bacteria transforms the effluent to a semi-treated condition.

(X) **Twin combination filters**
Water passes through two combination roughing, slow sand and carbon filters.

RECLAIMED WATER
Water is recycled 3 to 5 times.

(Y) **Reclaimed-cold-water tank** [1,200 L]
Supplies tub, laundry, showers and toilets.

(N) **Reclaimed-hot-water tank** [450 L]
Supplies tub, showers and laundry.

(Z) **GARDEN IRRIGATION**
Site gravel pack disperses overflow water under front garden (about 120 litres per day.)

Additional Sources—Homes & the Canadian Experience

For further information on Homes and the Canadian Experience , consider reading the following books and manuals, suggested by the chapters' contributors:

Books

The Healthy House: how to buy one, how to build one, how to cure a sick one, Third edition, John Bower, The Healthy House Institute, 1997.

The Smart Kitchen: how to design a comfortable, safe, energy-efficient and environmentally-friendly workspace, David Goldbeck, Ceres Press, Woodstock, N.Y., 1989.

Green Home: Planning and Building the Environmentally Advanced House, Wayne Grady, Camden House, 1993.

Heating Systems for Your New Home, Richard Kadulski, The Drawing-Room Graphic Services

Environmental by Design, Kim LeClair & David Rousseau, H&M, Vancouver, 1992.

The Super-Insulated Retrofit Book, Brian Marshall & Robert Argue, REIC, Toronto, 1981.

Build it with Bales: a step by step guide to straw-bale construction, Version Two, Matts Myrhman and S.O. MacDonald, Out On Bale, 1997

The Natural House Book, David Pearson, Simon & Schuster, Toronto, 1989.

Government Manuals

Canadian Home Builders' Association Builders' Manual, CHBA, 1994

Tap the Sun, passive solar techniques and home designs, CMHC, 1998 comes with CD-ROM. CMHC, 700 Montreal Rd., Ottawa, Ont. K1A 9Z9 1-800-668-2642

Healthy Housing: a guide to a sustainable future, CMHC, 700 Montreal Rd., Ottawa, Ont. K1A 9Z9 1-800-668-2642

Keeping The Heat In, EMR Canada, from NRCan, Energy Publications, 580 Booth St, 18th floor Ottawa, ON K1A 0E4.

Whole House Retrofit Techniques, EMR Canada and REIC, from NRCan, Energy Publications, 580 Booth St, 18th floor, Ottawa, ON K1A 0E4.

A Guide to Residential Wood-burning Systems, available from CMHC and NRCan.

Newsletters

Solplan Review, Drawing Room Graphic Services Ltd., Box 86627, North Vancouver, B.C. V7L 4L2—an independent journal of energy conservation, building science and construction practice

Environmental Building News, E Build Inc., 28 Birge Street, Brattleboro, VT 05301—a newsletter on environmentally responsible design and construction

CD Rom

The Green Home: A New "Green" Way of Building and Operating Your Home, available from The Learning Team, 84 Business Park Drive-Suite 307, Armonk, NY, 10504, USA or call 1-800-793-TEAM

Web Sites

Hearthnet, hearth.com/hearthnet.html

Pellet stoves, www.pelletstoves.com

Masonry Heaters, mha-net.clever.net/

Wood Heat Home, www.wood-heat.com/

CHAPTER 2

Power Systems

RENEWABLE ENERGY can easily supply all the heating needs of a home, especially one built using the principles described in the previous chapter. For a home some distance from the nearest power line, electricity generated by renewable energy may also be less expensive. This chapter describes home electrical power systems.

Living off-grid requires an awareness of energy use in the home, an instinct for conservation, and a basic understanding of the system. Although there are a few inconveniences to living off-grid—for example, electric water heaters or clothes dryers are not suitable for off-grid systems—users feel good about living with renewable energy and like their independence from the utility company. Rural users often find that renewable energy is more dependable than grid power.

Energy generated by a renewable energy system that is not required for immediate use is often stored in batteries. Many customers may also want the security of a gasoline, propane, or diesel generator, but such backup is not always essential and will add noticeably to the costs of a system. The price of a 4000-watt portable gasoline generator starts at $3500. A controller is also required and will cost about $600.

In addition, there are some heavy electrical load appliances that are manufactured as non-electrical appliances. For example, propane refrigerators, which start at $1500 for a capacity of 7.5 cubic feet, may be less expensive to purchase and operate than a larger home power system and a specialty DC refrigerator or a super efficient AC refrigerator. For further information on electrical appliances, see Chapter 1.

In remote areas, the use of renewable energy for home power generation can result in substantial savings. The cost of utility power line installation varies from $10,000 up to $60,000 per kilometre, depending upon the terrain. A renewable energy system for a cottage can typically be installed for $1,000 to $10,000, while a system for a year-round home may cost from $5,000 to $30,000+.

Many Canadian suppliers offer customized, complete power packages for off-grid homes. Multiple power sources, e.g. photovoltaic (PV) panels plus a wind generator, and/or a fuel generator back up, are usually preferred for year-round homes. A complete power system includes a variety of equipment—batteries, inverter, charge controller, fuses, switch gear, etc. Small systems can be installed by the purchaser but the use of a qualified installer is recommended. However, be aware that few electricians are familiar yet with the unique features of a renewable energy system and suppliers should provide installation schematics, a user manual, component information, and warranties.

> *What is a watt?* A watt is a unit of power measurement that, for electrical systems, equals volts multiplied by amps. Between 10 to 100 watts are required to provide power for an indoor light. For small systems in off-grid applications, the lighting would be low-wattage (e.g. 10- to 20-watt compact fluorescents).

Photovoltaic systems, microhydro generators, and wind generators can all be used to meet the modest electrical needs of a zero-energy house. Space heating is often accomplished off-grid by burning wood or pellets. Solar air heating panels can help maintain above-freezing temperatures in a ski chalet while it is vacant. With extra panels added to a solar water heating system, it can be used for space heating as well as for hot water.

Properly sized, a renewable energy system can supply any quantity of electricity. Equipment is available for custom installations, or as packaged systems in various sizes, to produce, condition, and store electricity. Microhydro generators and large wind generators can produce 120 volts AC and must conform to Canadian provincial electrical codes. However, as AC cannot be stored, it may be better to produce and store DC at either 12 or 24 volts. Photovoltaic systems and small wind turbines produce 12 volts DC and must conform to the relevant provincial electrical codes if they produce more than 100 watts.

What is . . . Biomass?

The Source

Biomass energy utilises the energy content in all forms of organic matter, including agricultural crops such as wheat and other grasses, harvest residue trees, forestry residue and wood waste, the methane extracted from the decomposition of human and animal wastes, and municipal garbage. Biomass is a versatile energy source that has local applications. In forested areas, wood or wood residues can be used. In agricultural areas, crop wastes, including corn cobs and crop residues, are readily available. In urban settings, trash can provide a great deal of energy as can construction wood waste, discarded plywood and garden refuse. Forests, both natural and managed, can supply most of the biomass required in Canada, as there is always wood, bark and logging leftovers from harvesting trees.

How it Works

All forms of organic matter store energy from the sun through photosynthesis, or the transformation of carbon dioxide and water into sugar and other organic compounds. When the bonds between the carbon, oxygen and hydrogen atoms are broken, by digesting, burning or decomposing the organic matter, the stored energy in the atoms is released. Biomass can be used in various forms: from a simple wood stove that heats homes or cottages to large-scale biomass conversion and from boilers that burn whole trees to wood chips that convert the biomass into renewable substitutes for petroleum products. Garbage from cities and towns, deposited in landfills is composed of about 60 % biomass[1], and consists of paper, yard waste, kitchen waste and industrial & commercial refuse. Some municipalities are burning garbage in waste-to-energy plants. As the garbage is burned, it gives off heat, which is then used to boil water, generating steam which turns the turbines and generates electricity. The dried sludge from sewage treatment plants can also be burned for energy. Some municipalities also tap into the landfills to extract methane, which is produced in the moist darkness of a landfill. Microbes present in the landfill slowly digest all the organic matter and leave behind their own waste product—methane gas—which is then sold to businesses or burned to generate electricity.

Types of Biomass

Biomass energy is utilised in many different types of systems and applications. The most common method is through the *direct combustion* of the energy crop. By burning the crop, the boilers produce steam to use in an industrial application or to generate electricity. Direct combustion also

[1] Nancy Cole & P.J. Skerrett, Renewables are Ready: People Creating Renewable Energy Solutions. (Chelsea Green Publishing, White River Junction, VT: 1995), 192.

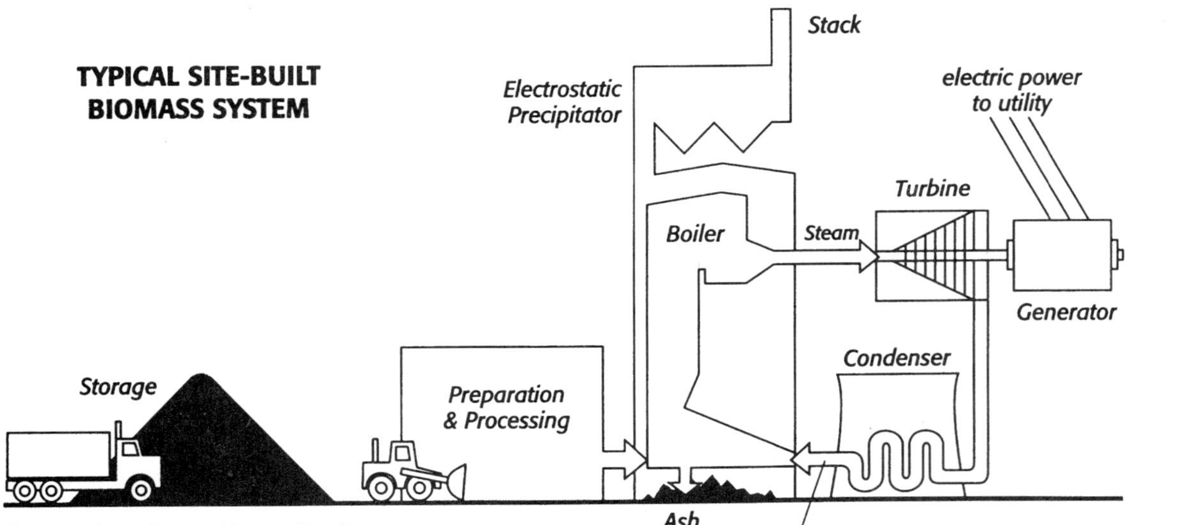

Diagram from *Renewables are Ready*

generates heat, which can then be captured and transferred to heat a building or greenhouse. It is important to design a system that will burn the fuel choice efficiently. Biomass can also be burned with coal, also known as *co-firing* in very small percentages in existing coal-fired plants and boilers.

A more recent application of biomass is *gasification*, which captures 65-70% of the energy in solid fuel by converting it into combustible gases. The gas is then burned like natural gas, generating electricity or powering vehicles. It works well in industrial applications and also can be converted into synthetic fuels.

Anaerobic digestion, another type of biomass, is best utilised where there is a great deal of human waste or animal waste, on cattle, hog and chicken farms. The process consists of fermentation, which partially removes harmful pathogens from the wastes. The resulting biogas can be used as a fuel, while the digested wastes are best used as a fertiliser, usually after months of storage in a lagoon.

Biomass seed crops containing high proportions of oils can be used directly as fuels in place of diesel, as *biodiesel*, or as a heating oil. The addition of methyl or ethyl alcohol to these natural oil fuels produces esters and the by-product glycerol. These biofuels are biodegradable and function efficiently over a long period in diesel engines.

Fuels and Biomass—Ethanol

Ethanol is produced by converting the starches of biomass feed-stocks, such as corn, potatoes, beets and sugarcane, into alcohol. Yeast and heat are used to break down the complex sugars in the feed-stocks into simple sugars. Ethanol can also be produced from other feed-stocks, such as trees, grasses and agricultural wastes, by utilising the cellulosic portion of the plants, and reducing it into a simple sugar.

The use of ethanol even at E10 (10% ethanol, 90% gasoline) levels can have significant environmental benefits. The use of E10 in cars would produce less carbon monoxide, sulphur dioxide and carbon dioxide. While E10 produces more volatile organic compounds, particulates and nitrogen oxide, these can be controlled with emission control devices such as catalytic converters.[2]

The Benefits of Biomass

The large scale use of biomass in Canada could result in job creation and the utilisation of land currently sitting idle. Jobs would be created in the agricultural sector, through the production and harvesting of the energy crops, in the industrial sector by the development and construction of conversion facilities and industrial feed-stocks and in the power industry. For example, in the United States, the Electric Power Research Institute estimated that the production of 5 quadrillion Btu (British Thermal Units) of electricity on 50 million acres of land would increase farm income by US$12 billion, providing farmers with a stable income, crop diversification and the revitalisation and strengthening of the local rural economies.[3] With a major movement towards ethanol, the United States Department of Agriculture has estimated that 17,000 jobs will be created for every million gallons of ethanol produced.[4]

The use of biomass would also provide many environmental benefits, including the mitigation of greenhouse gas emissions, the reduction of acid rain, soil erosion, water pollution, and landfill stresses. In the last 250 years, hundreds of millions of tons of greenhouse gases have been released into the atmosphere, mainly through the burning of fossil fuels. The release of such large amounts of carbon dioxide and methane, two byproducts of the burning of fossil fuels, have scientists concerned about disrupting the Earth's climate, by changing the whole biosphere of the globe.

Biomass technology captures the methane from landfills, wastewater treatment and manure lagoons and turns it into electricity. All plants, including the energy crops, absorb carbon dioxide into the plant and roots, including the carbon dioxide produced by the combustion of other energy crops.[5] Other byproducts of the burning of fossil fuels, especially coal, are sulphur and nitrogen oxides, which when released cause acid rain, affecting our lakes, our habitat and the habitat of wildlife. As biomass has no sulphur content, if it is mixed with coal, and burned a coal-fired power plant, it reduces sulphur emissions, and acid rain.

Through the planting of energy crops in marginal areas, in floodplains and between annual crops, soil can be stabilised reducing run-off into local water systems. In areas where there are large hog, cattle and chicken farms, pollution of rivers

---

2 Thomas L. Rosenberg, "Biomass Energy: America's Secret Renewable Energy Resource." Taken from the American Solar Energy Society World Wide Website, www.ases.org, 2.

3 Rosenberg, 2.

4 Rosenberg, 1.

5 Rosenberg, 2.

and streams can occur due to the flooding or leeching of large manure lagoons. Anaerobic digesters would capture the methane for energy and control some of the odour being released into the local area. Local municipalities could utilise the agricultural waste put into Canadian landfills each year, saving land space, having landfills pay for themselves, or even generating a profit.

What Is . . . Microhydro Energy?

Waterpower can be considered a solar—driven resource. Evaporation from oceans and lakes moves over higher land, condenses and rains. The land topography funnels run-off from large drainage areas into streams and the sun's energy is concentrated in the flowing water. Solar and wind are available nearly everywhere, but widely dispersed in nature. Hydro is site-specific; each stream is unique in its available head and flow, slope, seasonal variation, etc. Different types of equipment may be needed for different sites.

Microhydro is a small-scale renewable resource that can supply 100% of a site's energy requirement (including space heat) and possibly at a lower life-cycle cost per kWh than utility power. Where grid intertie is possible, hydro is even more cost-effective than other renewable energy sources because hydro generators operate efficiently at a constant speed producing clean AC electricity without the need for a power conditioner or expensive switchgear. This applies to both net metering and independent power producers.

Waterwheels (over or under shot) are not effective machines because they turn too slowly for the electrical generation desired in modern times and can freeze in cold climates.

Many microhydro projects are referred to as "run-of-river," a term which means that neither a large dam or water storage reservoir is built and the land is not flooded. The available streamflow at a given time is piped downhill, with little environmental impact. Obtaining a water license for power generation can be generally straightforward as water use is non-consumptive, water quality is not degraded and fish can be excluded with approved intake screens.

There are two basic categories of microhydro systems:

1) AC—consists of a hydro-turbine and a governor.

2) Battery-based—is necessary if less than 2kW of waterpower is available.

As with any hydroelectric system, an assessment of the loads to be powered is the first step. Power (watts) is produced in modern turbines via a gravity-feed pipeline with pressurized water ranging from a few hundred to hundreds of pounds per square inch (psi). Water-to-wire power output can be calculated using net head and flow information.

Head is the vertical distance from the surface of the intake to the turbine and is measured with a hand level, transit, altimeter, or the pressure in the existing pipe. Net head is the pressure remaining after pipe friction when water is moving.

Flow is the amount of water per unit time available to the turbine. Flow is measured with a container and a stop-watch, or the cross-section-times-average-velocity method, or by a measuring weir. Long-term and

seasonal knowledge of flow is important.
Low head systems for heads below 30 to 50 feet will need special consideration, and likely more money. The intake, pipe and turbines will be considerably more expensive due to the larger volume of water required, but the remainder of the system, generator, transmission and controls is the same as with a high head system.

System Considerations

System Sizing:

Because of the relationship between fixed and variable costs, as system (kW) size increases, it will generally be cost effective to install as much capacity as the water can provide, and heat with the excess water (hot water, house hot tub, pool, greenhouse, etc.) This recommendation has limits—there is no point in generating power without a useful load.

Pipe Sizing:

Pipe sizing is a trade-off between cost and power losses due to hydro-friction; large diameters have less pressure but cost more. For example, gravity-fed domestic water pipes are often 1.5" to 2" diameter and can deliver hundreds of watts if the necessary head and flow are available. In freezing climates, pipe damage may pose a risk, but there are several ways this can be dealt with and therefore should not be the only reason for rejecting a hydro-power option. Turbines are not at risk from freezing.

Turbines

Turbines fall into two categories. In an impulse turbine, a nozzle under pressure directs a jet of water at a rotating runner. In a reaction turbine, the runner is totally surrounded by water in the turbine housing and is turned by both the pressure (from gravity-feed) and velocity.

Impulse turbines (pelton, turgo and crossflow) are generally simpler, less expensive and have more predictable performance behaviour than reaction turbines. Impulse turbines are capable of operating efficiently over a wide range of flow, especially multi-nozzle designs. However, they require medium to high head. Certainly, the pelton turbine is best known, and dominates microhydro applications in North America.

Reaction turbines are needed at low heads, but Francis and propeller types are expensive at this scale, and therefore rarely installed. The use of industrial pumps in reverse mode is growing, as they are very cost-effective at low and medium head sites. A pump-turbine must have a constant flow available as there is no way to adjust the flow through it and a simple selection methodology is not yet available. Caution and expert advice is suggested.

Load Control

A transmission line (buried or overhead) is required if the powerhouse cannot be near the loads. Any voltage or power can be moved 1 to 2 kilometres or more. The limitations are due to costs, not physics. If necessary, voltage can be raised to reduce amperage, thereby reducing wire costs. Control of microhydro systems is accomplished by dumping excess power, usually to water or air heating loads. For a battery-based system, the controller dumps power to prevent battery overcharge, based on battery voltage. The inverter provides AC voltage and frequency control.

Meeting reasonable peak loads will require storing power during low periods of use (nighttime) to meet peak power periods. There are many more small streams that can be developed using battery-based hydro technology, where AC-only would be impractical.
Mixing these two basic systems is not practical as the specific equipment is entirely different and not complimentary. If stream power is between approximately 1 to 2 kW, the types of loads and patterns of power use will dictate the best system choice.

For an AC-only system, turbine-generator speed control is required and is based on frequency, which directly ties into RPM. An "electric-load governor" has computer-controlled electronic switches that turn auxiliary dump loads on and off as necessary to maintain power balance with the turbine's horsepower output and thereby maintain the correct speed. This is based only on frequency, therefore, the governor can be located anywhere in the system (the generator's voltage regulator controls the AC voltage to the loads).

What is . . . Photovoltaic Energy?

Photovoltaics(PV) literally means electrical power from light, usually sunlight. Today, photovoltaics are used in dozens of applications to bring safe, reliable electrical energy to sites where conventional hydro is either not available or too expensive.

There is a common misconception about solar power, where people confuse active solar power with solar electric power. Put simply, active solar energy is used to heat water, as in pool heating. The other type, solar electric energy or photovoltaics, generates electricity from sunlight. To better understand photovoltaics, it is essential to know exactly what happens in a crystalline silicon solar cell when it is exposed to a light source. This is known as the PV Effect.

The PV Effect

Crystalline silicon solar cells are made from two layers of silicon that are treated with negatively or positively charged elements, phosphorous and boron, in a process called doping. The two doped layers form a p-n junction. When light energy (photons) strikes the face of the cell, it excites the electrons within the cell. This flow of electrons (electrical current) from the negatively charged layer to the positively charged layer is what we call the PV Effect. Figure #1 illustrates this process. The electricity produced is then used to directly power a load or is stored in batteries for use later.

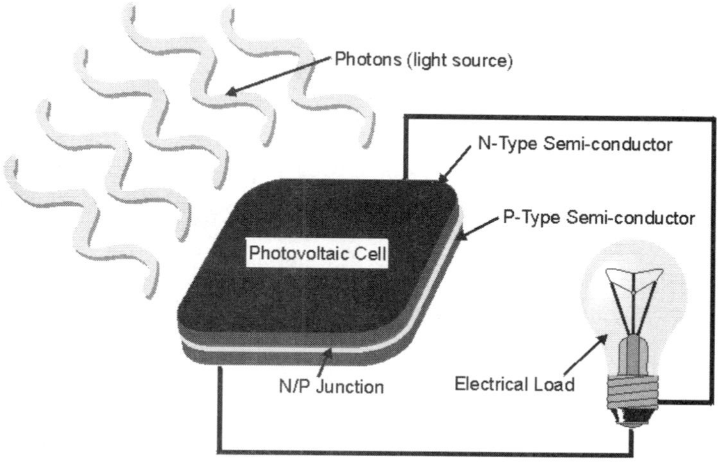

PV Technology

Because the amount of power produced by a single cell is relatively small, solar cells are interconnected to supply a more useful level of voltage, current and power. Each cell produces 0.5 volts, so several cells are connected in series to create a usable voltage. A standard crystalline silicon photovoltaic module consists of 36 interconnected in series to give approximately 17 volts. This higher voltage is necessary as 12 volt batteries need at least 15 volts to fully charge.

There are four factors to determine any photovoltaic module's output: load resistance, solar irradiance, cell temperature and efficiency of photovoltaic cells.

Solar PV Cell and Module Technologies

There are three main silicon based technologies used in the manufacture of photovoltaic cells and modules, amorphous, polycrystalline and single or mono-crystalline. Beside silicon based solar cells, there are other technologies available, including CIS, Cd-Te, In-P etc. Each type has different efficiency levels, can respond to the different parts of the spectrum and is suitable for different applications. The efficiency of solar cells is given as a percentage of electrical power resulting from a given amount of light (photons) and the area of the cell. Efficiencies of commercially available solar cells and modules range from 8% - 17%, depending on the PV technology.

Mono-crystalline Silicon Cells

Single-crystal silicon (Si) is the pioneer of solar cell technology. Cells and modules made from it have traditionally have had the greatest share of the market. Single-crystal silicon cells are highly stable and demonstrate high energy conversion efficiencies. Mono-Si cells are grown in a vacuum chamber keeping the molecular signature of the silicon intact throughout the cell. This results in a cell with an efficiency of 17% or more. Higher efficiency allows for better performance in low light resulting in more peak watt hours delivered throughout the day.

Monocrystalline Cells within a PV Module

Single-crystal solar cell production usually includes these stages, feed-stock purification, crystalline boule growth, wafer production, cell fabrication and mounting. The single crystal ingots must be cut and processed into wafers. The p-n junction is usually formed at this stage by doping the surface layer to give it an electrical conductivity opposite to that of the bulk material. At this point, metal contacts are attached to the cell to provide a current collection circuit; and an anti-reflective (AR) coating is added to increase the amount of sunlight entering the cell. Finally, the cells are inter-connected in a module and encapsulated in glass or plastic to protect them from environmental damage. The modules are then linked together to form arrays. Single crystalline cells are more expensive but are used in the same type of applications as their polycrystalline counterparts. Manufacturing of the solar module occurs the same way as polycrystalline panels.

A recent technological advancement in the crystalline Si family of solar cells is laser grooved buried grid type (LBGB). LBGB cells bury the gridlines directly into the silicon cell allowing the current to pass through more quickly and exposes more of the PV cells to the sun.

Polycrystalline Cells

Polycrystalline silicon solar cells are a lower cost alternative to single-crystal silicon. Although devices made of polycrystalline silicon are generally less efficient (approximately 11% - 14.5%) than single-crystal silicon, the polycrystalline material has certain advantages. It requires a less pure material and can be adapted to automated mass-production techniques because of the way the starting material can be prepared. These cells make use of abundant silicon material (computer grade silicon), have relatively high efficiencies, demonstrate good stability, and share the broad technology base associated with single-crystal silicon.

Polycrystalline silicon is a solid made up of many crystallites or grains. The grain size depends on the conditions under which the silicon is cooled from the molten state. The grain boundaries, however, make polycrystalline silicon cells less efficient than single-crystal silicon cells.

Current development in polycrystalline silicon focus on identifying and developing methods of fabricating lower cost high efficiency cells. In particular, researchers are investigating techniques that could reduce the destructive effects of grain boundaries. Similar to mono-crystalline Si, poly-Si cells are connected in series (for voltage gain) and sandwiched between a layer of glass and the substrate backing of the module. The module is framed by anodized aluminum and a junction box on the back allows for easy hook-up of the electrical connections.

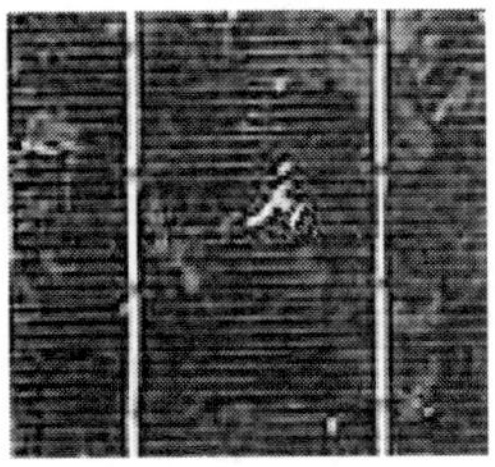
Polycrystalline Cell

Amorphous Silicon Solar Cells

Amorphous silicon solar cells, rated at 7% - 10% efficiency, are the least efficient of the three silicon technologies. They start out as a gas that is cooled and sprayed onto the glass or plastic panel surface. Amorphous silicon (a-Si) is disordered, unlike crystalline silicon which has a uniform lattice structure and poly-silicon which has a grains of crystal structure. The disorder in amorphous silicon limits the current more than grain boundaries in polycrystalline materials. However, alloying it with hydrogen, which provides acceptable conversion efficiencies, can diminish disorder deficiencies in a-Si. The standard amorphous silicon cell is made of three a-Si layers. Each layer is hydrogenated; the normal p- and n-type layers, plus an intrinsic (or i-type) layer between the p- and n- type layers. The p- and n- layers establish an electric field to separate the electron-hole pairs. These layers also determine the voltage of the device. The intrinsic or undoped layer is the active layer of the device where most of the cell's current originates. Each solar photon absorbed in the i-layer excites an electron from an energy band called the valance band (leaving behind a mobile, positively charged hole) into a higher energy band called the conduction band, (yielding a mobile, negatively charged electron). The internal electric field drives the hole to the p - layer and the electron to the n - layer, thereby building up a voltage across the cell. This is the essence of the photovoltaic effect in amorphous silicon. The polarity of the voltage is positive on the tin oxide and negative on the aluminum.

Amorphous Cells

Long term performance can be achieved by stacking two cells to form a p-i-n/p-i-n structure. Each cell then generates about 1.6 volts under open circuit conditions. Another design is based on a sophisticated triple junction multi-layer amorphous silicon thin film solar cell. This spectrum-splitting cell is constructed of three separate p-i-n types, amorphous semiconductor solar sub-cells, each having a different spectral response characteristic. In this way, the cell can convert the different visible and near infrared wavelengths of sunlight with optimal efficiency. This design is leading to higher

efficiencies, but the biggest advantage of the thin film a-Si is its low cost.

PV System Components

The standard components of photovoltaic systems are PV modules, charge regulators, deep cycle batteries, cable and mounting, and usually a DC to AC inverter. The PV modules produce the DC electricity when exposed to sunlight and it is carried down the cable (a positive and a negative conductor), to the regulator. The regulator's prime function is to protect the batteries from overcharge. Controllers often have several other features, depending on the specific application. If the batteries are not fully charged, the electrical current is then allowed to travel through the cable to the batteries. The deep cycle batteries store this power for use later when electrical loads, such as lights, need to be powered. If the system has 120-volt AC loads to be run, a DC to AC power inverter will be included in the system. The power rating on the inverter will depend on the total wattage of the AC loads in the system. Below is a drawing of a typical

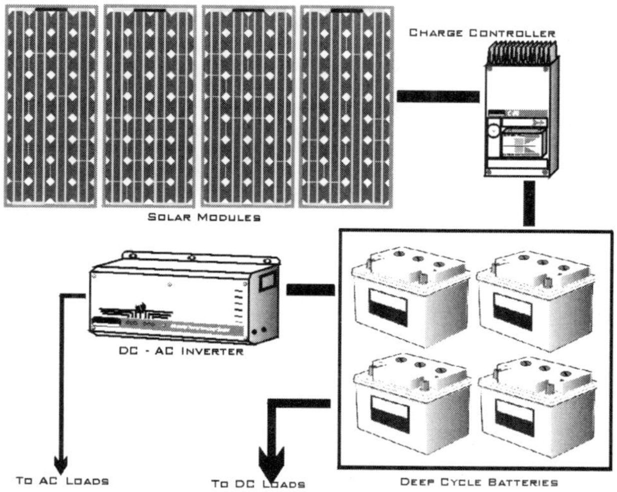

cottage solar system with an inverter, showing where each component fits into the system.

PV Modules

Photovoltaics modules are the basic building blocks of photovoltaic systems. The main role of the solar modules in the PV systems is either to charge batteries or to directly power loads. Modules are rated in terms of the total power they can produce in watts at a given radiation level (1000W/m^2) and at a temperature of 25°C This unit is known as the peak-watt (Wp). When considering the purchase of a solar system, it is important to first determine power requirements. This is accomplished by doing a load analysis listing all electrical loads in watts and the number of hours per day each will be on. This will give the watt hours per day necessary to power the PV system and determine the size of the PV module.

Charge Controller

The primary function of the regulator is to make sure the solar modules do not overcharge the batteries. It operates based on the battery voltage. Put simply, when the batteries need charging, the regulator allows current to flow to the batteries. When the batteries are fully charged, the regulator prevents this current from getting to the batteries. There are three types of PV charging regulators:

- A shunt type regulator shorts out the solar panel (safe & effective)
- A series type controller opens a relay, stopping the current flow (like a light switch)
- A pulse width modulated controller modulates the width of the charging pulse and allows full current flow with empty batteries and a very small trickle when the batteries are full. These are usually more efficient but more expensive.

Other common options with a regulator include low voltage load disconnect (LVD), temperature compensation for extreme climates and metering (showing array current and battery voltage). The LVD option will sense when battery voltage has dropped to a critical level and will automatically shut off the loads connected to that circuit. Because of increased internal resistance in cold weather, the temp-comp option will cause the regulator to stop charging the batteries at a higher voltage set point. The reverse is true in extremely

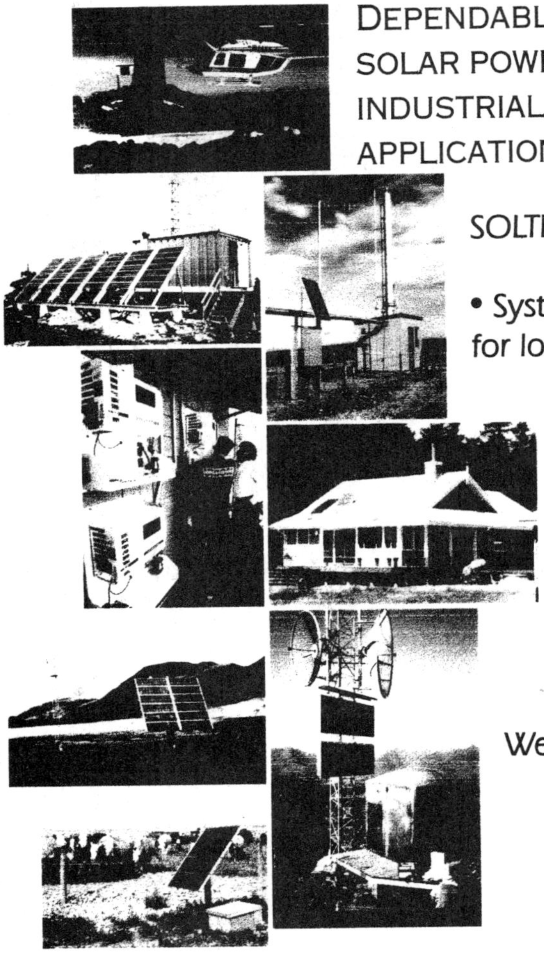

hot climates. Selecting what size controller to use is simple—add the amp rating of all solar modules in parallel and multiply by a safety factor of 1.25. For example, if the array totals 22 amps (5 x 75 watt modules), then the installation will need 22 x 1.25 = 27.5A or a 30 amp controller. Some new designs of PV charge regulators include maximum power point (MPP) tracking capabilities.

Batteries

The batteries are the heart of the system. They store the solar generated power that will run the loads when the sun isn't shining. To properly understand the battery bank, it is important to understand the difference between shallow and deep cycle batteries.

Shallow Cycle

These types, also referred to as cold cranking batteries, are found in cars. They have very thin lead plates, submerged in a lead-acid solution, and are designed to deliver large amounts of current over short periods. After starting the vehicle, the batteries are then recharged by the alternator. Therefore, the cycle (or discharge and recharge) is shallow. Cycling these batteries deeply will cause them irreparable harm (like leaving on car headlights). These batteries are not suitable for use in alternate energy systems.

Deep Cycle

These batteries are made of similar material, but the lead plates are thick, enabling them to deliver smaller amounts of current over a long period. This means the battery can be cycled deeply hundreds of times without being damaged. The number of cycles available from these types of cells depends on the depth of the cycle, the ambient temperature and the quality of the battery. If the batteries are regularly cycled down 50%, they will last longer than batteries that are continuously subjected to an 80% discharge. Therefore, the shallower the average cycle, the longer the life. All batteries sold for alternate energy systems are of the deep cycle type. The two main types of deep cycle batteries are FLA (flooded lead acid) and sealed. The sealed batteries use a gel or otherwise immobilized electrolyte, which will not spill or gas. Under normal conditions sealed batteries stand up better to colder temperatures.

Mount and Cable

Mounting structures for modules in a remote residential system are usually constructed from aluminum because it is lightweight and won't corrode. They generally have an adjustable mounting angle and use stainless steel hardware. A unique form of mount is the solar tracker. A tracker will increase the output of an array by tracking the path of the sun over the course of the day. The added expense of a tracker is not usually suitable for smaller systems as it will cost less to simply add more solar PV modules. There are passive and active trackers.

The cable used must be of a suitable gauge (thickness) to carry the system's voltage and current. It will either be UV resistant or run inside a conduit.

The Inverter

This component takes DC battery power and inverts it up to 120 or 240 VAC in order to run common household appliances or to allow grid-connection. Two different technologies are used, transformer based and high frequency switching. The transformer types are heavier, have higher surge capabilities and are usually more expensive for a given output. Many inverters have built-in battery chargers designed to assist the solar PV modules when an external AC power source is present. Examples of this are shore power for a boat or RV or a back-up generator in a residential system.

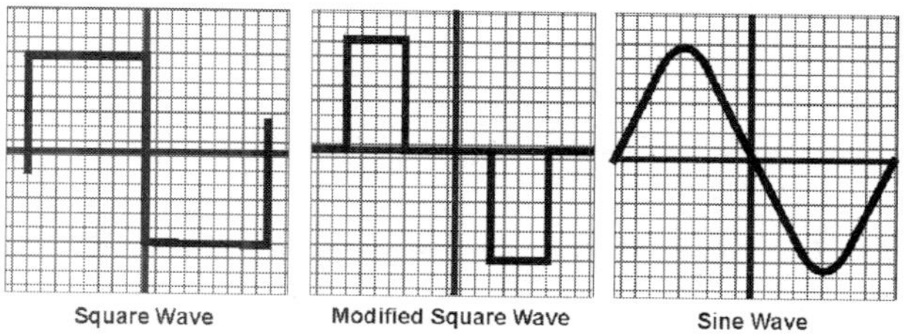

There are three different inverter waveforms which try to simulate the clean sine waveform put out by the local hydro utility. A square wave is inexpensive but has a dirty waveform so is effectively limited to running resistive loads, like incandescent lights and heaters. Square wave inverters are rarely used any more because of the harmonic interference produced, causing noise in radios, TVs and stereos.

The modified square wave is the most common type found on the market today. They are reasonably priced, highly efficient (around 90%) and will run most 120 VAC loads without causing any problems. The sine wave type inverter most closely simulate regular plug into the wall power. Any and all AC loads will run with no problems, assuming the inverter is large enough. The sine wave inverter is the most expensive but it is often more sophisticated although slightly less efficient than modified wave inverters.

The most advanced type of inverter are grid-connected inverters, which allow PV modules to be connected and feed back PV generated electricity directly into the utility grid.

For further information on photovoltaics contact:

Canadian Solar Industries Association
2415 Holly Lane - Suite 250
Ottawa, Ontari K1V 7P2 Canada
Tel: (613) 736-9077
Fax: (613) 736-8938
E-mail: cansia@magmacom.com
Web: www.cansia.ca

or the

Solar Energy Society of Canada Inc.
116 Lisgar Street - Suite 702
Ottawa, Ontario K2P 0C2 Canada
Tel: (613) 234-4151
Fax: (613) 234-2988
E-mail: sesci@cyberus.ca
Web: www.solarenergysociety.ca

What is . . . Wind Energy?

The Source

Heating of the earth's surface by the sun creates wind. Air warmed by the ground rises and is replaced by cooler air falling from higher altitudes. Also, large-scale convective currents set in motion by the sun's rays carry heat from lower latitudes to higher latitudes, including Canada.[6]

Causes of Wind Speed Variation

Wind speed is affected by a number of different factors including:

- Large lakes & oceans: as the sun heats land heats more quickly than water, wind will blow onshore. Conversely, if the land cools faster than water, the wind will blow away from shore, resulting in offshore wind. Winds also tend to speed up after traveling a long unobstructed path over the water.
- Land height: the greatest increase in wind speed occurs as air passes over hilly or mountainous terrain, and at the least rate over smooth terrain.
- Ground features: in attempting to get over barriers such as hills, tall trees, mountain ranges and tall buildings, the wind speeds up and causes friction as it passes over the obstacle
- Time of day: on average, there is a noticeable difference between wind speeds during daylight hours and night, as the effects of convective circulation have been set in motion
- Season: wind speeds in Canada may be higher in the winter months because weather systems are more intense

Site Selection and Surveying

In order to properly site a wind turbine in a rural, commercial or residential setting, it is important that the following factors are considered:

INCREASE IN WIND SPEED OVER A RIDGE

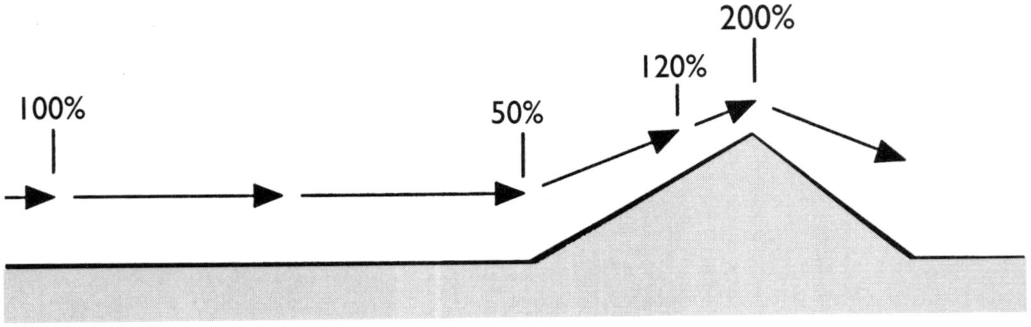

Variation in wind speed over a ridge. Wind speed increases near the summit of a long ridge lying across the wind's path. Paul Gipe, *Wind Power for Home and Business*

Space

There must be sufficient room to not only place the tower and attach the guy wires, if necessary, but also to install it safely

Turbulence & Exposure

All wind turbines should be located far away from trees, buildings and other obstacles in order to minimise the effects of turbulence, or rapid changes in wind speed and direction, and maximise the turbines exposure to smooth flow. In order to maximise output it is essential that the turbine is located far enough upwind or downwind to avoid the turbulent zone around all nearby obstructions.[7]

Average Wind Speed

Data on the average wind speed can be gathered in many ways: a) through the use of an anemometer, which measures both wind speed and direction b) observation of wind direction can also be achieved by looking at the vegetation around your potential site (strong winds and severe climate effects can deform trees and bushes and the level of the deformation can be used as a gauge of wind speed) c) checking the Atmospheric Environment Service of Environment Canada for data on wind speed, gathered from hundreds of locations in Canada.

Cable Routing

The owner must consider which method will be used to deliver the power from the turbine site to the house. The power lines can be either strung or buried. When determining which method will be used, the owner must consider the obstacles of the site, including gas lines, sidewalks, driveways or roads. The owners must also consult the Canadian Electrical Code and understand the restrictions and clearances that must be met in order to meet the code's requirements.

Zoning

In order to install the wind turbine in a rural, industrial or residential setting, the owners must ensure that the plans meet all local or municipal zoning ordinances or restrictions on land use. Contact the local building inspectors, planning office or board of supervisors and determine which rules and regulations are to be followed, which papers must be filed and who you must contact in order to proceed with your project. If there are problems with zoning, they will most often involve height restrictions, setback requirements and concerns about noise and safety.

How Wind Turbines Operate

Wind turbine blades are shaped like airplane wings, but instead of creating lift, the aerodynamic forces cause rotation. The blades are connected to a single rotor fitting into the turbine housing. When the "cut-in" wind speed is achieved, and a speed of 4-5 m/s maintained, the controls allow the rotor to rotate, producing power. As the wind speed increases, power output rises rapidly. As the rotor spins at a regular or constant speed, a magnet is turned through a coil of wire, which generates a flow of uniform-frequency electricity.

If the wind-speed continues to climb, eventually the control system will shut the wind turbine down to prevent damage to the machine. The "cut-out" wind speed is usually above 25 m/s.

Components

The major components of wind energy systems include the following:

Blades

The blades capture the wind and are most often made from fiberglass, polyester or epoxy resins. Some blades have a wood core. The diameters of blades range from less than 1 metre for a small turbine to over 50 metres in a large commercial turbine. Two or three blades make up a rotor in a Horizontal Axis Wind Turbine (HAWT).

[6] Paul Gipe, Wind Power for Home & Business, (Post Mills, VT: Chelsea Green Publishing, 1993), 22.
[7] Gipe, 264.

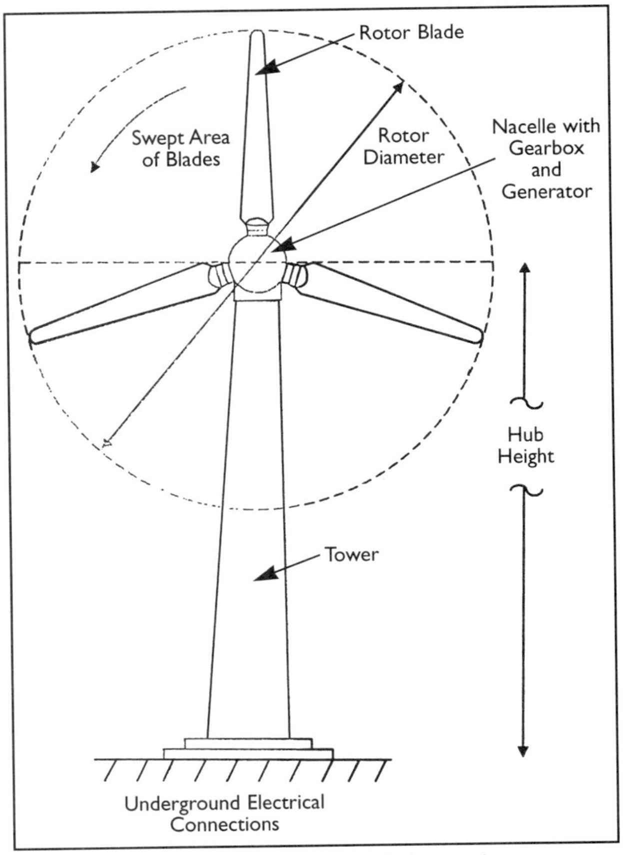

Wind Energy System Schematic
Source: RETScreen, Renewable Energy Analysis Software

Rotor

Blades attach to the rotor, called a center hub and the blades and the hub is then attached to the drive shaft.

Transmission

A series of gears, or the transmission, transfers the power through a spinning drive shaft and increases the RPM of the blades from 60 to between 1,500 and 2,000 RPM.

Generator

Once the high RPM of the blades are outputted from the transmission, the generator then produces electricity, similar to the way in which heat/steam systems generate electricity.

Controls

Many of the control systems are co-ordinated and monitored by a computer that is often accessed from a remote location. Pitch controls will twist the blades to improve performance at different wind speeds while yaw controls can point the entire turbine into the wind. The electronic controls maintain the voltage flowing from the generator as speed is changed. Smaller turbines may use mechanical controls, or less sophisticated electronic controls, according to their needs and price range.

Nacelle

A fiberglass casing houses the turbine components, protecting all of the components. It also allows maintenance to be performed safely from the attached platform.

Towers

There are two traditional types of towers, the "monotube," which is made of solid steel, or the "truss," which is a number of interconnected metal parts. All towers vary according to the rotor size, but generally range between 20 – 50 feet for a small system to 80 - 150 feet for a commercial system.

Foundation

A concrete foundation is needed to ensure the stability of the turbine in high winds. In some locations it may also be prudent to also use guy wires.

Sizes & Types

Generally, wind turbines come in two sizes, small and large. Small turbines are traditionally built for individual applications including home electricity production and water pumping in rural or remote areas, will produce 0.025 kilowatts to 10 kilowatts of electricity. Large wind turbines are used for utility-scale electrical generation and will generate from 50 kilowatts up to 1.5 megawatts.

The orientation of the drive shaft, the part of the turbine connecting the blades to the generator, decides the axis of the machine. Wind turbines come in two different types:

Horizontal Axis Turbines (HATs)

These are the most common types of turbines. They have 2 or 3 blades driving a horizontal shaft.

Vertical Axis Turbines (VATs)

The drive shaft of the turbine is oriented vertically, the blades are long, curved and attached to the top and the bottom of the tower.

Wind Turbine Applications

Off-Grid Applications

In off-grid applications, wind energy is used to charge batteries that store captured energy generated by the wind turbines and provide the user with electricity at a later time. Another important off-grid application of wind energy is water pumping for agricultural, community or livestock uses. Wind energy is competitive in off-grid applications where the energy demand is smaller, typically less than 10 kWh. The cost of wind energy is competitive when compared to the costs of electrical grid extension, primary (disposable) batteries, diesel, and gas generators.

Grid Connected Applications

Wind energy systems can also be used to feed electrical energy directly into the electric utility grid. In Canada, there have traditionally been two different types of grid-connected wind energy:

1) Isolated diesel-grid generation with additional wind turbine generation. The capacity will usually range from 10 kW to 200 kW. These wind-diesel hybrid systems are often used in remote communities, where the cost of electrical generation is high due to the costs of transporting diesel fuel to the remote sites. If the community has a good wind site, this type of system is installed to offset the communities' dependence on diesel generation.

2) Central-grid electricity generation, with additional wind turbine generation. This type of system is becoming more common as wind farms, or multi-megawatt projects. The capacity usually ranges from 200 kW to 1.5 MW. Wind farm benefits include the large-scale nature of the project and alternate use of land, such as livestock grazing and other agricultural uses.

A complete wind farm will consist of numerous wind turbines, access roads, electrical interconnections, a substation, a monitoring and control system and a maintenance building. To properly develop a wind farm, detailed project management is essential at both the planning and implementation stage. The developer must acquire all the proper local permits and authorisations, design and plan the civil, mechanical and electrical infrastructure, and oversee the layout and siting of each turbine, and the construction and commissioning of the wind installation. The construction of the wind farm will include site preparation, road grading, construction of the turbine foundations, installation of the electrical lines and transformers, construction of the substation and other buildings and installation of the turbines. In order to get accurate measurements of the wind resource, it is important to take measurements for at least one year at the exact location of the wind farm. Large wind farm projects may take up to five years to develop, as it may take four years to obtain and secure all the proper approvals, assuming that a comprehensive environmental impact study has been completed.

Environmental and Social Considerations

When planning the installation of a wind turbine on either a small or large scale, the impact upon the local environment must be addressed. It is important that the community does not perceive that you are asking for special treatment; what you are asking for is equal treatment. Over the last 20 years, many technical advances have been made to reduce the impact of wind turbines upon the general population and upon wildlife. The following impacts may need to be addressed at the local level before any turbine can be installed:

Aesthetics

In some communities, wind turbines are seen in a positive light. The community is working towards a more sustainable future - while in other communities, turbines and towers are perceived as "ugly" and an "eyesore." If your local community is less tolerant, it is important that you point out other similar structures that have been accepted. You may also consider using a "tubular" rather than a "truss" or "latticed" tower. Examples include utility poles and towers, radio station antennae and cellular towers. Certainly, you should not dismiss your community's concerns as they could make it very difficult to get your tower and turbine approved.

Noise

Noise can also be a cause for concern for some communities. When siting a turbine, it is important to take into consideration the distance to the listener, the repetitive nature of the noise, the pitch of the noise and the noise in context to the rest of the community's sounds around it. Ensure that the rated noise of the turbine is well within the limits allowed by the local zoning laws and try to ensure that objections are minimised by informing the neighbourhood of your plans.

TV and Radio Interference

Some objections to wind turbines may include a concern that the turbine will disrupt radio and television reception. When blades were manufactured using metal, that may have been the case, but today they are composed of fibreglass, plastic and wood, which do not cause disruptions.

Birds

Environmentalists have often raised concerns about wind turbines killing birds. A study of the 7000 turbine Altamont Pass Wind Farm in California, has shown that one turbine will kill a bird every 20-40 years. Putting that figure into perspective, it would take 1000 years for the bird mortality figures at the Altamont Pass to match the numbers killed by the oil spill from the Exxon Valdez in Alaska.[8]

For further information on wind energy contact the:

Canadian Wind Energy Association
Suite 100, 3553 31 Street NW
Calgary, AB T2L 2K7 Canada
Tel: 1 800 9 CANWEA (in Canada) or
(403) 289-7713
Fax: (403) 282 1238
E-mail: canwea@canwea.ca
Website: www.canwea.ca

[8] Gipe, 276.

Utility-Intertied Solar Electric Houses—Characteristics, Benefits, Limitations, and Potential Barriers

Contributed by Gordon Howell, P.Eng., Howell-Mayhew Engineering, Inc.

Introduction

INTEREST HAS RISEN SHARPLY in solar electricity generation technology (called photovoltaics, or PV) by large and small electric utilities and manufacturers around the world. This is fueled by the near-term promise of price reductions, the public's strong interest in solar energy, the interest in branding companies with a "green" image, and the restructuring of the electric utility industry.

One area of strong development is with utility-intertied PV systems installed on houses. The Sacramento Municipal Utility District in California has over 300 roof-top residential systems, Germany has over 2,500 systems, Japan is currently installing 70,000 systems, Holland is installing 60,000; Italy is installing up to 10,000 and the United States has started a programme to install up to 1 million.

In 1994, EPCOR Utilities, a utility owned by the City of Edmonton, approached Howell-Mayhew Engineering (HME), an engineering research and consulting company, for assistance in evaluating, and gaining regulatory and operating experience with PV technology. A project was initiated to design, assemble, install, operate and monitor a 2.3 kW PV utility-intertied system on the roof of HME's R-2000-certified energy efficient Cold Climate Demonstration House in Edmonton and to evaluate the potential barriers encountered in establishing and operating it.

The project goals included:

- gaining a better understanding of PV technology
- exposing and eliminating systemic barriers to the implementation of PV technology
- providing training to utility inspectors, technicians, and professional staff
- demonstrating the technical feasibility of this technology in the Edmonton region
- monitoring to evaluate design tools and PV system and component performance
- evaluating safety, commissioning, and field acceptance tests for PV systems.

Characteristics

Equipment Installed

A utility-intertied PV system is very simple. As shown in Figure 1, it consists of three major components: the solar PV array, a DC disconnect switch, a DC to AC inverter, plus DC and AC wiring. There is usually no battery bank. In the HME system, the array's output is rated at 2.3 kW DC in bright sunshine from 12 Siemens PC4-JF single-crystal 75 W modules and 24 Solarex MSX 60 poly-crystalline 60 W modules. The array was wired to normally operate at 205 VDC as required by the brand of inverter employed. An industrial-rated DC disconnect switch was used in this system because standard 120 VAC switches and breakers are not normally rated for DC voltages greater than 65 VDC. The switch needs to be able to interrupt the DC short circuit current produced by the array as well as be rated for the system voltage. Wiring sizes were selected to minimise wiring power losses. Size AWG #14 was used for inter-module connections, and AWG #12 for all other wiring from the array to the disconnect switch, inverter, and into the breaker panel. As a result, under conditions of maximum sunshine, calculations show that only 0.6% of the rated array power was lost due to total system wiring.

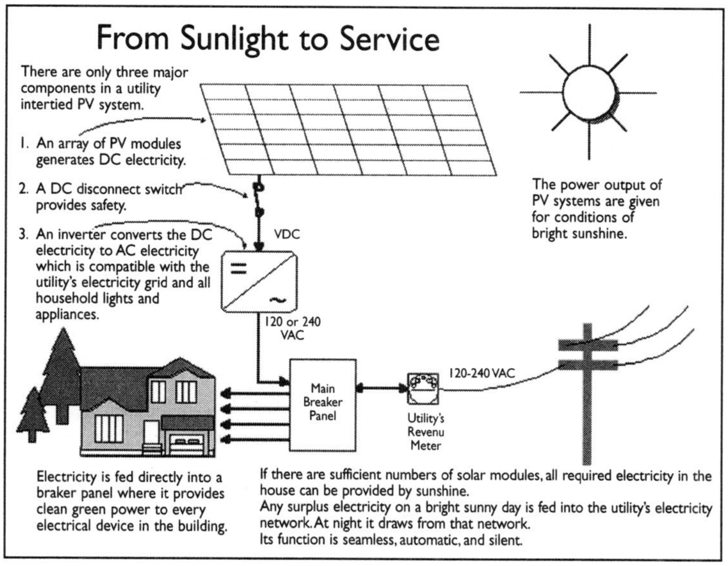

Fig 1. Typical schematic of a utility-intertied PV system.

Figure 2 shows the exterior of the house with the array installed. The array was placed flat to the roof at a tilt angle of 27°. The roof is a 6/12 pitch. This angle optimises the system's summer power production, but does not provide for

Fig 2. A PV array on the roof a house in Edmonton

automatic shedding of snow in the winter. Automatic snow shedding typically requires a tilt angle of at least 60°. Optimal tilt and azimuth (east-west) angles for flat-plate (non-concentrating) solar modules are dependent on the latitude of the location, (Edmonton is 53°N). Reductions in solar exposure due to moderately large variances in tilt and azimuth angle from the optimum, however, need not have an overbearing weight on system placement. This is shown in Table 1 and Table 2, where, for example, any tilt angle between 0° and 50° has only a maximum 6% reduction in solar radiation in the May to September months relative to the optimum angle of 30°. This is important to understand in order that array mounting costs and visual impact can be minimised without undue concern for loss of electrical productivity.

A StatPower PROsine GT inverter, rated at 3 kW and providing a 240 VAC 60 Hz pure sine-wave output, was used in the system. As is seen in Fig. 1, it is connected directly into the main AC breaker panel in the house. This inverter was from a production run being test marketed for the Japanese market and so did not have any labels nor documentation about its performance. It is currently being re-designed for the North American market by an Ontario inverter manufacturer under license. At the time of installation, the choice of inverters suitable for a residential utility-intertied system was very limited. Other inverters are now in the process of being produced for this market.

In this project, a data acquisition system was installed to measure the performance of the complete PV system and its components every five seconds and to store averaged and peak data every 15 minutes. The guidelines for this monitoring came from the PV system monitoring standard developed by the International Electrotechnical Commission.

System Installation

Installation of the system was simple. The array was assembled on-site by HME staff. The modules were assembled into panels by mounting them in groups of six onto aluminum rails. Each panel, weighing 42 kg, was in turn lifted by two HME staff onto the garage roof using ladders, and then onto the house roof. The rails of the array were mounted on aluminum feet bolted into 2"x4" pressurised wood sleepers lag-bolted into the roof trusses. The array protrudes from the house roof by 10 cm allowing air to circulate beneath the modules for natural cooling thus reducing the heat build-up that lowers summer operating efficiency.

HME installed the module interconnection wiring under the supervision of a local electrical

Period of the year	Optimal angle	Range of tilt angles	Reduction in solar exposure from optimum due to listed range of tilt angles
May to September	30º	0º to 50º	6%
October to April	60º	40º to 80º	8%
Year round	50º	20º to 70º	7%

Table 1. Optimal tilt angles for Edmonton and the reduction in solar exposure due to varience from optimum.

Period of the year	Optimal orientation	Range of azimuth angles	Reduction in solar exposure from optimum due to listed range of azimuth angles
May to September	South	SE to SW	3%
October to April	South	SE to SW	12%
Year round	South	SE to SW	7%

Table 2. Optimal azimuth angles for Edmonton and the reduction in solar exposure due to varience from optimum.

contractor. Under the direction of HME, an electrician installed the DC equipment downstream of the modules (consisting of the main feed cable from the array to the house plus the DC disconnect switch) and the AC equipment (the inverter and cable into the breaker panel). Final module connections were made in full sun under full shock hazard potential. HME supervision was required because this was the first time that this contractor had installed equipment which is an electrical source, as they normally only install equipment which consist of electrical loads. Once PV systems become commonplace and when installation training is common, an electrical contractor would normally do all the installation.

System Operation

When the PV system output is greater than house usage, the excess electricity is fed into the utility's electricity grid. When the inverter output is less than house usage, the shortfall of electricity is drawn from the utility grid in the normal manner. No relays are needed to control the direction of power flow. The system operates automatically, seamlessly, without noise, without light bulb flicker, and without causing motors or stereo equipment to hum.

The inverter is completely silent in its operation, except for three small computer-type cooling fans which cycle during hot summer weather. It operates automatically, turning on each day and going into sleep mode at night. It synchronises its AC power generation frequency with the utility's electricity grid, tracks the output of the array to ensure optimum efficiency, and, for important utility safety reasons, contains anti-generator-islanding technology which automatically turns it off during utility outages. Therefore this type of utility-intertied PV system is not producing electricity when the utility's system is out. It is not a fail-safe backup system for the customer unless the customer incorporates a battery bank with the system and uses a utility interactive inverter such as manufactured by Trace Engineering. That type of inverter isolates the utility grid from the house during outages.

There are two main methods of metering the electricity generated and consumed in such a system, either through a single meter measuring the net amount of electricity consumed, or through two unidirectional meters individually measuring what is exported to the utility's electricity grid and what is imported from the grid. Initially, the house's standard revenue meter continued to be used for billing—this amounted to net metering. This was changed to meet Alberta's electricity regulations which require that two unidirectional meters be used: a revenue meter measuring supply to the house and an export meter measuring supply to the utility. There are no special billing arrangements for the house. Normal consumer billing is based on the revenue meter. Credit to the system is based on the export meter. There are no free-for-all net-metering programs in Canada. Toronto Hydro, Ontario Hydro, and EPCOR are experimenting with them.

It is important to understand that, in a standard municipally sited system, it is much more expensive and dangerous to incorporate on-site electricity storage (such as a battery bank) than it is to use the utility electrical grid for storage. If a system is net-metered, then, in effect, the utility grid is being used as a storage device, usually without cost. In addition, the utility supply is always available in the rare instances that the PV system cannot meet the electrical demands of the house due to malfunction or maintenance, if there is an unanticipated cloudy winter month or heavy snow cover, or if the electrical demands increase due to expanding numbers of household electrical devices.

Performance

Continual improvement in component and system performance, testing, standards, and simulation is occurring throughout the industry. This is resulting in steady increases in actual system performance. The performance of this PV system, as measured by the monitoring system, showed some surprises. The array was rated at an efficiency of 11.1% according to the manufacturer specifications. The simulation software, WATSUN PV, suggested that the system would operate at an efficiency of 10.2%. Measurements showed that the average operating efficiency in the summer was 7.9%. The reasons for this have not been confirmed but could arise from excess module mismatch, poor array peak power tracking by the inverter, higher module temperatures due to its roof-top location, and errors in the simulation models.

The system simulation indicated that the PV system would meet 77% of the house loads or approximately $220 worth of electricity annually. The average inverter efficiency was 83%. Typical new inverters can reach efficiencies of over 90%.

Maintenance

The system has now been operating for three years. Required routine maintenance has been virtually nil. Array cleaning in the summer is not required as normal rainfall keeps the modules very clean. Between rain storms, a very slight layer of dust accumulates, the effects of which cannot be accurately measured. Since the system is designed to operate by natural cooling mechanisms, and since the roof tilt does not provide for automatic snow shedding, snow does accumulate and can fully block any sunlight penetration. Normal winter maintenance can optionally include snow removal at a rate of up to five times per winter. Snow removal has been useful to the project to evaluate winter performance.

Inverter operation has been reasonable although operational improvements in the latest technology are welcomed. This particular inverter is sensitive to noise on the utility's electrical grid and usually ceases to operate when a lightning storm passes by. This requires manual resetting of the inverter by turning it off and then on. In addition, an inverter component failed early in its operation and was replaced by HME staff using a simple field upgrade kit.

Benefits

The ultimate value of utility-intertied PV systems depends on the value that the owners place in their characteristics and products. These include:

- generating electrical energy
- providing secure, uninterruptible power (if the system includes energy storage)
- feeding electrical power distributed throughout the utility network
- reducing stress on utility electrical lines and equipment
- shaving peak daytime electrical loads
- reducing greenhouse gas emissions
- providing a image green for buildings
- fostering green values in the community
- providing highly visible opportunities to reinforce personal, community and corporate commitment to environmental protection and sustainable development
- providing significant prestige which can sell lifestyle, building projects and environmental programmes
- nurturing utility-customer relationships
- providing tools for learning about the characteristics, values, opportunities, and technical and commercial issues surrounding utility-intertied PV and distributed generation systems
- providing opportunities to identify technical, economic and other barriers blocking its market penetration
- providing tools for documenting and resolving codes, regulations, permits, and power pool issues
- providing tools for assessing and identifying niche markets and growth potential
- providing tools for displaying and marketing PV technology
- providing tools for targeting markets and facilitating efforts to accelerate retail efforts.

Since most of the above characteristics are of greater value to utilities, communities and industry than to residential customers, it is likely that utility-intertied PV systems will be developed and owned by commercial and industrial organisations before they are widely owned by families.

Limitations

If the PV system does not include on-site electricity storage, then the system will shut down during utility power outages, for safety reasons. This means that such systems will then not provide any more secure a supply of electricity than the utility company.

If the utility's electrical load profile has a winter evening peak, then the PV systems will not lower that peak (because there is no sunlight) unless on-site electricity storage is included in the design.

Potential Barriers

An objective of this project was to expose potential barriers to the implementation of utility-intertied PV technology. To catch as many potential barriers as possible, the approval, permit, and inspection process was approached from the perspective of an ordinary person applying for miscellaneous house renovations.

Thirteen potential economic and regulatory barriers, itemised below, were encountered in developing this project. Some were expected as this was the first grid-connected PV system in the province. Others were discovered in the process. Problems and confusion in securing necessary approvals, permits, design, equipment selection, installation, inspection, and final authorisation for start-up were significantly minimised by expressing, presenting, and discussing the technology and its characteristics in a confident and expert manner. As knowledge and experience with the technology are gained, more wide-spread solutions are expected to be found. Some solutions to remove or lower the barriers are suggested below.

1. **Provincial electricity production regulations:** Alberta's Energy and Utilities Board (EUB) required that the PV system be subject to all the regulations of the Electric Utilities Act because it generated electricity which was exported across a property line. In Alberta, the Power Pool administers the market-based pricing of electricity and a utility-intertied generator must

sell its excess electricity into the Power Pool. HME was required to apply to the EUB for permission to construct and for permission to operate a power plant, at a fee of $550 each, and was required to become a member of the Power Pool (www.powerpool.ab.ca) for an annual fee of $150. Neither HME nor the utility had understood this before the system was installed and connected. Only when the EUB became aware of the project because of its extensive publicity was HME approached and requested to follow regulations. Initially the EUB suggested that an environmental impact assessment and a neighbourhood impact and noise assessment would be required for the system.

- *Interim solution*: At this time, the cost of application and membership cannot be waived. Assessments of noise, environment impact and neighbourhood impact can easily be answered by a simple letter.
- *Long-term solution*: Simplify EUB's requirements for residential utility-intertied PV applications and waive Power Pool membership. This, however, needs to be assessed very carefully as such a solution needs to consider the impact of large-scale distributed generation, and address distributed-systems' generation capacity and type of technology. New residential generation technologies such as micro-gas turbines, fuel cells, and wind turbines are going to become widely available.

2. **Building permits:** The City of Edmonton's Building Inspection Branch required a Residential Building Permit and a Building Development Permit. Initially, two sets of site plans, elevations, and engineer-stamped construction drawings were to be submitted with the application. In the end, the permits were granted, without any problems, based on module layout and mounting drawings, and a photograph of the site elevation, none of which were stamped.
   - The City treated the application in the same way as someone who was adding a deck or satellite dish, and was specifically concerned with roof loads and mechanical fastening. The zoning in the area restricted the height of the system to no more than 7.5 m from street level. This was not a problem.
   - No neighbours appealed the development. No community impact studies and information meetings were required.
   - *Interim solution*: To reduce the possibility of a permit being denied, the PV system project can simply be treated as a home renovation. Regulatory people can overstate requirements because of unfamiliarity with a PV system. Discussing the project with understanding and with confidence can help in reducing any fear of public risk. Examples and pictures from other PV systems across North America can help.
   - *Long-term solution*: Regulatory and development staff need to be trained to understand what PV systems are and how they perform.

3. **Electrical permit:** The electrical contractor applied for an electrical wiring permit in the standard manner. It was approved without delay. When the inspection department was questioned about the approval, the inspector wanted to know what "PV" was and what would happen during a utility grid outage. He felt that this system was conceptually not much different, in regulatory terms, from systems such as uninterruptible power supplies. He also required that an AC contactor, and not just a silicon controlled rectifier, be employed to fully isolate the inverter from the grid. (The StatPower inverter contains contactors on both the DC and AC sides.) The utility's Customer Services Department required a letter describing the project, its relay protection, a single-line diagram, inverter data sheets, and equipment sizes. This was provided.
   - *Interim solution*: Reasons for denying an electrical permit need to be placed into three categories: jurisdictional, electricity quality and public safety. Tens of thousands of utility-intertied PV systems now installed world-wide, having met the concerns about safe, high quality operation in hundreds of electrical jurisdictions thus show that widely-held reasonable standards of quality and safety can be addressed by proper system design and component selection. Jurisdictional reasons for denying a permit then underlay a lack of understanding which can be met by proper staff training in order that systems operation can be understood, or a threat to financial stability which then underlies something quite more substantial.
   - *Long-term solution*: Inspectors and utility staff need to be trained to understand what PV systems are and how they perform.

4. **Roof strength:** The array weighs 300 kg (661 lbs) and covers 21 m^2 (225 ft^2) for a load density of 14 kg/m^2 (2.9 lb/ft^2). This was well within the

roof trusses' live load design of 107 kg/m^2 (22 lb/ft^2), as reported by the manufacturer, and did not present a problem.

- *Interim solution*: If the roof's strength is insufficient, then the PV system's mounting may need an engineered design to overcome this.
- *Long-term solution*: Work with truss manufacturers and house builders to ensure that roof designs are sufficiently strong. It is not expected that roof strengths are typically a problem.

5. **Equipment availability:** An inverter with a much higher than usual array operating voltage (205 VDC) was selected in the expectation of encountering additional safety and equipment rating barriers. Sufficiently rated DC switchgear were much more difficult to source. Common AC breakers and switchgear could not be utilised as they are typically only rated to 65 VDC. The choice for utility-intertied inverters is also very limited.
   - *Comment*: Equipment availability is becoming much better. DC switchgear rating is not always fully understood, even by inspectors!
   - *Equipment approval:* The utility required all equipment to be approved to Canadian Standards Association (CSA) or Canadian Underwriters Laboratories (CUL) standards. At the time of purchase, no PV modules came with this approval. Instead the local CSA office was contracted to perform their simple Special Inspection Service on each module. This was easily obtained since each module showed UL and FM certification. Warnock-Hersey in Vancouver was contracted to test and approve the inverter.
   - *Interim solution*: Special Inspection Service for each component without a CSA or CUL label.
   - *Long-term solution*: Get manufacturer-based component approval.
6. **Installer training:** Electrical installation was sub-contracted to a residential electrical contractor with no previous experience in PV systems. The contractor needed to obtain the 1994 Canadian Electrical Code (CEC) with the new Section 50 on PV systems. In installing the system, the electricians had to re-focus their thinking and procedures to account for the fact that the system was an electricity source instead of a load as in normal house wiring.
   - *Interim solution*: Explain the operation of the PV system and make sure the electrician knows Section 50 and Section 84 (on Interconnection). Standby to ensure correct installation.
   - *Long-term solution*: Electricians need to be trained to understand what PV systems are, how they perform, and how they need to be installed.
7. **Inspection:** The electrical inspectors first had to familiarise themselves with Section 50 of the CEC before inspecting the system since this was the first PV system they had encountered. Other utility safety and operating personnel enquired about the performance of the array under short-circuit conditions and large house loads, and about the inverter's frequency synchronising, grid isolation, grid detection, and anti-islanding methods. There was a question whether the CEC permits parallel generation.
   - The inspector noted the following deficiencies in the system:
   - 1. Power cannot be fed into the utility's system. (In fact this did not appear to be correct as no utility nor city regulation was found which stated this!)
   - 2. The breakers in the main AC breaker panel are not allowed to be reverse-fed.
   - The final inspection report indicated that:
   - 1. Acceptance is subject to approval of the utility's Customer Services and Engineering Departments, and
   - 2. A warning sign shall be installed on main panel stating "CAUTION, TWO POWER SOURCES".
   - Although not noted by the inspectors, CEC Rules 14-414(3), 50-004, 50-014(2), 84-026(1), and 84-032 also indicate requirements for additional signs.
   - *Interim solution*: The chief inspector of a jurisdiction makes the final decision about whether the system can be operated and interconnected to the utility. However, if the PV system has been designed, sourced, certified and installed according to the CEC, then any technical and safety objections to it become questionable, considering the large number of similar systems now installed and accepted in North America.
   - *Long-term solution*: Inspectors and utility staff need to be trained to understand what PV systems are, how they perform, and how they need to be installed.
8. **Integrating into the utility's operations:** A separate transformer with fused disconnect was installed on house's service pole to isolate

the PV system from the neighbours' service feeds. This was to integrate the system into the utility's operations and to enable them to conduct outage tests without interfering with neighbours. This cost was borne by the utility.

- *Interim solution*: It is acceptable if the utility needs to do this, though the cost needs to be theirs because it is a matter of training and experience.

- *Long-term solution*: Utility operations staff need to be trained to understand what PV systems are and how they perform.

9. **Insurance:** For insurance purposes, the PV system was deemed to be a part of the house, and the increase in annual premium was $25. The only problems in obtaining insurance were due to the insurance staff being uninformed. Initially, the insurance agency declined to insure it because they did not understand what it was. They then required stamped engineering blueprints, which was declined because we did not feel that such blueprints should be required for this project. In the end, all they wanted to know was that the system was CSA and utility approved. Since CSA approval was already a standard requirement, however, by the electrical permit, they should have just asked for a copy of the permit (to prove that it was obtained).

   - *Interim solution*: Work through the questions asked, show pictures of other systems, be confident and knowledgeable about the technology and about what is reasonable for them to require.

   - *Long-term solution*: The insurance industry and staff need to be trained to understand what PV systems are and how they perform. The insurance industry are important allies of the renewable energy industry.

10. **Residential property taxes:** A reassessment of the house for residential property taxes has not yet been done.

    - *Interim solution*: Since the PV system is only affordable for high-end consumers, it could be argued that there is no market value to it as installed on a middle-income house and thus it should not be taxed at its fully-installed value.

    - *Long-term solution*: The real estate industry and tax assessment staff need to be trained to understand what PV systems are, how they perform, and how they are valued.

11. **Industrial property taxes:** In Alberta, all electricity exported across a property line comes under the jurisdiction and price-control of the EUB. Any equipment which produces such electricity becomes "linear property" and is subject to an annual property tax based on the municipality's mil rate. The mil rate for linear property in Edmonton in 1996 was 26.96, which would have attracted a $755 annual property tax on the PV system. This has not been executed by the Alberta government for unknown reasons.

    - *Interim solution*: Apply for the tax to be waived because this is residential property not industrial property and so should not be exposed to industrial taxes.

    - *Long-term solution*: The tax policy department needs to be trained to understand what PV systems are, how they perform, and how they are valued. Such training needs also to include other upcoming distributed generation technologies such as fuel cells and micro-gas turbines in order that policies are in place to consider them.

12. **System cost:** PV technology produces high-end, high-value electricity. The system's purchase and installation cost was $28 000 in 1995 or $12/W of capacity. System costs are expected to be reduced by 60% by 2005. This capital cost results in an electricity cost of 37 ¢/kWh if it is spread interest-free over 25 years, and 93 ¢/kWh if a 6% interest rate is included. This compares to an Edmonton residential retail rate of 7.15 ¢/kWh. While this may be a burden to some home-owners, the economic viability of it is no poorer a choice than what most people make for the brand of automobile they own!

    - *Interim solution*: If costs are a concern, then consider that the cheapest electricity is the electricity that is saved through energy conservation and efficiency, and the second cheapest is that which is purchased from the utility. Make choices now for energy conservation and efficiency so that your electricity bill is reduced immediately in the manner which shows the most environmentally and sustainably green leadership to your community, your family, and yourself.

    - *Long-term solution*: Make choices for energy conservation and efficiency so that a smaller

PV system will be sufficient for all your electricity needs and thus be more affordable, sooner.

Acknowledgments — This project is sponsored by EPCOR Utilities, the Energy Diversification Research Laboratory of Natural Resources Canada/CANMET, Howell(Mayhew Engineering Inc., and the City of Edmonton departments of Public Works, Transportation, and Parks and Recreation.

Designing A Home Power System

The power source for off-grid homes will likely be PV, wind, or microhydro operating "stand-alone" or in some combination. Storage batteries, power controllers, and power conditioners may be required to complete the system. It is always important that before any decisions are finalised, a renewable energy installer is consulted for the most current and reliable industry information.

There are two main steps to designing a photovoltaic, wind or microhydro energy system. First, the owner must determine how much energy should be produced by the renewable energy systems to meet expected electrical loads. An "energy budget" can assist the owner in assessing energy demands. Second, assess the amount of energy that is available from the renewable energy source and select the correct equipment to harness the energy and the proper methods to store it for later use.

System Design

Solar, wind and small hydro electric systems are used in thousand of homes world wide. The standardization of the components and design of these systems has changed renewable energy from an experimental hobby to a practical alternative to fossil fuel generators. These systems can easily pass electrical inspections when installed with UL or CSA approved equipment.

Purchasing an independent electrical power system is not as simple as buying a mass produced consumer appliance. These systems usually operate in remote locations where the owner is responsible for installation, maintenance and service. An understanding of the equipment and how it functions will save money and frustration.

Some of the reasons people purchase renewable energy systems include becoming more self-reliant, saving money and reducing their impact on the environment. In the right location, a renewable energy system is the most cost-effective method of generating electricity and it produces virtually no pollution.

CASE STUDY:

The Kitto Residence
Tara, Ontario

THIS RESIDENCE is completely independent of the electrical grid, using photovoltaic power in combination with solar water heating. Electrical power is stored in a 24 volt system, the house is wired to Hydro standards, and a 4-kilowatt inverter provides 120-volt AC power. Propane is used for the refrigerator, freezer, clothes dryer and cook stove. The lifestyle enjoyed by the occupants is typical of a modern Canadian household. The washing machine, water pump, water softener, and many other small appliances operate on 120-volt AC. The office includes a fax, photocopier and computer. Low wattage lights are used in areas where lights are left on for extended periods.

There are 32 industrial quality storage batteries in the basement wired for 24-volt. There is a 100 amp control centre that is CSA approved. A 5000 watt Honda generator provides back-up power and will come on automatically if the batteries get down to 23.7 volt, charge the batteries and turn off. There is a wood stove for heating with a closed loop running through the fire box. It also provides enough hot water to heat the office through a water loop that runs under the floor. The only energy purchased is wood and propane at a cost of about $900 per year.

The garage is a separate system with a power-star 1300 watt inverter and 6/6 volt golf cart batteries, a 30 amp charge controller, and a 75-amp Todd charger. 3/75 Seimens solar panels charge the system, and if the generator is on it will charge this system as well as the house. This system runs electric garage doors, outdoor lights and tools in the shop.

System Specifications for the House

12 Solarex VLX53 PV panels fixed & 6 Kyocera 50 watt on a WATSUN tracker
APT control centre
Trace SW4024 inverter 4 kilowatt
32/8L-16 batteries
Thermomax ETC solar water heater
Honda 5000 generator

Contact: **Judy Kitto**
Solar Power Systems
RR#4
Tara, Ontario N0H 2N0
Tel: (519) 372-6555

Power System Design

There are two main steps to designing a photovoltaic, wind or microhydro energy system. First, the owner must determine how much energy should be produced by the renewable energy systems to meet expected electrical loads. An "energy budget" can assist the owner in assessing energy demands. Second, assess the amount of energy that is available from the renewable energy source and select the correct equipment to harness the energy and the proper methods to store it for later use.

Energy Budget

An energy budget is used to assess the electrical loads in a residential renewable energy system. It is important that electricity is viewed as a finite commodity like flour, firewood or money. An energy budget is a list of all your electrical appliances and the energy they use.

Conservation

An owner should revise an energy budget a few times to see if the home's energy need can be reduced. One way of conserving energy is by using energy efficient lighting. A 15-watt compact fluorescent tube will produce the same light intensity as a 60-watt incandescent bulb. The energy saved by using the 15-watt compact fluorescent tube for three hours is equivalent to the daily output of one 45-watt solar module. Energy conservation is less expensive than energy generation.

Rated Watts

To complete the energy budget, the owners must know what the rated watts of their appliances are. This information will be listed on the nameplate of the appliance.

If the appliance lists amps instead of rated watts, multiply the amps by the voltage to determine the rated watts. For example, if the blender says 4.5 amps and 120 volts AC on the back, then multiply 4.5 by 120. The rated watts of the blender is 540 watts.

Loads and Operating Time

The concept of load and operating time is essential to assess the viability of large, intermittent or small constant loads. The energy consumed by an appliance is the size of the load in watts multiplied by the run time. For example, operating a large load like a microwave at 1000 watts for six minutes is the same as operating a 50 watt light bulb for two hours.

Fine Tuning

If the electrical system is used only on weekends and holidays, the energy budget will have a much lower average daily load than a full time system. In addition, the system will have a whole week to charge and store in preparation for weekend use.

There may be some loads that should operate from a generator, such as a washing machine or large power tools. These loads should not be listed in the energy budget because they will not be drawing energy from the storage battery.

Appliances that have build in clocks or instant-on features will continue to draw power even when it appears that they are turned off. These loads are called phantom loads.

Energy Budget Calculations

1. Write down all loads and the **rated watts** of those loads in the energy budget chart shown.
2. Determine the **hours per day** and **days per week** you will operate each load and enter these values in the appropriate place in the energy budget.
3. Calculate the **Rated watts** x **hours per day** x **days per week** which equals the **weekly watt hours** of each load.
4. Add together the **weekly watt hours** for every load.
5. Divide the **total weekly watt hours** by seven days to obtain the **average daily load** in watt hours.
6. Take the **average daily load** in watt hours from the energy budget. Multiply the **average daily load** in watt hours by 1.20 for DC only systems and 1.30 for AC systems only. This factor accounts for battery and inverter inefficiencies and gives the **corrected average daily load** in watt hours.
7. Divide the **corrected average daily load in watt hours by the system voltage.** This gives the **total daily load** in amp hours. The **total daily load** allows the determination of the size of the battery bank and charging system.

Load	Rated Watts	Hours Per Day	Days Per Week	Weekly Watt Hours

Total Weekly Watt Hour s ____________

Average Daily Load in Watt Hours ____________

x 1.2 or 1.3

Corrected Average Daily Load in Watt Hours ____________

System Voltage

Total Daily Load in Amp Hour ____________

While the energy budget is a valuable tool in designing an energy system, accurate estimation of multiple variable loads is difficult. Avoid underestimating the energy budget as it will result in an undersized system. This is especially critical when choosing the battery. PV modules can always be added later, but a battery bank is much less flexible.

Typical Appliance Loads

Appliance	*Watts*
Toaster	1150
Blender	700
Freezer (Chest 15 cu. ft.)	*350
Refrigerator (stand. 14 cu. ft.)	*300
Refrigerator (frost free 14 cu. ft.)	*360
Vacuum Cleaner	500-750
Microwave Oven	700-1500
Washing Machine	450-800
Gas Clothes Dryer	250
Small Water Pump	300
Deep Well Pump	1000-1700
Computer Monitor	50-85

Appliance	*Watts*
Laser printer	600-800
Bubble Jet printer	24-50
Hair Dryer	1500
Colour Television	75-150
Satellite Receiver	25
3/8" hand Drill	250-400
Small Drill Press	600-700
Router	700-1000
Orbital Sander	300
Belt Sander	700-1000
Circular Saw	700-1200
Jig Saw	300-500

*Very High Efficiency Appliances

The information on Home System Design was kindly provided by Bob Mathews of Energy Alternatives Inc., from his company catalogue, "Energy Alternatives: Solutions for Remote Power Supply, Design Guide & Catalogue, 7th Edition.

CASE STUDY:

Microhydro/PV Hybrid Power System
Gambier Island, British Columbia

IT IS NATURAL to think of the west coast of British Columbia as wet—rainforest is a common expression. But the mountain ranges of Vancouver Island actually create a relatively dry region between their eastern side and the Coast Mountains of mainland BC. The islands are largely bedrock with little soil to hold precipitation, so when it rains, the streams run, but when it stops raining, the streams stop running. If the streams ran more regularly, there would be little or no need to supplement a microhydro system with photovoltaics. In the Strait of Georgia, there are numerous islands that are ideal for cabins, but there is no utility supply due to the difficulty and expense of running cables over or under a water barrier.

Solar PV provides a reasonable cost/benefit during the sunnier periods in the region, especially for cabins, which tend to concentrate their electrical demands only on a part-time, weekend basis, so the batteries can charge all week long. But as with all PV, there is little output on the cloudy, shorter days of fall and winter. But the use of a microhydro system during the rainy season provides sufficient power and eliminates the need for a back-up fuel generator.

John Cosulich lives and works in the Vancouver area and has a large cabin on Gambier Island, where he intends to retire. Like many, he would prefer not to live with the noise, maintenance and repairs of a generator. Over a two to three-year period, he has installed a hybrid microhydro/PV system with battery storage and DC to AC inverters.

The head (elevation difference from intake to turbine) is 37 metres (120 ft.). By changing the size of the nozzle for the Turgo-type turbine, he can utilise flow rates from as little as 0.13 to approximately 4 litres/second, with approximate proportional power outputs of a few to 700 watts, 24 hours/day. At 700 watts, the microhydro output is roughly two to three times more kWh/day than the total of both PV arrays on a sunny day. Excess power from hydro is dumped to a resistance heater.

For additional power in the dry season, the wind resource is also being assessed.

System Specifications

8 Solec 80 W PV modules, mounted on a Zomeworks TrackRack Tracker
8 Solec 75 W PV modules, mounted on a Zomeworks TrackRack Tracker
1 Trace C40 Controller, Diversion type
12 2-volt cells (lead-acid motive-power type) for 24-volt battery, 1465 amp-hours
1 Trace Power Panel with dual 4 kW Sine Wave inverters for 120/240 VAC, 8 kW total
1 Steam Engine by ES & D, single nozzle Turgo turbine with Permanent Magnet brushless generator

The Big Picture

The natural world provides energy resources at different times of the day and/or week, hence the need for energy storage, which allows the energy to be retained for later use. The battery storage with DC to AC inverter is a power subsystem unto itself, with re-charging possible from many different resources.

The natural world also may provide us with different energy resources in different seasons. Each renewable energy system (microhydro, wind, photovoltaics) adds to overall capital cost, but if this significantly reduces fossil-fuel generator run-time, noise, pollution, maintenance, etc., it should be considered cost-effective.

Contact: **Bob Mathews**
Energy Alternatives Ltd.
825 Laval Crescent
Kamloops, British Columbia V0E 1M0
Tel: (800) 265-8898 Fax: (250) 374-3943
E-mail:
energyalternatives@bc.sympatico.ca

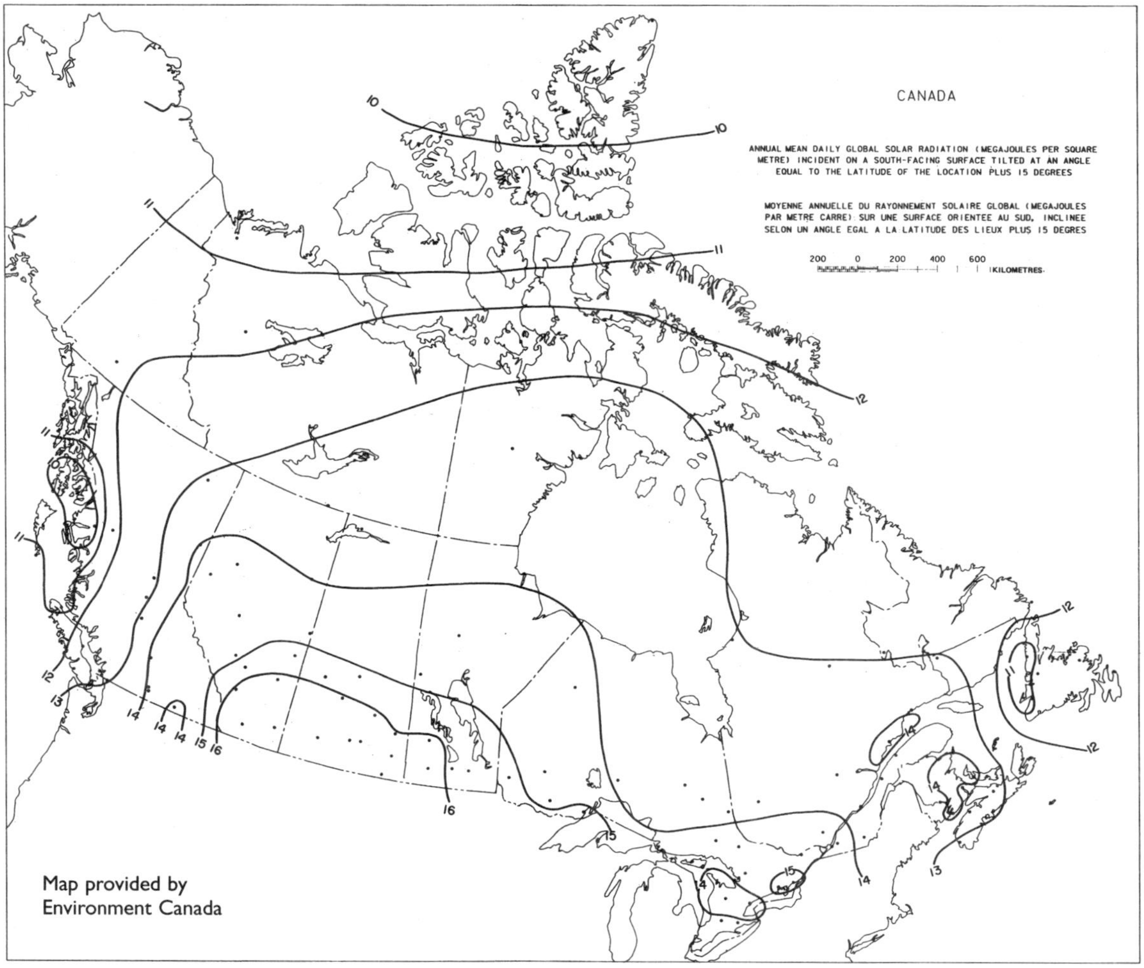

Map provided by Environment Canada

Photovoltaic System Design

When designing a solar electric system two factors determine the size and cost of the system - the daily power consumed by your electrical loads, as determined in the energy budget, and the sunlight levels or insolation available at your particular location.

Insolation

Insolation or sunlight intensity is measured in equivalent full sun hours. A full sun is equal to the amount of sun striking the earth in one hour when the sun is directly overhead in a clear sky.

The lower the sun in the sky, the more atmosphere the light must pass through. Water molecules and other gases in the atmosphere reflect and absorb some of the light passing through it, reducing the insolation. Most of the sun's energy is delivered between 9:00 a.m. and 3:00 p.m., when the sun is highest in the sky. Shorter days, cloud cover and the sun's position in the sky reduce the amount of energy available in the winter.

See Solar Insolation Data Map in this Chapter.

Array Size

To calculate the size of a photovoltaic array, the amount of sunlight at the location must be determined. Sunlight varies considerably throughout the year. If the system is being designed for a summer cottage, then the owner should use the figures given for the summer months. If the system is being designed for a year round residence, with a back-up generator, then use the yearly average or the lowest value for the months when the generator will not be operating (see Step 2).

1. From the energy budget find the **total daily load** in amp hours.
2. Determine the isolation in full sun hours from the chart. Divide the **total daily load** in amp hours (from the Energy Budget) by the full sun hours to give the **array current** in amps.
3. Select an appropriate number of modules to supply the **array current**.

What Can Be Done With Solar Power?

Lighting

In order to reduce energy demands associated with room lighting, use of compact fluorescent bulbs, which use less than one-quarter the power of regular bulbs, will better serve a renewable energy system.

Appliances

120 volt AC appliances, like vacuum cleaners, kitchen blenders or power tools are all compatible with solar energy. These high wattage appliances only use power for a short time so total energy consumption is low enough for a solar electric home.

Refrigeration

Refrigerators and freezers are often the largest electrical load in a solar powered home. Special, energy efficient models are the only practical way to refrigerate with solar power.

Electronics

Any colour TV, VCR, satellite dish, stereo, computer may be used.

Water Pumping

Some pumps are powered directly from a solar module, if the well is far from the dwelling. Deeper wells need jet, submersible or jack-type pumps which may be AC powered by an inverter or generator.

What Should Not Be Used?

Major Heating Appliances

Space and water heaters, cook stove and clothes dryer account for 50% to 80% of typical monthly electric bills. It is not practical to operate these major appliances with solar electricity.

Conventional Refrigerators and Freezers

Standard refrigerators have very long hours of operation and poor insulation. Frost-free models and those with special butter drawers have heating elements in them. So-called energy saving models usually save only 20%. None is practical in a solar powered home.

The information on Photovoltaic System Design was kindly provided by Bob Mathews of Energy Alternatives Inc., from his company catalogue, "Energy Alternatives: Solutions for Remote Power Supply, Design Guide & Catalogue, 7th Edition.

CASE STUDY:

Solar Panels Replace Hydro Quebec's Umbilical Cord
Montreal, Quebec

WHEN THE overhead power line to this summer island residence not far from Montreal fell down during a 1997 ice storm, Hydro Quebec and its client, Mme. Croisier, decided to look for alternatives. The cost to install either an over the water or submersed power line was excessive and posed a number of environmental and safety concerns. The noise, pollution and maintenance of a generator were also deemed unacceptable.

While the 2400 sq. ft., five bedroom residence, complete with inground swimming pool is used full time from April through to November, it is also used as a weekend getaway when the ice is thick enough to allow the owners to walk or drive to the cottage. Since propane was used for air and water heating, electricity was needed for the rest of the household appliances, lighting, entertainment, and water and pool filter pumps.

A 1080 watt solar array consisting of 12 90-watt photovoltaic modules was used, nine of which were installed on the existing hydro pole with three installed on a tree nearby. This temporary installation is being used until the roof is repaired where upon the entire array will then be roof mounted. A power panel consisting of a 2.5 kW sine wave inverter, DC disconnect, four 30 A controllers, AC bypass and DC distribution panel, was installed in a separate building adjacent to the house. This building also serves as the enclosure for the 1600 amp hours lead acid battery bank.

Mme. Croisier adores her new system and the quiet autonomy she now enjoys. "There's nothing I do without. Even when my whole family including two sons, their wives, six grandchildren and friends, come over for a week or two, the solar system is so reliable I forget where the energy comes from."

Manufacturer/Installer:

Matrix Energy Inc.
16807 Hymus Boul.
Kirkland, Quebec H9H 3L4
Tel: (514) 630-5630
Fax: (514) 426-9123
E-mail: matrixenergy@sympatico.ca

The Big Picture:

Thousands of cottage owners are using photovoltaic systems generate their electrical energy. From simple single module systems charging one battery for a few lights and a radio, or larger systems such as the one mentioned above for a complete home power package, solar systems continue to prove their reliability, cost effectiveness and practicality in a wide range of applications.

Microhydro System Design

Two factors determine the power available from a stream, head and flow. Head is the vertical distance from the water surface at the pipeline inlet to the turbine nozzle. It is a measure of pressure. Flow is the volume of water per unit of time available to the turbine.

Head

Static (or gross) head is the measured head with no flow. Net head is the pressure remaining after pipe friction is deducted. It is the actual head measured while the water is flowing.

Head is the most important measurement to make. The following are different methods of calculating head, but whichever one is chosen, always take multiple measurements to ensure accurate results.

a. Use an altimeter to measure the difference in elevation. Altimeters measure atmospheric pressure so you must make your measurements within a short time and preferably on a day when the weather is not changing rapidly.
b. A pressure gauge installed on an existing pipe is the most accurate method of measuring head.
c. A pressure gauge on a length of garden hose may also be used. Run the hose downhill as far as possible, take a reading and mark the spot. Move downhill and repeat the process. Air bubbles in the hose can cause large errors, so be sure the hose is filled.
d. A hand-held sight level (inclinometre) is effective if there is not a lot of dense bush. Measure the distance between the intake and the proposed turbine site. Measure the slope and calculate the head using trigonometry.
e. A surveyor's transit may be used or one may be approximated with a carpenter's level and a measuring stick. The latter method is slow and prone to error.

Flow

Flow varies seasonally and may vary along a creek's length if tributaries flow into it. Measuring the flow at different times of the year helps to estimate maximum and minimum usable flow. If the water source is not year-round, you may need some other source of power during the dry season. It is best to make at least one measure of flow during its lowest season. The following are different ways of measuring flow. Again, no matter which one is chosen, take multiple readings to ensure accurate results.

a. The easiest method of measuring small flows is to channel the water into a short pipe and fill a container of known volume. Measure the time it takes to fill the container and calculate the flow.
b. If the creek has a known name, it may be possible to obtain flow information from the provincial Ministry of the Environment.
c. The weir method may be used for measuring higher flows. A rectangular or triangular opening of specific dimensions is cut into the top edge of the weir. The opening must be large enough to allow all the water to flow through it. Measuring the height of the water flowing through the opening allows one to calculate the flow.

Equations

1 cubic ft/sec = 450 USGM = 28.3 litres/sec
1 psi = 2.31 feet of head = 0.7 metres of head

A good approximation of the power available to a battery charging microhydro system is given by:

Power (Watts) = Net Head (ft) x Flow (USGM)/14

Hydro turbines and generators for AC only microhydro systems are more efficient, so the above power equation needs to be modified:

Power (Watts) = Net Head (ft) x Flow (USGM)/9

Battery System Example

The total distance between where to draw the water from the creek and where the turbine will be located is 600 feet. Thus the pipeline (L) must be 600 feet long. Along this 600 foot pipeline there is an elevation drop of 80 feet. Thus the static head (H_S) is 80 feet. The flow (F) available from the creek is measured to be 48 US gallons per minute.

L_P = length of pipeline = 600 feet
H_S = static head = 80 feet
F = flow = 48 USGM

From the pipe friction table, friction losses for 2 inch PVC pipe at 48 USGM are 2.7 feet per hundred feet.

H_f = friction head = 2.7 feet/hundred feet x 600 feet = 16 feet
H_N = net head = H_S - H_f = 80 feet - 16 feet = 64 feet

$$\text{Total Power} = \frac{\text{64 feet x USGM}}{14} = \text{219 watts}$$

Power generated would be approximately 219 watts. Over 24 hours, 5 kW of energy would be produced.

In a nominal 12-volt system the turbine and generator would deliver approximately 15 amps continuously or 360 amp hours per day.

The information on Microhydro System Design was kindly provided by Bob Mathews of Energy Alternatives Inc., from his company catalogue, "Energy Alternatives: Solutions for Remote Power Supply, Design Guide & Catalogue, 7th Edition.

CASE STUDY:

Buck Creek Hydroelectric
Lillooet, British Columbia

BUCK CREEK is a 200 acre ranch located far enough west of Lillooet, BC to be off the electrical grid. It has a gravity water system for irrigation and domestic service. This consists of 500′ of 5″ PVC pipe, buried below the frost line. The static pressure on the system is 49 psi.

This site has been generating electricity for many years. The latest upgrade uses an Energy Systems and Design 4″ turgo runner on a vertical axis, directly coupled to a 12 kw brushless alternator. Three one-inch jets use 525 gallons per minute of water to produce 3.2 kilowatts. Irrigation, a rambling, mile long system using a big gun sprinkler, runs from the same pipe. Power production is reduced 10-15% while irrigating.

This system replaced the older 600 watt AC system. It runs year round, with the cleaning of trash racks the only maintenance. The new design has no other scheduled servicing between bearing changes every ten years. It provides excellent electrical service—refrigeration and freezing are difficult off grid—as well as excess power which provides all domestic hot water (washing machine, dishwasher, soaker tub, etc.) and a significant amount of space heating as well. The system is governed by a Thompson and Howe electronic load controller.

The ESD turbine was used originally for smaller, battery charging applications, but is a rugged design, capable of generating several kilowatts. It produces excellent service at very low cost. Today the turbine would cost $1300, and a similar alternator would be about $3000 plus mounting and assembly.

Contact: **Scott Davis**
Yalakom Appropriate Technology
Box 1509
Lillooet, British Columbia V0K 1V0
Tel: (250) 256-0268

Bob Mathews
Energy Alternatives Ltd.
825 Laval Crescent
Kamloops, British Columbia V0E 1M0
Tel: (800) 265-8898
Fax: (250) 374-3943
E-mail:
energyalternatives@bc.sympatico.ca

The Big Picture

People tend to overestimate how much AC hydro is required to give good service. In fact, small machinery like this can fit into existing infrastructure on farms and ranches to give excellent performance at low cost.

CASE STUDY:

Case Study: Nielsen "AC-only" Residence
Kamloops, British Columbia

LIKE MANY PEOPLE in rural British Columbia, Jan Nielsen produces his own electricity. But unlike most people, he does it without burning fossil fuels. By diverting some of the water from a creek near his property east of Kamloops, BC, he has 120/240 volt AC power 24 hours a day.

Two dwellings, a workshop and a barn containing 300 pigs are supplied with hydroelectricity and Nielsen's 14kW microhydro power plant generates sufficient power for all of their demands. "It would have cost the same to bring BC Hydro in," state Nielsen, "but I don't have to pay the bill every month." He estimates he would be paying at least $300 a month to the utility, just to operate the farm. The system is presently operating at a fraction of its capacity, but it will still pay for itself in six or seven years.

The barn contains 1500 watts of heat lamps and two 3/4 hp feed augers. Four large freezers, a meat saw and two fridges round out the main electrical loads. Electric heat in the house is controlled by a computerised load controller, which turns non-critical loads on and off to maintain a constant load on the generator. The generator and the turbine operate at a constant output, which is set by turning the needle valve. The speed and frequency of the generator is regulated by the load control governor, which monitors frequency and diverts excess power into resistive loads, such as the baseboard heaters and hot water tanks. An additional 6000 watts of baseboard heaters are located outside the main house to provide a load dump when no other loads are available. Modern load control governors are a critical component of microhydro systems, making them cost effective.

A Lima generator directly coupled to a pelton turbine was installed and commissioned in 1996 by a local contractor and Bob Mathews of Energy Alternatives. 1430 feet of five inch PVC pipe was run down a hillside to provide 136 psi of static pressure at the power house. The pressurised water in the pipeline squirts through a nozzle and the jet of water strikes the buckets of the pelton turbine. The turbine is essentially a shaft with spoon shaped buckets arranged around the perimeter. The energy in the water is transferred to the turbine and provides the shaft power to drive the generator. The flow rate and power output is manually adjustable via a needle nozzle, like a carburetor. About 575 litres flow through the turbine every minute to generate 5kW to 7kW.

An AC-only hydro system, like the Nielsen's, is virtually maintenance free. The intake must be inspected and the screen cleaned a few times a year, especially during spring runoff. Smaller systems, which utilise batteries, will require additional attention.

Nielsen admits it took a while to learn how to live with the system. "The first time we ran out of power we were having a party. I hadn't adjusted the nozzle and both feed augers came on at the same time." The power plant has a frequency sensor which shuts down the system if the generator is overloaded or underloaded.

System Specifications

Generator: Lima SER 19 kW, 1800 rpm, brushless
Turbine: Canyon Industries Pelton , bronze, 7.5" pitch diameter
Controller: THES A2 Computerised Load Control Governor
Penstock: 1430 feet, 5" PVC Class 100 to Class 160, solvent welded
Tailrace: 540 feet, 6" PVC Class 63, solvent welded
Gross Head: 315 feet (136 psi)
Flow (max): 400 US gallons per minute
Power (max): 14 kW @ 270 feet net head

Contact: Bob Mathews
Energy Alternatives Ltd.
825 Laval Crescent
Kamloops, British Columbia V0E 1M0
Tel: (800) 265-8898
Fax: (250) 374-3943
E-mail:
energyalternatives@bc.sympatico.ca

Wind System Design

Although wind generators may cost less per watt than a comparable PV array, finding the right location is critical. The determining factor is the amount of wind available at the site.

Measuring the Wind

There are three primary ways of determining how much wind is available:

- Installing an anemometer.
- Using existing data.
- Visual observations—Griggs-Putnam Index.

Anemometer

An anemometer is the most accurate way to determine wind speed. However, for a large installation it usually takes two to three years of readings, an anemometer, a data logger and a computer to compile accurate data. Unless you plan on spending $15,000 or more on a wind energy system, this is not a cost effective choice.

There are inexpensive anemometers that can be read manually on a daily or weekly basis. For a small residential wind generator, a minimum of three months of data should be recorded and correlated with another source of wind data. Some people use a small wind generator like the Air 303 as an anemometer.

Existing Data

Existing data can be obtained from local airport or meteorological stations. Other places to contact include universities, colleges and radio stations. The Canadian government compiles wind data in catalogue form and information is available on the internet. Wind forms can vary a great deal within a small area, so this information should be correlated with another method, such as the Griggs-Putnam Index.

Griggs-Putnam Index

The Griggs-Putnam Index is the simplest and quickest method of determining average wind speed. Flagging of foliage (deformation of branches away from the prevailing wind) is correlated with the average annual wind speed. If you have a flag at your site and the flag flies stiffly in the wind for a few hours a day, then the site may be a good candidate for wind power.

See Wind Speed Data in this Chapter.

How much wind is enough?

As a general rule, an average annual wind speed of at least 8 to 10 mph is necessary for a cost effective wind power installation. If wind power is being used as seasonal back-up for a solar electric system, or if less power is required during calm periods of the year, this rule can be modified. Use the average wind speed for the season in which you need the power.

How much power?

Average annual wind speed is used to predict annual energy output. There is some variation in annual energy output because some locations have gusty, intermittent winds, and other locations have steady trade winds. Different locations may have the same average wind speed yet produce very different amounts of energy.

To determine how much power is available on a daily basis, the **average energy output** must be divided by 365 days to give the **daily energy output**. For the purposes of a battery based system design, this daily energy output must be divided by the **system voltage** to give the **daily amp hours**.

Example

In a 12-volt system where the average annual wind speed is 10 mph, a wind turbine such as the AIR 303 will produce 140 kWh.

Daily Energy Output = 140 kWh/365 days = 383 watt hours/day

Daily Amp Hours = 383/12 volts = 32 amp hours per day.

Therefore, if you have an energy budget of 32 amp hours per day, the AIR 303 will be able to supply the power you require at an average wind speed of 10 mph.

The information on Wind System Design was kindly provided by Bob Mathews of Energy Alternatives Inc., from his company catalogue, "Energy Alternatives: Solutions for Remote Power Supply, Design Guide & Catalogue, 7th Edition.

Batteries

Storage batteries are the heart of independent power system. They store electricity for use at a later time when the charging sources, such as the sun, wind or water, are not available. They also provide a reserve of available energy to run loads that require more power than that provided by the charging source.

Batteries wear out and must eventually be replaced. Regular monitoring and maintenance will extend the life span of your battery bank and save you money. Batteries contain toxic materials and should be disposed of accordingly when they wear out.

Choosing a Battery

There are many types of batteries available but many are unsuitable for an independent power system. Choosing the right type of battery is the first step in designing a reliable renewable energy system.

Engine Starting Batteries

Engine starting batteries (car batteries) are designed to provide a brief, high current for ignition, not for deep discharge applications. The "Cold Cranking Amps" rating does not indicate the battery's storage capacity. The thin plates in these batteries are good for a quick release of energy, but deep discharging will cause rapid deterioration.

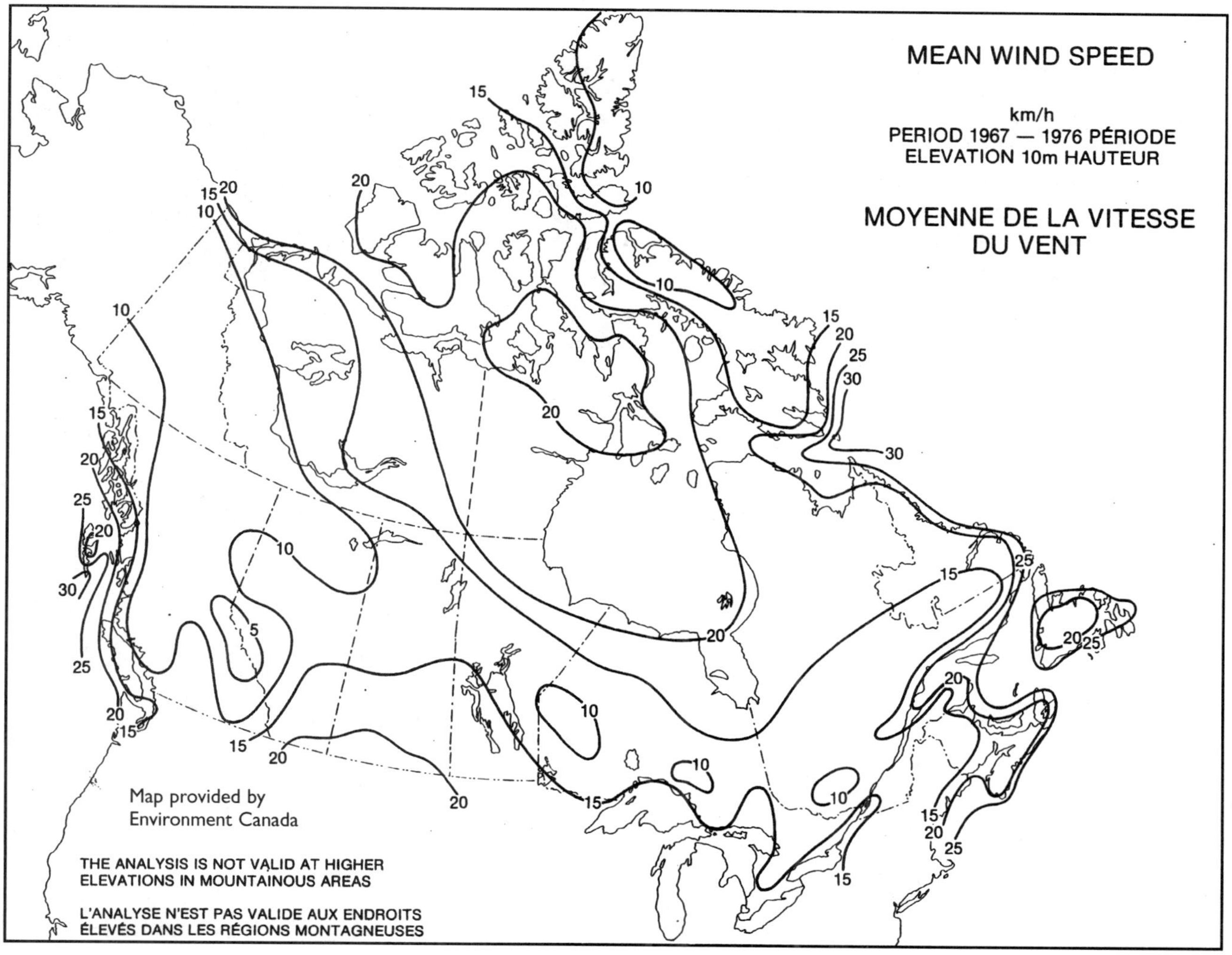

RV and Marine Batteries

RV and marine batteries are a compromise between engine starting batteries and deep cycle batteries. The plates are similar to the engine starting batteries and will not withstand repeated deep discharging. They are adequate for small (under 200 amps-hours) systems and might last two to four years. These are not a true deep discharge battery.

Motive Power Batteries

Motive power batteries (Alco 2200, Global-Yuasa) are deep cycle batteries used to provide energy for electric vehicles such as golf carts and forklifts. They have thick plates that will withstand many deep discharge cycles. These are used for most independent power systems as they are durable, have good storage capacity and are cost effective. These are usually in a 2V or 6V casing, or occasionally in a 12V casing. Typically, the golf cart batteries will last four to seven years, while the Global-Yuasa forklift batteries will last 10 to 20 years.

Stationary Batteries

Stationary batteries (utility back-up cells) are used by telephone companies for back-up power supply systems. They are often designed with calcium alloy plates, which are not for deep cycling, so they are poor batteries for a renewable energy system. However, they may be suitable for use in systems where there is no requirement for autonomy, such as microhydro installations. When batteries are used this way, also referred to as "float service," they are storing reserve energy for peak load demands. Stationary 2V cells frequently come in a clear casing.

Gell Batteries

Gell batteries are filled with an electrolyte in the form of a gel or sponge. It is not possible to service this type of battery or replace the electrolyte. They have a much shorter life and cost considerably more than liquid filled batteries. These batteries require no maintenance, tolerate low temperatures, do not spill and do not produce corrosive gases. They are good at remote sites where maintenance is not possible and cold weather prevails. Precise charge control is critical.

Temperature

Ideally a battery bank should be protected from cold and be maintained at a constant temperature. Battery capacity is significantly reduced as temperatures fall. For example, if a battery is rated as 600 amp hours at 60°C, its rating will drop to 300 amp hours at -20°C. Batteries must be derated when operated in cold climates. Protecting your battery from cold temperatures by encasing them in a protective cover will improve the performance of your electrical system.

Capacity

The capacity of deep cycle batteries is measured in amp hours. The rate of discharge affects the total capacity. The slower the discharge, the higher the capacity. A 20-hour discharge is often used when labeling a battery for renewable energy use. However, 72-hour rates are sometimes used. A battery rated at 800 amps hours at a 20-hour rate can sustain a discharge of 40 amps for 20 hours before it is completely discharged. When purchasing a battery bank, one should consider price, quality, battery bank size, length of expected service, and weight.

Battery Hydrometer

A hydrometer measures the specific gravity of the acid in the battery. It is a reliable indicator of battery state of charge and is a valuable tool for locating weak or damaged cells in a battery bank.

Battery Installation

In the colder parts of Canada, many people place their batteries in an underground insulated box. This is a good way to protect them from freezing weather but the pit must be well drained.

Batteries may be installed inside a home, if they are vented to the outside. When batteries are charging they give off hydrogen and other explosive gases which are also corrosive and foul smelling. To minimize the risk of fire, batteries should be in a vented compartment, separate from any other electrical components. If battery cables to the inverter are run in conduit it is necessary to seal the conduit at the battery box to prevent gases from rising up the conduit and reaching the other electrical components. Even gel cells, which are sealed, should be kept away from sources of flame, sparks and heat. Batteries should be installed so children do not have access to them.

Battery Care

- Keep terminals clean and tight.
- Keep batteries charged. Do not store them in a discharged state.
- Check equalization periodically, especially with deeper discharge cycles.
- Avoid totally discharging your batteries as it greatly shortens their life.
- Inspect cells every three months.

The information on Batteries was kindly provided by Bob Mathews of Energy Alternatives Inc., from his company catalogue, "Energy Alternatives: Solutions for Remote Power Supply, Design Guide & Catalogue, 7th Edition.

CASE STUDY:

Minus 40 Battery Enclosures for Canadian Winters, Canada

WHEN BATTERIES are cold, their usable capacity is reduced. At a temperature of 0˚C, only 70% of their rated capacity can be used. If battery temperature reaches -20˚C, less than 30% of the room temperature capacity is available before it fails. To maintain overall system reliability, the batteries must be kept in a heated room, or a larger bank installed to compensate for the effects of low temperatures.

Minus 40 enclosures are a third option to maintain system reliability. These advanced thermodynamically designed battery containers maintain battery temperature above 0˚C regardless of ambient outdoor temperature. They do this passively, without fuel. By keeping batteries warm, more capacity will be available, so system reliability is increased. Minus 40 enclosures also allow smaller battery banks to be used, without lowering battery reserve autonomy simply by ensuring batteries stay above 0°C.

Minus 40 enclosures are custom designed for each site, and take into account climate data, battery bank mass and the required days of reserve autonomy needed. The enclosures use advanced insulation materials and select thermal ballast materials to ensure batteries stay warm even after prolonged periods with freezing temperatures. With a powder coated, welded aluminum outer shell, they are impervious to the effects of weather.

Applications:

Oil fields, science and research, and parks applications.

The Big Picture:

Minus 40 battery enclosures ensure that batteries located in remote, hostile locations will be protected from the elements and provide reliable service. Costs are minor when compared to risks associated with equipment failure, and site visits.

Contact: **SOLTEK Solar Energy Ltd.**
#2 - 745 Vanalman Ave.
Victoria, British Columbia V8Z 3B6
Tel: (250) 727-7720
Fax: (250) 727-2135
E-mail: soltek@soltek.ca
Web: www.soltek.ca

Inverters

An inverter must have the capacity to handle the appliances you intend to use. List the appliances that could be running simultaneously and determine their power rating. If the rating is listed in amps, multiply the value by the appliance voltage to calculate the watts. Inverters are rated according to the continuous power they can produce; however, they are also designed to deliver large amounts of power for short periods. This is important for some AC motors that draw many times their normal operating current when starting. Refrigerators and pumps generally have the highest surge requirements.

Twelve-volt systems are ideal for small applications because of their simplicity and the availability of DC appliances for entertainment, lighting, and communications. Twenty-four volts is ideal for larger systems. The current at 24-volts is half that at 12-volts and makes wiring, fuses, and circuit breakers smaller and cheaper. Larger inverters are available at 24-volts and are becoming cheaper and more powerful than 12-volt counterparts.

Wire sizing in DC circuits

The current is much greater with low-voltage circuits so larger wires are required. The wire size increases as the gauge number decreases. For example, for a 5% voltage loss in copper wire carrying 6 amps, #12 wire is good for up to 29 feet; #10 wire is good for up to 46 feet; and #6 wire is good for up to 118 feet.

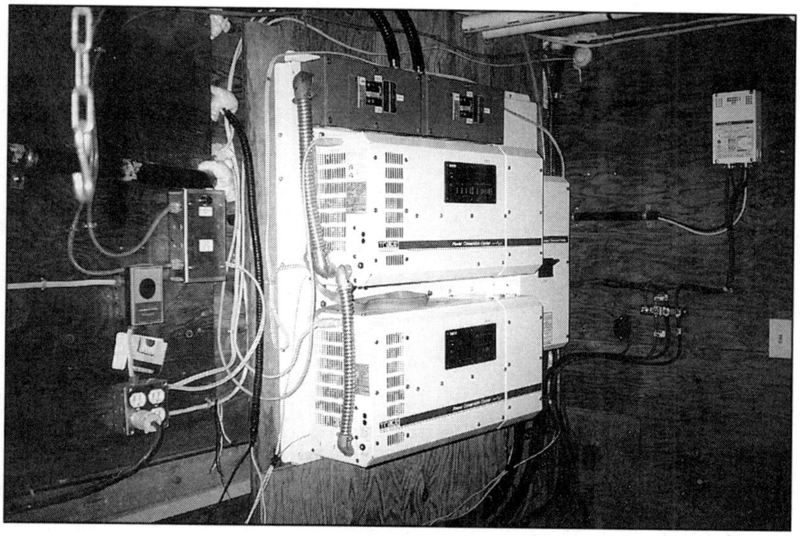

CHAPTER 3

Commercial and Industrial Applications

Renewable Energy in the Industrial & Commercial Sectors

THE INDUSTRIAL SECTOR accounts for the largest share of absolute energy use in Canada, with 39%, followed by the residential sector and then the commercial sector. The industrial sector can be defined as including all Canada's manufacturing industries, as well as forestry, construction and mining. In Canada, almost 28% of total energy use in the industrial sector comes from the following areas; pulp and paper, mining, petroleum refining, iron and steel, smelting and refining.[1] The energy demand in the commercial sector comes from any activity related to trade, finance, real estate, public administration, education and commercial services.[2] There are two major building types in the commercial sector: commercial buildings; offices, retail stores, restaurants, and institutional buildings; schools, colleges, universities, hospitals and religious buildings.

The utilisation of renewables in the commercial and industrial sectors comes in many forms, including the implementation of photovoltaics in a building facade or on the surplus, unused rooftops of commercial buildings, the use of solar hot water systems to heat water for

[1] Natural Resources Canada, "Energy Efficiency Trends in Canada"(Ottawa: Ministry of Supply & Services, 1996), 26.
[2] Natural Resources Canada, "Energy Efficiency...", 20.

hotels, motels and hospitals, municipal swimming pools and to sterilise jars for canning. For many companies or governments, where the "bottom line" is crucial in the decision making process, energy costs are significant and any way of reducing them should be attractive. If the industrial and commercial sectors began to use renewable energy on a larger scale, the Canadian environment would benefit as greenhouse gas emissions would be reduced, as would air and noise pollution.

Recently, Natural Resources Canada (NRCan) has begun to financially support the installation of commercial and industrial renewable energy installations with the Renewable Energy Deployment Incentive (REDI). This $12 million initiative was developed to stimulate market demand for reliable, cost-effective renewable energy systems. NRCan has targeted four types of systems to be developed. They include active solar hot water systems, active solar air heating systems, ground-source heat pumps, and high-efficiency/low-emissions biomass combustion systems.

Residential and Commercial Applications of Earth Energy

Solar energy is the sole source of heat from an earth energy unit, which transfers the warmth found in the soil below the frost line into the building. In the commercial sector, many buildings incorporate a heat pump with a conventional boiler and cooling tower or condenser (to the ARI 320 standard).

A ground-coupled loop can easily replace the boiler/tower with passive operation, thereby dropping operating costs and increasing efficiency by up to 30% over typical cap-tube type units. Commercial buildings can install an open-loop unit (ARI 325) or closed-loop (ARI 330), and units can handle entering water from -5°C to 40°C. Units are insulated and designed to fit in areas with limited mechanical space, with options such as ECM variable speed fan motors and scroll compressors.

Earth energy is ideally suited for commercial applications. Surrounding property and parking lots provide an excellent heat sink, and are easy and inexpensive to install. Decentralized control allows simultaneous heating or cooling of different zones, allowing areas of the building to set different temperatures.

Basic Heat Pump Cycle

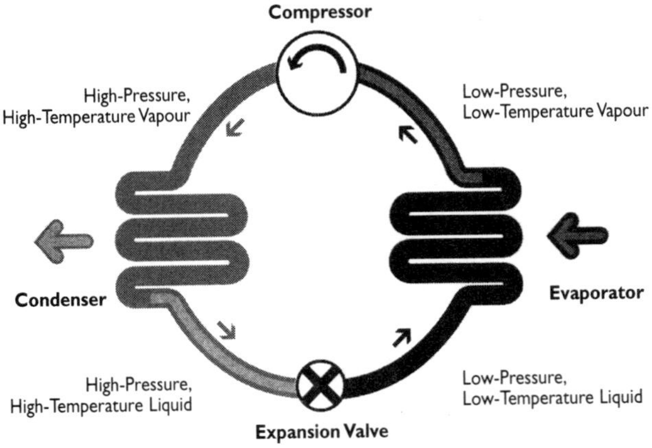

Source: *Heating & Cooling with a Heat Pump*

Locating the heat pump inside the building reduces damage from adverse weather conditions and vandalism, and eliminates the need for roof penetrations (further securing the building envelope and allowing enhanced design with non-horizontal rooftops). Occupant health and safety is increased by the elimination of combustion heat and the need for fuel storage, as well as the absence of possible carbon monoxide leaks and superior levels of indoor air quality. The smaller mechanical space requirements allow all available space to be dedicated to work area, which reduces the effective cost of construction, and modularity allows reduced inventory for boilers and chillers.

Studies on systems in schools indicate that lifecycle costs are as much as 19% lower than conventional combustion-heated systems, and maintenance costs at one secondary school have been quantified at less than 4¢/ft^2 per year.

To capture the earth's heat, there are four basic loop designs available:

1) Open-Water, Open-Well or Open-Loop

This configuration takes water directly from a drilled well or a body of water into the heat pump where the heat is extracted. The cooled water is then returned directly to the same lake or well. If a lake is used, it must be large enough to provide a sufficient heat sink capacity; the same is true of a river. The water cannot have high levels of chlorides and other mineral contaminants as these may damage the pumps. Government regulations may restrict the use of water from Crown lands or the placement of loops in navigable waterways. If a well is used, follow the regulations concerning the extraction and re-injection of water, which are designed to protect natural aquifers.

Advantages of open-loop systems:

- higher efficiency of heat transfer because there is no transfer fluid to serve as middleman
- the installed cost is usually lower, especially if there is already a well drilled for the building's domestic hot water (DHW) supply
- there is no need to use any chemicals in the transfer fluid

Disadvantages:

- the disruption of water tables and aquifers could affect the environment
- any fluctuation in water temperature will degrade system performance

2) Lake Loop Or Pond Loop

A closed-loop system of durable polyethelene pipe is positioned on the bed of a lake, and weighted to avoid shifting. The loop extracts heat from the water that circulates in the area.

Advantages of lake loop:

- usually less expensive to install than conventional land-based trenching
- relatively easy to find any leaks that may occur in the loop

Disadvantages:

- potential environmental damage to aquaculture if there is a leak, depending on the heat transfer fluid used and the degree of leak
- if the temperature of the lake water drops during spring runoff, system efficiency will drop

3) Horizontal Closed-Loop

This is the most common configuration. A trench is dug, in which polyethelene pipe is buried in a continuous parallel loop (the length of pipe and amount of trenching depends on the size of the heating unit). The loop must be buried at least 600 mm (2 feet) below surface, but most contractors dig to at least 1 m. The trench must be backfilled properly to avoid air pockets that reduce heat transfer and to ensure that the pipe is not damaged by large rocks. A heat transfer fluid is then pumped through the pipe.

Components of a Typical Ground-source Heat Pump

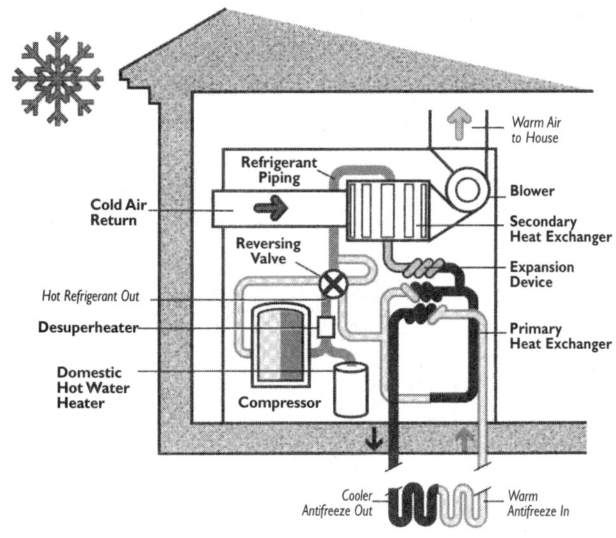

Source: *Heating & Cooling with a Heat Pump*

Advantages of horizontal closed-loop:

- relatively easy to install
- high flexibility of design
- greater control over the temperature of the fluid circulating in the loop

Disadvantages:

- high cost of trenching in rocky ground
- more difficult to find a leak in a buried loop
- landscaping may be required on retrofit installations

4) Vertical Closed-Loop

In this system, a deep hole is drilled and laid with a loop of pipe. A heat transfer fluid is pumped through the pipe. This is an efficient configuration due to the stability of the heat source at greater depth. Due to the cost of drilling, it can be an expensive option, but is viable when surface property is limited. The boreholes must be grouted and sealed after drilling.

Advantages of a vertical closed loop:

- the most energy efficient closed-loop system
- less loop required than comparable horizontal closed-loop installations, resulting in lower pipe costs
- a smaller footprint, demanding less ground area
- secure from any accidental post-installation damage

Disadvantages:

- potentially high-cost option if drilling is expensive
- potentially environmentally damaging to underground aquifers if the boreholes are not installed and sealed properly

Hot Water Heaters

Domestic water heaters (desuperheaters) are a popular option on earth energy units. Desuperheaters can be 'low' or 'high' priority. They are high priority if the solar heat is diverted first to water demand, or low priority if they heat the water after the space heating requirement has been satisfied. Both DHW and space heating require back-up systems. These systems kick in to bring heat levels up when earth energy is depleted by one of the applications. In summer, the concurrent demand for hot water and space cooling makes an earth energy unit highly efficient in supplying both.

Other Applications

Earth energy units are frequently used for the de-humidification of indoor swimming pool areas, where the heat pump can reduce humidity in the air and provide condensation control with a minimum of ventilation air. The heat recovered from the moisture is used for heating domestic or pool water, or for heating the inside space. They are also used as HRV units for exhaust air from buildings or wastewater from an industrial process. The recovered heat is supplied at a higher temperature at which it can be more readily used for heating air or water.

Efficient heating performance makes earth energy a viable option for heating or cooling of commercial and institutional buildings, such as offices, stores, hospitals, hotels, apartment buildings, schools, restaurants and a number of penitentiaries.

Important Factors

The design and installation of an earth energy system (and its subsequent performance) relies heavily on a number of important factors:

Heat Loss

The important first step in the design of a ground-coupled system is to determine how much heat is required to satisfy the comfort level inside the building. The national installation standard for residential systems (CSA C445) states that a heat loss must be calculated in accordance with the F280 program, which uses insulation levels, geographic location, soil type, and other factors to determine the total annual heat loss. With this calculation, the unit can be sized to provide heat in the winter, cooling in the summer and domestic hot water, if desired.

Sizing

Earth energy units can be sized to 60% of the building heat loss if there is an auxiliary heating source for backup and emergencies. Most of a residential heat load can be met by a unit sized to this level, where an electric plenum provides the balance of heat on cold winter mornings. This allows installation costs to be reduced considerably due to the smaller furnace and ground loop required. Over-sizing is expensive and can result in problems, but under-sizing will force the unit to extract excessive heat from the ground, causing it to fail. Most contractors now design for at least 75% to 80%, which is a good compromise between installation costs and slightly higher operating costs.

Air Flow

An earth energy unit provides a small temperature rise. This means the temperature of the air as it comes through the floor register is lower than that from a combustion furnace. As a result, the unit must move more air to supply the same amount of heat, and the size of the ductwork must be larger than normal. If an earth energy unit is installed during a renovation, there may be additional cost for ductwork.

Soil Type

Loose dry soil traps air and is less effective than moist packed soil for the transfer of heat from the ground to the buried pipe. Manufacturers provide specifications on the relative merits of each soil type. Low-conductive soil may require twice as much loop as a high-conductive one, for the same energy transfer. This means higher installed cost.

Loop Depth

Stable underground temperature is the key to heat transfer, so, the deeper the loop is buried, the more efficient it will be. Trenches should be 1 m to 2 m deep, not deep enough to require installation scaffolding. Vertical boreholes are very efficient, but drilling is expensive.

Loop Length

The longer the piping in a loop, the greater heat can be transferred from the ground to the building. 'Short-looping' for the selected heat pump will produce less heat and place excessive strain on the system. However, too much piping can create other problems, as well as increase installation cost. Each manufacturer provides specifications for the amount of pipe required for any specific site.

Loop Spacing

If the distance is greater between buried loops, there is more heat to be extracted from the ground between the pipes. If the pipes are separated by 3m, they can collect heat without interference from any neighbouring loops. It is also possible to install up to six horizontal loops in one trench, in order to reduce the amount of land required for trenching.

Type of Loop

Most pipe has a polyethelene base, a durable and long lasting plastic also used in natural gas pipelines. Two coiled loops (the Svec Spiral and the Slinkey) require less trenching than conventional straight pipe and, thereby, reduce the footprint required and installation costs.

Water Quality

Open-loop systems require a source of water that is adequate in temperature, flow rate and mineral content. The efficiency of an earth energy furnace will drop markedly if the entering water temperature drops below 10°C or if the flow rate is constricted. An official water well log must be completed to ensure a sustainable yield for open-loop configurations. Some units may be susceptible to contaminants in the source water, such as chlorides and metals, which can damage the heat exchanger.

Water Discharge

Environmental regulations govern how the water that has circulated through an open-loop system must be returned to the earth. A return well must put the water back into the same aquifer or level of water table.

Balance Point

This is the outdoor temperature at which an earth energy unit can fully satisfy the heating needs of a building, and is usually -10°C in most regions. As long as the outdoor air temperature is above the balance point, the unit can satisfy the demand for heat inside. When the ambient temperature drops, the unit runs continuously, and activates the auxiliary heater (second stage heat) to meet the demand.

Heat Transfer Fluids

Provincial environment officials authorize which fluids can be used in closed-loop units, and manufacturers specify which are acceptable to their units. The most common fluids are denatured ethanol or methanol (the latter is not approved for use in Ontario).

Solar Heating for Industrial and Commercial Applications

There are two different types of solar heating in industrial and commercial applications: active solar hot water systems and active solar air heating systems.

Solar Air Heating

Since 1992, over 25 solar air heating systems have been installed in Canada. In most systems, fans draw air through south-facing solar collectors, mounted on the building's wall, and the air is preheated before being introduced into the building. A Canadian company, Conserval Engineering, has developed the SOLARWALL® which is rapidly becoming the Canadian and international standard for commercial and industrial air heating.

The system works by drawing ventilation air through a perforated aluminum absorber. The unglazed absorber is mounted directly on the building wall and replaces conventional wall cladding. The absorber is heated by the sun, and transfers the heat to the air as it passes through the perforations. The system will function as a preheater, even on cloudy days, reducing the fuel used by the conventional heater. In the summer, the system is bypassed to prevent overheating.

A major advantage of the SOLARWALL® system is that it can be integrated readily into new building designs, and can be supplied in a wide variety of colours to ensure an attractive appearance.

Solar Water Heating

In Canada, there are over 300 industrial/
commercial solar hot water systems currently installed. The majority of the systems are used for institutional pool heating, with the remaining systems being used in commercial systems, such as car washes and canning plants. Many of these systems were installed in the early 1980's, when fuel prices were high and government programs existed to defray installation costs.

Some factors to look for when prospecting for solar water heating opportunities include:

- large volumes of hot water used
- load evenly spaced through daytime
- load spaced evenly by month, or biased to summer load
- relatively high cost of fuel used for water heating

The following three examples illustrate some of the features of commercial water heating applications:

1. Commercial Hot Water

Commercial hot water systems typically use collectors, and storage tanks, along with a pump and controller. Likely solar applications include fitness centres, hospitals, and hotels. While the components do not vary much from SDHW, the size of the systems do vary. Commercial systems

may have collector areas of over 60m^2, while a typical SDHW system will have no more than 6m^2.

Because many commercial sites have more regular maintenance regimes, commercial solar water heaters can be more complex than residential systems, and closed loop glycol based systems are more common. While these systems require more care, the site requirements may preclude the opportunity to use a drainback strategy.

While glazed flat plate collectors are traditionally used for commercial systems, unglazed collectors can present a more economical option, though the solar fraction may be lower.

2. Aquaculture

Aquaculture is a growing field that is an interesting, niche application for solar water heating. In a study released by the Canadian Aquaculture Industry Alliance, "Canadian Aquaculture Industry Profile and Labour Market Analysis" concludes that there is significant growth potential from the industry in Canada. In British Columbia alone, by 2010, the industry could be a billion dollar business with 20,000 full-time jobs.

As fish are cold-blooded, their metabolism and growth rate are dictated by the surrounding water temperature. Consequently, an increase in water temperature of only a few degrees can have a dramatic increase in the fish growth.

Solar heaters for aquaculture use unglazed plastic panels because they outperform glazed panels at the temperatures used, as well as costing less. Additionally, as copper is quite toxic to fish, plastic panels are safer than metal panels for the aquaculture operation.

The heating requirements for this application differ from a commercial hot water system in several important ways. The volume of water being processed is much greater, often in the millions of litres per day, but the temperature increase required may be quite small, perhaps just a few degrees. The precise delivery temperature of the water is very important; it is usually required to be within 0.1°C.

Consequently, solar heaters for these sites tend to be relatively large, with collector areas of over 100m^2. Additionally, unglazed panels are used as they were found to be both more efficient, and less expensive than glazed collectors at the required temperatures. Because the water flow is continuous, there is a reduced requirement for storage. The hatchery water is passed directly through the panels and accurate controls are used in conjunction with a backup heater to ensure that the exact water temperature is maintained.

3. Municipal Pool Heating

As with residential pool heating described in Chapter 1, municipal and other public pools are excellent applications for solar. With virtually every town in Canada having its own public pool, this is one of the first places for Canadian municipalities to look for cost-effective renewable energy installations.

Solar heaters for commercial pools can use either glazed panels designed to contribute significantly on a year-round basis, or use unglazed panels, with the strategy to have a lower solar contribution, but on a more cost-effective basis. The appropriate strategy will vary from pool to pool, depending on the local climate, cost of energy, and economic requirements of the pool owner.

Seasonal pools may use solar on a stand-alone basis, in which case the pool temperature will fluctuate with the prevailing weather conditions. If a minimum temperature is required regardless of the weather, an auxiliary heater is needed. Pools operated year-round will require an auxiliary heater, and the solar will simply reduce the amount of energy used by it.

Photovoltaics in Commercial and Industrial Applications

Since the late 1970s photovoltaics have been mounted on the rooftops of commercial and industrial buildings still connected to the grid. Recently, the numbers of building integrated photovoltaic (BIPV) buildings has grown rapidly in North America, Europe and Japan with over 10,000 BIPV systems with 25 MW aggregate generating capacity being installed annually.[3]

PV systems are connected to the existing electrical grid with the use of an inverter, which converts the PV-generated DC electricity to the AC electricity used on the grid. The grid is a ideal storage component for the generated electricity and also ensures that the building will maintain a

---

[3] J. Nodelman, T. Tupper and D.S. Shugar, "Performance and Value of Building-Integrated Photovoltaics: A Case Study of EPCOR's 13kW PowerGuard System," RETCCC/SESCI'98 Conference Proceedings, 4-6 May, 1998, Montreal, Quebec.

regular supply of electricity. Regulations differ widely from province to province with respect to the interconnection requirements.

Some photovoltaic systems are conventionally mounted to the building's roof which requires that additional structural materials, such as structural steel, are adhered to the roof, penetrating the existing seal. The penetration of the roof allows for water to seep into the roof, damaging it and reducing its lifespan.

BIPV systems are installed as part of the building envelope, replacing a conventional facade or traditional roofing material. BIPVs can be installed without any additional structural materials and no penetration of the building envelope. As a facade system, BIPV's may provide shade and reduce noise. As a roof mounted system, BIPVs can perform the additional roofing functions of displacing heavy roofing rock ballast, providing extra R-value insulation and providing ultra-violet light protection. BIPV roofing systems can be installed on either flat roofs or moderately sloping roofs and may be installed during new construction, re-roofing or a retrofit.

The PowerGuard system, which was installed by the Edmonton Power Corporation, is manufactured by the PowerLight Corporation of Berkeley, California, and combines PV panels backed with two inches of extruded polystyrene foam. The system is installed, unadhered over the fully adhered roofing membrane. The panels are then electrically interconnected to an inverter, which feeds AC power into the building.

RETScreen™:

Renewable Energy Analysis Software

Contributed by CANMET Energy Diversification Research Laboratory

AS PART OF Natural Resources Canada's Renewable Energy for Remote Communities (RERC) Program, the CANMET Energy Diversification Research Laboratory (CEDRL) has developed, with the help of industry, a Renewable Energy Technologies (RETs) project assessment "tool" called RETScreen™. The RETScreen™ pre-feasibility analysis model is a computerised generic tool developed to help an energy project proponent prepare a preliminary evaluation of the annual energy performance, costs and financial viability of potential RETs projects located throughout the world.

Five months after the official launch in May 1998 by the Hon. Mr. Ralph Goodale, Minister of Natural Resources, there were over 1200 users of RETScreen™ in 70 countries. These results indicate that RETScreen™ is positionning itself to become not only a standard in Canada but also around the world.

The RETScreen™ tool includes computer software, detailed user's manual and an accompanying information database. Version 98 of the RETScreen™ software has been developed for use with Microsoft Excel, a spreadsheet program for PC compatible computers. Currently, the tool can be used to evaluate projects based on the following commercially available and reliable technologies:

- Wind energy
- Small hydro
- Photovoltaics
- Solar ventilation air heating (VAH)
- Biomass heating

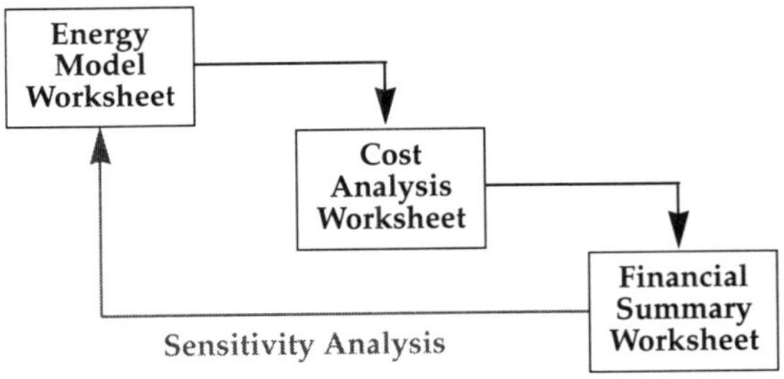

Three additional models are currently under development and will become available in Spring 1999. The upcoming models deal with:

- Ground source heat pumps
- Solar water heaters
- Passive solar designs

The models are supplemented by the manual, which features a database facilitating the evaluation of projects in Canada's 300+ remote communities. The database includes estimates of fuel costs, electricity demand and renewable

A 20 kWp rooftop photovoltaic array and a solar ventilation air heating system are installed at the CANMET Energy Diversification Research Laboratory.

energy resource availability. An international database of solar radiation is also expected to be released in Spring 1999.

RETScreen™ is intended to be used by technical and financial personnel from consulting engineering and planning firms, electric utilities, government and development agencies, financial institutions, R&D organisations, private power developers and product suppliers. Some of the main benefits obtained by using the tool include:

- reduces the costs, time, risks and errors associated with conducting project pre-feasibility studies
- facilitates the consideration of RETs by decision-makers who perform community, government and utility energy resource planning
- provides a low-cost preliminary design approach for project developers and industry;
- facilitates training and information transfer for RETs
- increases the initiation of project studies which help identify the best opportunities for successful RETs project implementation

The tool will also be useful to government agencies for policy development activities and R&D organisations to "benchmark" the project cost and/or performance impact of their research and development efforts.

RETScreen™ can be downloaded free of charge from the RETScreen™ web site at cedrl.mets.nrcan.gc.ca/retscreen/. Other inquiries on RETScreen™ should be directed to :

Natural Resources Canada
CANMET Energy Diversification Research Laboratory
1615 Lionel Boulet, Varennes, Québec, J3X 1S6
Tel: (450) 652 4621
Fax: (450) 652 5177
E-mail : rets@ncan.gc.ca

Waste Treatment

Municipal solid waste (MSW) generated in one locale and disposed of in another has increasingly become a political issue. Each of us must address the question "Why do we generate so much waste?" With modest efforts, including composting and recycling, we can reduce our solid wastes. For a time, garbage was incinerated to reduce its bulk, but studies have shown that this produces dioxin and other cancer-causing toxins. Moreover, recycling waste recovers three to five times more energy than incineration produces.

Approximately 25% of MSW is compostable wet waste, such as disposable diapers, which can be converted by anaerobic digestion to produce methane gas, which can be used as a fuel to produce electricity. See the following Gazmont case study.

Anaerobic Digestion Plants

Most of us are familiar with composting, a natural process in which microbes convert organic wastes, such as leaves and grass clippings, into compost, or humus. The process can be either aerobic (with air or oxygen) or anaerobic (without air). A facility on a scale appropriate for a municipality can provide an efficient, effective, environmentally appealing, and commercially viable way to convert organic wastes generated by residential, industrial, commercial, and institutional sources. See the following Newmarket Composting Plant case study.

Aerobic Digestion Plants

Waste water and sewage is usually treated by exposure to the atmosphere, an aerobic process. The aerobic process is accelerated by injecting oxygen into the waste stream. A recently developed sewage treatment system converts the variety of organisms—from bacteria and algae to higher plants, snails, and fish—that reduce the nutrients and pathogens by breaking down and digesting them in a continuous food chain. See the following Body Shop case study.

CASE STUDY:

Production of 25 MW of Electricity From Landfill Gas (Gazmont)
Montreal, Quebec

ON JANUARY 1, 1988, the City of Montréal took possession of the Centre de Tri et d'Elimination des Déchets (CTED) at the Miron quarry, a landfill site which has received approximately 30 million tons of waste since 1968. The main objective of Montréal officials was to improve the environment and the quality of life in the neighbouring communities. It was from this perspective that the City of Montréal accepted the offer by Gazmont to construct and operate a plant that would produce electrical energy from the treatment of landfill gas produced by buried waste. If landfill gas is not controlled, it becomes a source of environmental problems. On the other hand, if it is effectively managed, it can generate a fair amount of electrical energy. A 25 MW power plant was built at the cost of C$ 37 million. The electricity generated by the landfill is sold to Hydro-Quebec. The main benefits have been a reduction of nuisance caused by landfill gas and the production of energy from a waste product.

System Specification

- **Power plant**
  - Gross capacity: 25 MW
  - Landfill gas maximal inlet flow: 17 383 Nm^3 /h
  - Landfill gas heating value: 16,54 MJ/Nm^3
  - Heat rate (net): 12 600 Btu/kWh
  - Output voltage: 13,8 kV
  - Transmission voltage: 24,94 kV
- **Boiler**
  - Quantity: 1
  - Inlet energy: 309 500 MJ/h
  - VOC destruction efficiency: 99%
  - Steam flow: 100 000 kg/h
  - Steam temperature: 955 °F
  - Steam pressure: 1270 psia
- **Turbine/alternator**
  - Output power: 25 MWe
  - Net power: 23,3 MWe

- **Electrical supply (substation)**
  - Substation supply voltage: 25 kV
  - Equipment operating voltage: 600 Volts
  - Substation: 1

Ownership

Gazmont is owned by SNC, Biothermica and Novergaz.

Other application

Landfill gas valorization technology can be applied to landfill sites as an improved and worthwhile solution to change waste to energy.

The Big Picture

An energy source for 8,200 households.
Less than five-year payback period.
Decrease in atmosphere pollutants and greenhouse gas emissions.

Contact: **Mr. Guy Drouin, President,**
Biothermica International Inc.
3333, Cavendish Boulevard Suite 440
Montreal, Quebec H3B 4M5 Canada
Tel: (514) 488-3881
Fax: (514) 488-3125
E-mail: biodir@total.net.
Web: www.biothermica.com

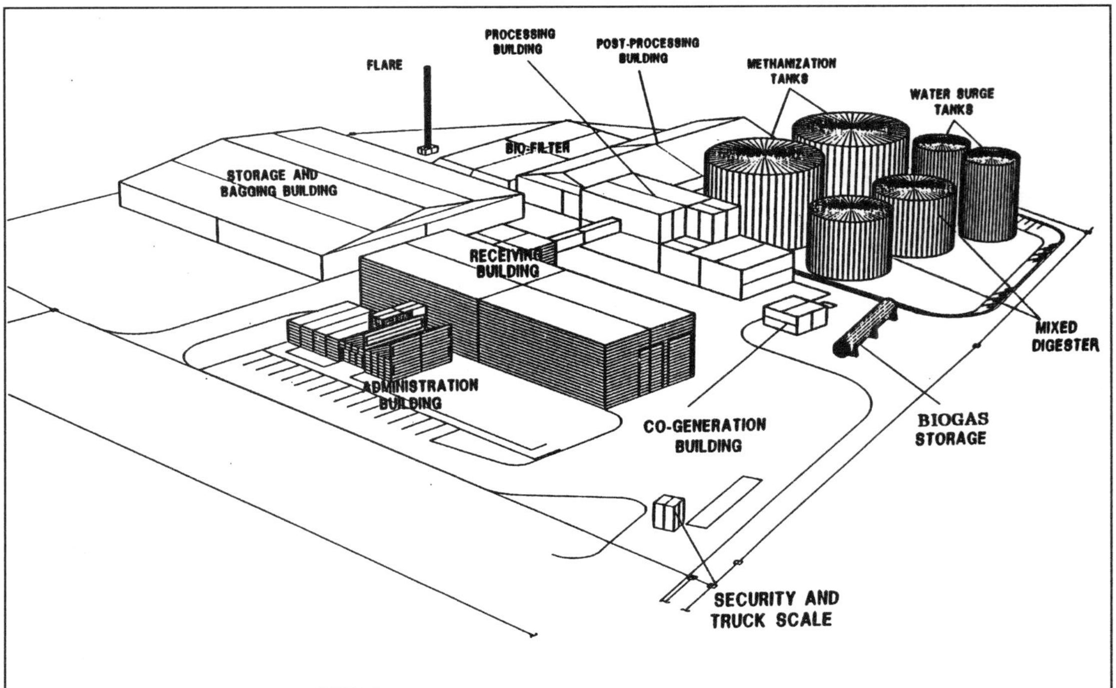

CANADA COMPOSTING INC. - NEWMARKET PLANT DESIGN

CASE STUDY:

Newmarket Composting Plant
Newmarket, Ontario

A COMPOSTING PLANT using a commercial process developed by Biotechnische Abfallverwertung Gmbh (BTA) of Munich, Germany, is under development in Newmarket, Ontario. The technology, licensed to Canada Composting Inc. (CCI), converts the biodegradable elements of municipal solid waste (MSW) into two primary, high quality, end products—compost and biogas. The $2 million Newmarket facility will be the first plant of its kind in North America, will process more than 120,000 metric tonnes of MSW annually and will occupy less than three acres of land. This plant, a plant in Gloucester, Ontario and another composting plant being built for the City of Toronto are due to open by the end of 1999.

The plant will produce 13 million cubic metres of gas per year, which can be converted into 2 megawatts of energy. The retention time for the waste in the plant is a matter of a few days, rather than the 30 days typical of other processes, and will operate 24 hours a day, 365 days a year. The process is completely enclosed to eliminate contaminants, such as leachates, from entering the water table, noise, dust, and odours. CCI expects to produce compost similar to the quality soil enhancer produced by BTA in Europe. The biogas, which has a methane content of 73% with 27% carbon dioxide and trace amounts of hydrogen sulphide, will be used to produce electricity in the plant in a co-generation facility. The tipping fee to receive MSW will remain in place and will contribute revenue to ensure the process is profitable.

System Specifications

120,000 metric tonnes MSW annually

Supplier Contact:

Paul Blanchard or William Faulds
Canada Composting Inc.
Newmarket, Ontario L3Y 7V1
Tel: (905) 830-1160

Client/Customer

Town of Newmarket, Ontario

Other Cities

Three BTA plants are operating in Germany and one in Denmark; eight others are at various stages of development in Europe.

The Big Picture

A process that turns MSW into compost and usable methane gas is significant. This plant will lead the way for municipalities to process their solid wastes.

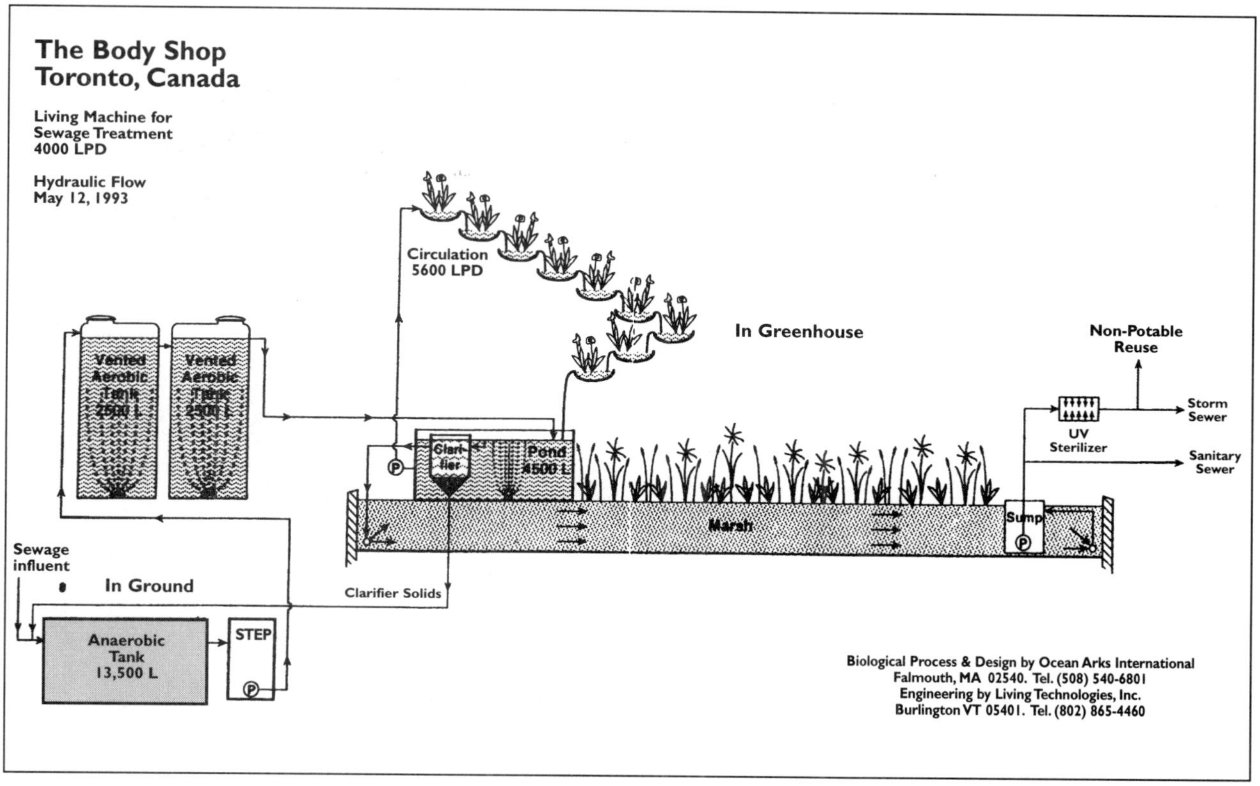

CASE STUDY:

The Body Shop Sewage Treatment
Toronto, Ontario

FOR SEWAGE treatment the Body Shop factory in Toronto, Ontario, uses a Living Machine designed by Living Technologies Corporation in Toronto. Living Technologies Corporation is affiliated with Living Technologies in Vermont and with John Todd (who designed the ARK in PEI some twenty years ago), at the Centre for the Restoration of Waters at Ocean Arks International. The Living Machine uses an engineered biological system to treat waste water and sewage.

The system consists of an anaerobic digester, aerated bioreactors, rock biofiltres, a clarifier, and a constructed wetland, all enclosed within a greenhouse. All stages, except for the anaerobic digester, incorporate higher plants, including many of the plant species used in the Body Shop's products—dwarf banana, bay laurel, camellia, comfrey, eucalyptus, evening primrose, horsetail, St. John's wort, juniper, passion fruit, peppermint and rosemary. Using solar energy to heat the system and to energize photosynthesis, the Living Machine closes the cycle, going from pure water to contaminated water and back to clean, potable water.

System Specifications

8, 000 litres of sewage are treated daily

Supplier Contact:

Vera Lusney
Living Technologies Corporation
Don Mills, Ontario M3C 3N6
Tel: (416) 445-0647

Client/Customer:

The Body Shop
Toronto, Ontario
Tel: (416) 441-4189

Other Cities

Boyne River Ecology Centre, Orangeville, Ontario; Muncie, Indiana; Frederick, Maryland; Stirling, Scotland; San Francisco, California.

The Big Picture

This important process has great potential for treating sewage close to the source while growing useful plants and resulting in potable water.

CASE STUDY:

Solarwall™ on the Windsor Senior Citizens Building
Windsor, Ontario

WINDSOR HOUSING AUTHORITY'S Ouellette Manor Senior Citizens Building in Windsor, Ontario, needed some renovations. As with many older brick apartment buildings, the exterior of the 24-storey building was deteriorating from moisture penetration. Freezing of the moisture caused the bricks to crumble. Recladding was necessary to prevent further damage. An unglazed, brown perforated panel solar collector was chosen to cover the facade.

The SOLARWALL® offers an ideal solution. It will prevent further moisture damage, has an attractive appearance and provides free heating of air in the building. 335 m^2 of solar panels were installed on the south side of the building at a small incremental cost over traditional cladding. The extra costs cover the additional supports to mount the solar cladding out from the main wall to form an air cavity and a summer bypass damper on the roof to connect the SOLARWALL® to the existing fan. In winter, the solar panels heat the ventilation air and in summer the SOLARWALL® panels shade the south wall from heat gain reducing the building's cooling requirements.

It looks like conventional metal cladding except that it is mounted 250 mm from the wall and is perforated to allow the outside air to travel through the metal where it picks up the solar heat. The heated air is then directed to the fresh air intake of the ventilation system and supplying 28% of the ventilation heating requirements.

System Specifications

Collector Type: Unglazed brown SOLARWALL®
Collector area: 61m high X 5.5m wide, 335 m^2
Ventilation: 23,000 m^3h
Solar system output: 195,700 kWh/a
Dollar savings: $4,860/year

Client/Customer:
Windsor Housing Authority

Other Applications:

The SOLARWALL® can be installed on any building that uses ventilation air during theheating season and is especially beneficial for industries that produce heat and fumes during their manufacturing process.

Supplier: Conserval Engineering
200 Wildcat Rd.
Toronto, Ontario
Tel: (416) 661-7057
Fax: (416) 661-7147
E-mail: conserval@globalserve.net
Web: www.solarwall.com

The Big Picture

Both the building superintendent and the manager of the housing authority have been interviewed on television stating "This system works! We are saving energy because the gas heaters haven't been on. The system even works on a cloudy day."

CASE STUDY:

Solar Lumber Kiln
Canadian Prairie Provinces

A SOLAR LUMBER KILN designed for the Canadian climate is effective 10 months of the year, even in the cold prairie provinces, and will accept auxiliary heat for year-round operation. The original kiln has been operating since 1987, a second was finished in spring 1994; and the third, in Manitoba, is under construction. Each of the first two has twin 25,000 board-foot capacity chambers, but solar lumber kilns can be built to any size.

Unique solar collectors incorporate heat dispersion fins to rapidly transfer solar energy and efficiently reduce the relative humidity of a large volume of air, creating an ideal drying environment. Deciduous woods respond beautifully to the natural daily stress-release cycle. The result is exceptionally stable lumber that does not warp or cup when worked.

System Specifications

Lumber capacity: any size
Operating season: 10 months (year-round with auxiliary heat)
Payback potential: first year
Moving parts: fans

Supplier: **Vic Ellis**
Ecology Products International Inc.
P. O. Box 510
Pilot Butte, Saskatchewan S0G 3Z0
Tel: (306) 781-4561
Fax: (306) 781-4858
E-mail: radiant@sk.sympatico.ca

Client / Customers

- D & A Halland, Love, Saskatchewan
- Solar-Dri Lumber Specialties, Rocky Mountain House, Alberta
- Woodworkers Sawmill & Kiln, Headingly, Manitoba

The Big Picture

In a Forestry Canada-funded study of drying technology appropriate for the small independent woodlot owner, the consultant chose this kiln design for his demonstration project. The work is in cutting the wood, the profit is in drying and grading it.

CASE STUDY:

First Solarwall® System Installed Under The REDI Program: Tapis Coronet

Farnham, Quebec

TAPIS CORONET was building a 150,000 ft^2 expansion on its existing carpet manufacturing facility located in Farnham, Quebec. In addition to this space, the firm also needed to supply 60,000 CFM of ventilation air to compensate for the negative pressure already existing within the premises.

Originally a roof mounted, gas fired make-up air system was specified, however, the company was also interested in demonstrating its "green" philosophy. With the added benefit of Natural Resources Canada's REDI program, a 12,700 ft^2 SOLARWALL® providing 66,000 CFM of fresh air proved to be cost competitive versus conventional steel cladding and the conventional make-up air system.

The SOLARWALL® performs like a solar collector but looks like a conventional metal wall. The concept is simple. The dark colored wall heats up in the sun and the heated air is drawn off as quickly as possible through small holes in the cladding by axial flow fans located along the top of the wall. The SOLARWALL® system also recaptures all heat losses through the existing wall and destratifies hot ceiling air for additional energy savings.

Tapis Coronet is now enjoying the benefits of its solar air heating system. Not only is the system providing the needed fresh air, it is also helping the environment by reducing the plant's natural gas consumption by 250,000 m^3. Producing over 3,500 GJ per year in useable solar energy is saving the firm $35,000 annually.

Quotes Sylvain Roy of Tapis Coronet, "all existing customers were satisfied with the product and the predicted energy savings proved accurate . . . which was a key factor in the company's decision."

Manufactuer/Installer:

Matrix Energy Inc.
16807 Hymus Boul.
Kirkland, Quebec H9H 3L4
Tel: (514) 630-5630
Fax: (514) 426-9123
E-mail: matrixenergy@sympatico.ca

The Big Picture:

Solar heated ventilation systems such as this one are suitable for any building that needs ventilation or make-up air. This same system is being used around the world in cold climates for space heating and ventilation, while in hot climates the same product is used for crop drying.

CASE STUDY:

Power Purchase Solar Pool Heating
Lillooet, British Columbia

THE LILLOOET RECREATION CENTRE operates its indoor 25 m pool only three months per year—May, June, and July. The pool building was originally built with a low slope, south-facing roof in the hopes that solar heating would one day be installed.

In May 1998, the recreation centre comissioned Taylor Munro to install a new system using 63 4′x12′ plastic collectors which generate a peak output of approximately 250 kW, and are expected to last 15 to 20 years. All plumbing is PVC and a booster pump is used to circulate water through the collectors. The solar heating system operates automatically, requiring attention only at start up and shut down.

Using innovative local financing, the recreation centre did not pay for the capital cost of installation. Instead, the system is heat-metred, and the recreation centre pays the Energy Service Company (ESCO) only for the energy used.

This system cost $38,000 to install and the solar energy delivered during its April through July operating season is worth approximately $4,300.

System Specifications

Installation Date: May 1998
Solar Collector: 63 - 4′x12′ FAFCO unglazed plastic flat plate
Solar Collector Area: 280 square metres
Solar Collector Mount: south facing, roof mount
Fraction of Heating Load Met by Solar: 90+ %

Other Applications

Hundreds of municipal pools would benefit from solar heating to reduce operating costs and improve their environmental performance. Solar collector technology is efficient and well-proven for swimming pools. To limit financial and technical risk, some may prefer to purchase solar energy under an energy service agreement rather than make an outright purchase.

Customer:

Lillooet Recreation Centre
Lillooet, British Columbia

Supplier Contact:

Taylor Munro Energy Systems
4504 45A Street
Delta, British Columbia V4K 2L8
Tel: (604) 878-1588
E-mail: thwaites@direct.ca
Web: www.direct.ca/taylor-munro

Yalakom Appropriate Technology
Box 1509
Lillooet, British Columbia V0K 1V0
Tel: (250) 256-0268
E-mail: sldavis@uniserve.com
Web:
www.yalakomappropriatetec.bc.ca

CASE STUDY:

Solar Heating at Rosewall Creek Salmon Hatchery

Fanny Bay, Vancouver Island, British Columbia

THIS SOLAR WATER HEATING SYSTEM at Omega Salmon Group's Rosewall Creek Hatchery in Fanny Bay, British Columbia, was installed in the summer of 1997. It works by circulating the hatchery water directly through solar collectors mounted on building roofs. The heated water from the solar collectors is returned to the insulated tanks which allows energy to be stored overnight. The heated solar water was designed to raise the temperature of the 700 litre per minute flow from 8.5°C to 14.5°C, but if desired, it can raise the temperature to as high as 18°C. Highly accurate controls and a mixing valve are then used to blend the heated solar water with cold ground water, ensuring the fish are not exposed to stressful temperature fluctuations.

Sixty 4'x12' unglazed collectors provide a peak output of approximately 240 kW. Ultra-violet inhibitors in the solar collectors help ensure an expected life of 15 to 20 years.

Hatchery manager Dave McLellan, who operates the solar heater March through December, is delighted with the performance and hands-off automatic operation. An extensive and ongoing energy and environmental monitoring program, conducted by Natural Resources Canada (NRCan) has shown the system to perform as predicted, providing about 25% of the annual heating requirement.

System Specifications

Solar Collector: 60-4'x12' FAFCO unglazed plastic flat plate
Solar Area: 266 square metres
Solar Collector Mount: south facing, roof mount
Solar Storage Tanks: 20 m^3

Fraction of Annual Heating Requirement Met by Solar: 25%
Cost to Install: $66,000
Annual cost of heating fuel avoided: $13,000

Other Applications

The growing aquaculture industry has expressed interest in purchasing solar power. Many fish hatcheries are located in remote areas and rely on propane for heating; this makes solar heating more competitive. Not only does solar reduce operating costs, it reduces greenhouse gas emissions thereby improving environmental performance. Recent NRCan feasibility studies for other salmon hatcheries have shown solar to be cost-competitive with conventional heating.

The Big Picture

Low temperature solar water heating is often the most economical since inexpensive unglazed collectors can be used. Agricultural and commercial applications which use large volumes of heated water stand to gain from solar energy. In some circumstances, it may be possible for customers to purchase solar energy under an energy service agreement, rather than make an outright purchase of equipment.

Customer:

Omega Salmon Group
Campbell River, British Columbia

Supplier Contacts:

Taylor Munro Energy Systems
4504 45A Street
Delta, British Columbia V4K 2L8
Tel: (604) 878-1588
Fax: (604) 946-3804
E-mail: thwaites@direct.ca
Web: www.direct.ca/taylor-munro

Gerald Van Decker
Energy Technology Branch
Natural Resources Canada
13th Floor, 580 Booth Street
Ottawa, Ontario K1A 0E4
Tel: (613) 996-3630
Fax: (613) 996-9416
E-mail: gvandeck@nrcan.gc.ca

CASE STUDY:

Geothermal Heating and Cooling for Montreal's Biosphere

Montreal, Quebec

THE MONTREAL BIOSPHERE is a large interactive museum designed to educate its visitors about the St. Lawrence River and the Great Lakes Basin's ecosystem. The Biosphere is located in a large geodesic dome on Sainte-Helene's Island and was initially used as the United States Pavilion for Expo'67.

When the new owners of the building were transforming the dome into the Biosphere in 1995, they were looking at an efficient, yet environmental way to heat and cool the 76m diameter, 63m high dome. Preliminary quotes showed natural gas was the most economical system for heating and cooling the dome, but they decided to use a geothermal system. To reduce its energy demands, the Biosphere also incorporated other energy-efficient measures, including energy-efficient lighting, high-efficiency motors, high performance heat pumps, and fresh air preheated or pre-cooled by a heat-exchanger.

Geothermal systems use the consistent temperature of the ground, or an underground source. In the Biosphere, a semi-open loop hydrothermal system was installed. There is an underground water source capable of providing enough energy to meet the needs of the building, which are 610kW (175 tons) in cooling and 360kW (1.23 Btu/h) in heating capacity. The main components of the system are:

- the use of submersible pumps sunk 91m (300 ft) below ground level to carry a total flow of 22.1 l/s (350 gal/min) water from an underground source at a constant temperature of 9°C (49°F) to two plate heat exchangers
- the transfer of ground-water heat to a semi-open loop through heat exchangers before returning it into the earth through an injection well
- the use of plate heat-exchangers injecting heat into the water loop, depending on heating or cooling needs in each room of the museum

The temperature extremes in Montreal, fluctuating from 31°C (88°F) to –26°C (-16°F), made the planning of the system important. There were definite concerns about freezing, especially as the thermal loop has a low supply temperature (around 4.5°C) and no anti-freeze is used in order to maintain the Biosphere's high environmental standards. To protect the underground piping from freezing and frost, a high-density rigid insulation was installed four feet deep.

Estimated investment cost and payback period:

Energy Source	Estimated Investment Cost C$	Annual Energy Cost C$	Energy Cost C$/m^2	Payback Period (years)
Geothermal	525,000	124,155	0.24	0.41
Electrical	508,375	163,330	0.31	-
Natural Gas	532,600	109,426	0.21	0.44

Contact: **Mr. J. Lagacé, ing**
Bouthillette Parizeau & Associés
9825, Verville
Montreal, Quebec H3L 3E1
Tel: (514) 383-3747
Fax: (514) 383-8760
E-mail: bpa@sympatico.ca

Mr. Michel Lamanque
Natural Resources Canada
580 Booth Street, 13th Floor
Ottawa, Ontario K1A 0E4
Tel: (613) 947-3812
Fax: (613) 947-1016
E-mail: michel.lamanque@nrcan.gc.ca

CASE STUDY:

Building-Integrated Commercial PV
Edmonton, Alberta

EPCOR Utilities, Inc., Edmonton's electricity, water, and electrical services utility, is evaluating the technical, economic, and regulatory aspects of solar generated electricity.

In October 1996, EPCOR developed a commercial 13.4 kW DC utility-intertied, building-integrated PV system located on the roof of their 22-storey corporate headquarters in downtown Edmonton in partnership with the building's owner's, Lehndorff Tandem Group, and PV consultants, Howell-Mayhew Engineering. The system is the third largest PV system in Canada and, when built, was the tallest building-integrated PV system in the world. Its height had a considerable effect on design, equipment selection, and installation.

Simulations show that the system as mounted horizontally will produce 13.6 MWh of electricity annually. The horizontal angle reduces its generation by 26% from the optimised-annual 40° tilt angle, and by 5% from the optimal summer angle. The horizontal also allowed the system to be spaced closely to reach the desired generating capacity from the available roof area.

System Specifications:

Modules: 112 poly-crystalline Solarex 120 W.
Rated PV Power Output: 13.4 kW DC, 10 kW AC
PV Operating Voltage: ±218 volts DC
Tilt Angle: Horizontal
Mounting Characteristics: Building-integrated PowerGuard™ roof tiles from PowerLight Corporation of Berkeley, California. No roof penetrations. Lightweight. Able to withstand severe wind loads and uplifts found on high buildings.
Inverters: 2 Omnion Series 2400 inverters
Inverter Output Rating: 6 kW AC each
Inverter Output Voltage: 120 volts AC
Inverter Characteristics: Automatic silent operation. Maximum power point tracking. Automatic synchronising with the utility grid. Shuts down when the utility grid is not available.
Total Cost of Equipment: $150,000 installed.

Other Applications

Many countries are developing roof-top PV programmes. There are over 25 utility-intertied PV systems in Canada and tens of thousands in Japan, Holland, Germany, Italy, and America. The biggest market for these types of systems focusses on their integration into society's infrastructures: building facades, windows, roofs, awnings, car ports, sound barriers, *etc.*

The Big Picture

Utility-intertied photovoltaic systems can be installed on any building or ground surface. They provide high-value, economically-viable electricity in many niche markets.

Contact: Joel R. Nodelman, P.Eng.
EPCOR Utilities, Inc.
Environment Affairs and Sustainable Development Department
16th Floor, Capitol Square
10065 Jasper Avenue
Edmonton, Alberta T5J 3B1
Ph: (403) 412 3487 Fax: (403) 412 3346
E-mail: jnodelma@epcor-group.com
Web: www.epcor-group.com

CASE STUDY:

AC-BIPV Module Installation
Vancouver, British Columbia

THE BRITISH COLUMBIA Institute of Technology (BCIT) recently formed PEARL (Photovoltaic Energy Applied Research Lab). PEARL is part of BCIT's Technology Centre which conducts technology development projects for industry clients. Since 1997, when the lab was formed, the PEARL research team has been working on a steady flow of projects for industry involving photovoltaics and other renewable energy technologies. PEARL has a staff of well-qualified and experienced researchers and engineers, and also draws on part-time help from BCIT instructors and students. The PEARL research team has specialized experience and is involved in building integrated photovoltaics (BIPV), PV system design, engineering and prototyping, R&D in PV, and PV training.

One of the most exciting projects involves integrating 1 kW of the AC photovoltaic modules into the south facade of the BCIT Technology Centre building and connecting it to the BC Hydro main utility grid. The AC PV modules were mounted vertically for several reasons: to minimize the cost of mounting structures, to avoid penetration of the roof membrane, to take advantage of the surrounding area reflectance, to decrease snow, dirt and debris built-up, and to preserve the building's aesthetics. Powered by the sun, the PV system is producing electricity and feeding it back into the utility grid. There are no batteries. For safety, the PV system automatically shuts down if the BC Hydro system is not available (anti-islanding protection feature of the inverters). A data acquisition system was also developed to continuously monitor the installation and allow further research.

The PEARL research team will study and evaluate the interactive compatibility of the PV system and Hydro's utility grid, and investigate future applications for grid-connected BIPV systems in the urban environment.

Other Applications

This installation is the first AC PV module installation in Canada.

The Big Picture

BIPV technology has rapidly gained utility grid-connected market acceptance worldwide. Even more exciting technologies are AC PV modules allowing individual PV modules grid connection. Some of the advantages of AC PV modules include:

- no DC interconnections between modules (only regular AC wiring)
- each PV module is individually optimized (they have their own Maximum Power Point Tracker, so no mismatch losses, which allows

installation on curved roofs and facades)
- ease of installation and incrementability
- built-in data acquisition allows monitoring and fast false PV module detection in the array

BIPV System Specifications

Rated Power: 1 kW
Number of AC PV modules: 10 single-crystal silicon Solec 100W
Inverter: 10 MicroSine PV module Inverters—MS100 (grid connectable), OKE/Trace
108-132V AC, 59-61 Hz output; 24-50 V PV input; 93% peak efficiency; MPPT; automatic synchronization with the grid; built-in data acquisition.

Contact: **Ljubisav Stamenic**
BCIT Technology Centre
Photovoltaic Energy Applied Research Lab (PEARL)
3700 Willingdon Avenue
Burnaby, British Columbia V5G 3H2
Tel: (604) 451-6934
Fax: (604) 436-0286
E-mail: lstameni@bcit.bc.ca
Web: www.bcit.bc.ca/~tc

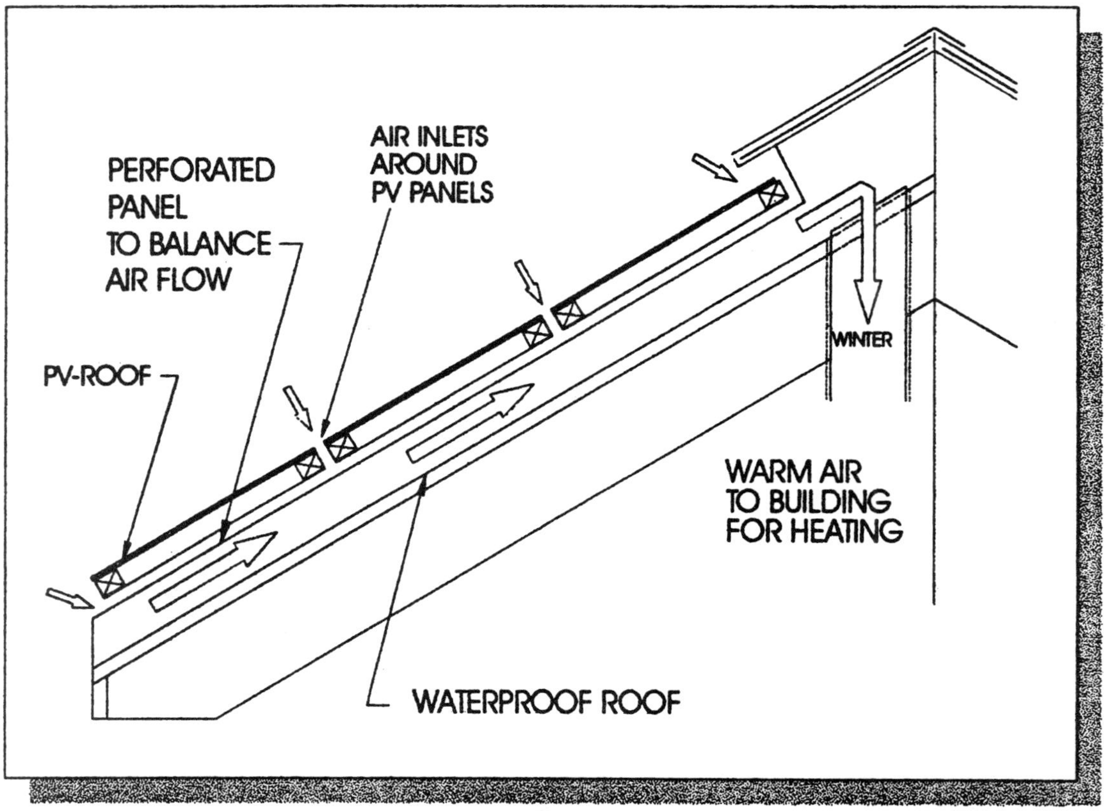

CASE STUDY:

PV Solarwall™ Hybrid System
Toronto, Ontario

PHOTOVOLTAIC SOLAR PANELS actually produce more energy in the form of heat than electricity. SOLARWALL® panels are designed to remove this heat and also provide solar heated fresh air for use in the building's ventilation system. Portable classrooms use electricity for heating resulting in high energy costs. Indoor air quality is also an issue and PV SOLARWALL® addresses both of these concerns.

The two amorphous photovoltaic panels, integrated into the SOLARWALL®, power two variable speed, 340 m^3/hour (200cfm) fans that will bring solar heated fresh air into the classroom. The air is then distributed in the classroom by two 150 mm. diameter ducts. The 15 m^2 SOLARWALL® panel system provides heated fresh air that is as much as 30°C above ambient.

System Specifications:

Area of portable classroom: 72m^2 x 2.5 m high
SOLARWALL® area: 15m^2 plus 120 watt PV panel mounted along the top surface of the SOLARWALL®

Ventilation rate: varying from 0 to 340 m^3/hr., controlled by temperature sensors.

Supplier: Conserval Engineering Inc.
200 Wildcat Rd.
Toronto, Ontario
Phone: (416) 661-7057
Fax: (416) 661-7146
E-mail: conserval@globalserve.net
Web: www.solarwall.com

Client/Customer: Toronto District School Board.

* Funds for the demonstration were provided by Natural Resources Canada, CANMET, Toronto Hydro and Ontario Hydro.

The Big Picture:

As more is learned about indoor air quality and air tight buildings, the need for a cost effective, low energy consumption solution becomes important. The PV SOLARWALL® could be the ideal solution.

CHAPTER 4

Remote, Northern & Rural Applications

THE APPLICATIONS of renewable energy are well suited to remote, northern and rural locations in Canada and throughout the world. The most common application of renewable energy in remote locations is the generation of electricity through the use of photovoltaics, microhydro and wind energy. The use of active solar energy for space heating is also occurring more regularly.

The diverse requirements for electricity in remote locations are similar to populated areas. They include cottage and home systems, water pumping, computers, research stations, telecommunications stations, navigational buoys, and oil and gas well monitoring systems. As remote areas are less populated, often physically inaccessible and politically removed, the cost is prohibitive for installing a complete infrastructure, including an electric grid, a complete telecommunications network and water and sewer systems. In examining the costs of introducing electricity to a location, many factors must be taken into account including the initial costs of installing a line over a long distance and the ongoing costs of servicing the line.

There are four types of renewable electricity systems that can provide electricity in rural or northern remote locations.

1. *Stand-alone system without battery* will power a load that does not need an electric storage system like ventilation or circulation fans, which only needs to operate during daylight hours. Another application is water pumping for agricultural purposes. The water is pumped during a sunny or windy period and is stored in a water tank or reservoir.
2. *Stand-alone system with battery* is the most common application of renewable energy systems. During the period of electrical generation, on sunny or windy days, the excess electricity charges a bank of batteries. This allows the system owner to draw upon the stored energy at night, or during periods of little or no sun or wind.
3. *Stand-alone hybrid system* ensures access to electricity at all times by supplementing the renewable energy source with an additional source of electricity, such as a diesel, gasoline or gas generator or even another type of renewable energy. As the system owner depletes the available electricity in the batteries, the generator may be started to recharge the batteries.
4. *Grid-connected system* ensures that the system owner has access to electricity at all times. Instead of using batteries as storage, the

Solar powered radio controlled hazard beacon, TRigolet Airport, Labrador

renewable energy system feeds its excess generated power into the electric grid. When the power demands are higher than the renewable energy supply, the extra demand is met directly from the electric grid. A grid-connected system can be connected to either a very large grid, such as the North American grid, or to a smaller, regional grid, such as those found in remote communities.

The application of wind, micro hydro and photovoltaics for household and cottage use is more common in rural, remote and northern settings. The renewable energy systems for homes and cottages are relatively small, from several hundred watts to over 2kW. Coupled to a battery system and with efficient energy use, these systems can provide sufficient energy for all domestic uses.

Monitoring systems and telecommunications stations in remote and northern locations need an electricity supply that is both reliable and low maintenance, which wind or photovoltaics can provide. Telecommunication repeater stations relay telephone calls, radio-frequency and microwave signals. To ensure the clarity of the signal, they are often located along inaccessible ridges and mountain tops. It is in these types of locations that renewable energy is a viable solution.

An interesting application of photovoltaics in the telecommunications industry is the PV-powered distress telephone. This system has proven to be cost-competitive when grid-electricity and telephone lines are not readily available. These safety telephones have been installed along remote sections of highways and on university and college pathways and in parking lots.

1.2 KW Solar powered lightstation. East Light, Sable Island, Nova Scotia.

Renewable Energy Applications in Remote Areas

Visual or radio navigational aids are often located where it is difficult to supply electric power. Small airports and the Canadian Coast Guard have begun to utilise photovoltaics in their more remote locations. Marine navigational aids, either land-based or buoy-mounted, require a constant electrical source as the flashing light on the marker warns mariners of dangerous waters or marks the navigable channels. In a PV-powered buoy, a single 10- or 20-watt PV module regularly charges the sealed battery, whereas with a traditional marker buoy, the Coast Guard has to perform regular maintenance on the buoy and replace the drained battery. Some of the new marker buoys have also been designed to withstand complete encapsulation in ice. They are able to resume normal functions after being in ice for over 100 days.

Photovoltaics are increasingly being used at infrequently-used airstrips. Rather than having the landing lights on at all times, a pilot preparing to land at a remote location turns on the PV-powered landing lights by a radio signal. He is then able to turn off the lights on take-off. PV systems are also used to power ground radio transmitters to pilots who are flying in remote locations.

There are some systems that do not need electricity at all times, including fans for air circulation and cooling and water pumping. In many Canadian homes, attics and homes often overheat during the summer months. With the installation of a photovoltaic-powered fan in the attic, a homeowner may avoid complicated rewiring of the house. The fan will only operate when it is necessary to cool the space—that is, when the sun is shining.

In remote rural areas, such as agricultural locations in the prairie provinces of Canada, it would be costly to extend the electrical grid to remote locations to pump water for livestock. Photovoltaics are a reliable, affordable solution to this problem. A PV-powered water pump is capable of pumping sufficient water from deep levels into a trough for animals to drink. The water remains clean and bacteria-free as it is

constantly refreshed. It is when electricity is required intermittently that renewable energy applications are the most cost-effective and cost-competitive.

Recently, wind energy and photovoltaics have been installed in greater numbers in the remote areas of Canada's national and provincial parks. Traditionally, the electricity demands of these parks have been met through the use of diesel generators, as the extension of the electrical grid into those areas is cost-prohibitive. However, diesel generators pose their own

problems, including continuous noise and air pollution in environmentally sensitive areas, the continued threat of oil or gas spills affecting the sensitive ecosystems in the parks and the high cost of delivery to remote locations. In Canadian parks, wind energy and photovoltaics have been used in rangers' or wardens' cabins, in campground lighting systems, and in information kiosks.

In remote communities of Canada, especially in the Yukon and Northwest Territories, wind, micro hydro and photovoltaics are effectively providing some local grids with clean electricity during the long summer days. Many of these communities cannot afford the high capital costs to become connected to the North American grid and have become dependent on diesel-generated electricity. They have consequently experienced the growing costs of maintaining noisy diesel systems. These communities have become concerned with the ever-present environmental risk of an oil spill or slow leaks from the storage containers. Also, northern communities pay at least ten times as much for their electricity as the grid-connected locations in the rest of Canada. With the assistance of the Canadian government, especially Natural Resources Canada, some communities are learning that a cost-effective way to generate electricity involves the installation of a wind, micro hydro or PV system to supplement diesel generation.

CN Telecommunications sites, Northern Quebec, Canada

CASE STUDY:

PV Powered Highway Beacons
New Brunswick

IN 1996, the Government of New Brunswick was looking for a number of photovoltaic powered highway beacon systems to reduce project infrastructure costs on a new highway. They wanted to save on the usual costs that are associated with running extensive distances of hydro lines from the main power grid to the individual median warning beacons.

Cimat Power Systems Inc. had proposed initially that the entire individual system be mounted on one pole. The client requested that the PV panels be situated at the side of the road as there was some concern about replacement costs should a vehicle hit the pole carrying the entire system assembly. Although most of the battery enclosures were buried to ground level, a few were positioned above ground level and attached to the beacon pole. Burying or pole mounting the battery enclosure appeared to be best as at least one ground level enclosure was found with a truck bumper still attached—where someone had tried to pull the enclosure from the pole.

System Specifications

The key components per system include the pole mounted photovoltaic panels, charge and flasher control assembly, lightning protectors, low wattage beacon assemblies and freeze tolerant, sealed lead acid batteries with heavy duty battery enclosure.

The Big Picture

It is difficult to say exactly how much money was saved by using the independently powered beacon assemblies instead of running grid power cables the length of the highway. The same logic can be applied to most remote site, low power requirements such as area and sign illumination.

Client/Customer

Province of New Brunswick

Contact Information:

Ray Lalonde
Cimat Power Systems Inc.
P.O. Box 422
Pointe Claire, Quebec H9R 4P3
Tel: (514) 695-1546
Fax: (514) 630-8938
Web: www.cimat.com

CASE STUDY:

Microwave Telecommunications Site Northern Ontario

THE Ontario Ministry of Transportation has installed a turn key solar power system to replace prime diesel generators at a microwave repeater site in northern Ontario. The twin diesel system consumed over $50 per day in fuel, and service requirements amounted to over $3000 per year. From the time the government placed the solar system order to complete installation took less than six weeks. The project included complete engineering design along with all materials and supervision of the installation. It was commissioned in the spring of 1996. The diesel generators have since been removed.

System Specifications

38 Siemens SP-75 PV modules
APT control centre using 60 amp mercury contactors for redundancy
2100 AHr lead acid battery bank

Other Applications

All types of remote telecommunication repeater stations

The Big Picture

A solar photovoltaic system can be installed anywhere there is suitable exposure to the sun and a demand for electricity. In remote areas, the costs of installing and maintaining a complete PV system are significantly lower than the costs associated with installing and maintaining diesel generator systems. PV systems will also improve overall system reliability.

Client: **Ontario Ministry of Transportation**

Contact: **SOLTEK Solar Energy Ltd.**
#2-745 Vanalman Ave.
Victoria, British Columbia V8Z 3B6
Tel: (250) 727-7720
E-mail: soltek@soltek.ca
Web: www.soltek.ca

CASE STUDY:

Solar Power for VHF Radio Repeaters
British Columbia

PROVIDING telephone communications to remote northern B.C. communities is a difficult job. With networks of mountaintop repeaters accessible only by helicopter, site reliability is paramount, as operating costs can skyrocket with each maintenance visit. Historically, site power was provided by disposable, caustic potash batteries which had to be replaced annually. Rising battery costs and environmental issues associated with disposal of used cells have led utilities to install solar electric systems.

A solar power system was designed to operate the VHF repeaters that service the Public Mobile Telecommunications System (PMTS). The goal was to reduce the $10,000 annual service costs and the frequency of site visits.

The power system consists of four Siemens SM-50 solar modules, providing up to 12.4 amps of charge current. These modules were selected because of their high power density and narrow profile. The solar modules are secured to the curved fibreglass communications shelter with SOLTEK's CM-50 mounting kits. These include special low profile panel brackets and a hood to deflect falling ice from damaging the panels. A bank of sealed, maintenance free batteries provide up to 60 days of reserve power and were installed to ensure operation during the winter.

System Specifications

Four Siemens SM-50 PV modules
Temperature compensated electronic charge controller
1150 Ahr GNB Absolyte IIP battery bank

Applications

Radio telecommunication repeater sites.

The Big Picture

A solar photovoltaic system can be installed anywhere there is suitable exposure to the sun and a demand for electricity. With an average payback of less than two years, conversion of potash battery-powered repeater sites to solar is a sound financial decision.

Client: NorthwesTel

Contact: **SOLTEK Solar Energy Ltd.**
#2-745 Vanalman Ave.
Victoria, British Columbia V8Z 3B6
Tel: (250) 727-7720
E-mail: soltek@soltek.ca
Web: www.soltek.ca

CASE STUDY:

Hybrid Solar PV/Wind System for Full-time Year-round "Off-grid" Home
Silver Islet, Ontario

SILVER ISLET is a summer cottage community with a small number of year-round residents, approximately 100 km from the city of Thunder Bay, Ontario. The Ontario Hydro grid is several miles away and most cottagers use woodstoves, propane appliances, kerosene lanterns, and gas generators when they need electricity. However, many are now turning to solar panels for some electrical conveniences.

The Thunder Bay area is typically a good solar regime. Even in winter when days are short, the skies are typically clear and the sun bright. This particular location on the shore of Lake Superior is also ideal for wind energy. A renewable energy system that incorporates both solar and wind can make year-round living feasible and comfortable without the necessity of constantly running fossil-fueled generators.

Cam and Brenda Snell live year-round in this remote community. Cam is the superintendent of the nearby Sleeping Giant Provincial Park and Brenda is an accomplished artist. They have chosen to provide power for their log home with a hybrid solar PV/wind system. This energy system gives them the ability to live with comfortable modern conveniences amidst the incredible beauty of this northern wilderness.

System Specifications

- Four 75-watt (4.4 amp) Siemens PC4 solar panels, configured for 24 volts, for an 8.8 amp array
- Wattsun single axis tracker
- Custom-built pole mount for array on deck
- Whisper 1300-watt 2-blade wind generator (high voltage model) on mountain behind house, 125m of teck cable
- 15m tubular steel tower with gin pole to facilitate lowering for maintenance

In Utility Shed:
- Trace 4024 SW inverter with fused DC disconnect in insulated compartment

- 820-amp/hour battery bank (four 6-volt Surrette dual-case) with hydrocaps in sealed vented insulated compartment, with 40-watt light bulb in explosion-proof fixture for heating compartment when necessary
- Enermaxer 60-amp diversion control with custom built 1850-watt resistor bank and shield
- Transformer, rectifier and brake for the wind generator, solar disconnect switch
- Custom built manual override for tracker
- Back-up propane 4.5 kW generator

In House:

- Remote Photron MPC-3m to track battery voltage, battery compartment temperature, amp consumption, and system amp production

The Application

This renewable energy system was installed in the fall of 1995. Cam and Brenda have learned to live with the system, adjusting their consumption habits to suit the weather. The system provides power for most conventional household needs, such as lighting, stereo, TV/VCR, small kitchen appliances, microwave, vacuum cleaner, power tools, hair dryer, submersible water pump (120 V AC), washing machine, and an ultra-efficient Vestfrost two compartment two-compressor fridge/freezer.

The home is heated by a woodstove and two propane heaters; the kitchen range and water-heater use propane. If there are extended periods of cloudy, calm weather or greater than average load requirements, the back-up generator provides power and charges the battery bank via the inverter's standby feature.

The advantage of a hybrid system in this climate and latitude is how well solar power and wind power complement each other. A solar system can provide for most needs in the summer when daylight hours are long and power requirements are less. In winter, daylight hours are short, but typically the wind is stronger, colder, and therefore more powerful. A system of this type can be used to provide power for household needs as long as energy conservation is part of the homeowners' routine. The wind regime at Silver Islet is particularly good. In other locations within the Thunder Bay area, the solar array would need to be larger because of reduced wind.

The Client:

Cam and Brenda Snell, Silver Islet, Pass Lake, Ontario

Contact Information:

Frank's Alternate Energy
RR # 3
Thunder Bay, Ontario
P0T 2V0
Tel/Fax: (807) 964-2050
Toll Free 1-888-SUN-WIND (786-9463)
E-mail: sunwind@norlink.net
Web: www.sunwindwater.com

CASE STUDY:

Tundra Ecosystem Research Station, PV/Wind/Generator Hybrid System
Daring Lake, Northwest Territories

THE TUNDRA Ecosystem Research Station is a semi-permanent tent camp situated in Canada's Arctic barrenlands, about 300km north of Yellowknife and 50km west of Canada's first diamond mine, at 64°52'N and 111° 35'W. With extreme northern weather conditions and fly-in only access, this camp is only managed on-site from mid-May to mid-September. It is available for use year-round by various schools and other groups, accommodating a maximum of 25 people. Clearly, in this beautiful remote region with an abundance of wildlife in its natural habitat, the noise, costs and pollution associated with a generator-only system were unwanted and prohibitive. With high sunlight hours for the camp's peak season and good wind speeds, a hybrid power system utilizing mostly renewable energy proved to be the ideal solution. The PV/wind/generator hybrid system is used successfully to power all loads for this research and educational facility, including an electric bear fence to protect the camp and its inhabitants from some of the wildlife they study. The gas generator is used periodically in the winter months only.

System Specifications

8 Siemens 75W solar panels
2 4-Panel ground mounts
1320-Ahr battery bank
Trace 40-amp charge controller with digital monitor
1 Trace 2500W inverter/charger with safety/disconnect assembly
1 MSE power panel
1 Air 303 wind generator with tower
1 5kW gasoline generator

Other Applications

Hybrid power systems work beautifully and economically for a variety of off-grid applications world wide, such as mining exploration camps, remote residences, lodges and remote communities. Generally, a properly sized PV/wind/generator hybrid system installed in a region with adequate sun and wind conditions reduces generator running time by 80-95%. It is a very elegant, cost-effective solution for many remote power applications where a reliable, clean, quiet and low-maintenance power system is desired.

The Big Picture

Compared with running solely a diesel generator at this fly-in camp where fuel costs are at a premium, this hybrid system reduces operating costs by approximately $16,000 for each summer season alone. In addition to the huge cost savings it provides, the system enables the camp to fulfill its mandate for demonstrating the simplicity and other benefits of alternative energy use. The camp also facilitates various research projects for some of the country's leading biologists, archaeologists and other scientists.

Contact: **Midnight Sun Energy Ltd.**
5 Ellesmere Drive, Box 1683
Yellowknife, Northwest Territories
X1A 2P3
Tel: (867) 873-8760
E-mail: midsun@internorth.com
Web: www.midnightsunenergy.com

CASE STUDY:

Solar/Generator Hybrid System
Queen Charlotte Islands, British Columbia

A TURN KEY solar/generator power system is being used to service the electrical needs at the Huxley Island Park Warden Station. Located in the Moresby Island archipelago, Huxley Island serves as a base camp for park staff in the region and as a registration office for visitors arriving by boat. It is also used by visiting scientists performing studies in the area. The camp has a 75 m^2 building with full kitchen, bunks for four people and an office equipped with satellite telephone VHF radios and computers.

To facilitate installation, all electrical components were preassembled and wired on a 1.3m^2 power board before shipment. Also provided was a waterproof, aluminum outdoor cabinet for the batteries and power equipment which was installed directly behind the living quarters. A battery state-of-charge display and a remote control for the inverter and generator was installed on an interior wall of the building for the convenience of park staff.

One of the key features of the Gwaii Haanas integrated power system is its completely automated operation. The solar system supplies the majority of the power to the loads, with the generator available for backup charging using the 100-amp charger in the inverter. In the event of extended poor weather or excessive loads, the generator is programmed to start automatically when 50% of the battery capacity has been removed. In this way, the batteries are charged before a potentially damaging low battery condition is reached.

System Specifications

Eight Siemens SP-75 PV modules
Electronic charge controller with metering
4 kW Trace Sine Wave inverter
38 kWhr lead acid battery bank
5 kW gasoline generator

Other Applications

All types of marine navigation and obstruction lighting systems.

The Big Picture

A solar photovoltaic system can be installed anywhere there is suitable exposure to the sun and a demand for electricity. In areas where utility power is not readily available, a solar and generator hybrid system can provide enough power to run almost any load.

Client: Gwaii Haanas National Park

Contact: **SOLTEK Solar Energy Ltd.**
#2-745 Vanalman Ave.
Victoria, B.C. V8Z 3B6
Tel: (250) 727-7720
E-mail: soltek@soltek.ca
Web: www.soltek.ca

Renewable Energy in Canada's Remote Communities

Contributed by: Kim Ah-You and Greg Leng, Renewable Energy for Remote Communities Program, CANMET Energy Diversification Research Laboratory, Natural Resources Canada

Background

CANADA HAS over 300 remote communities with a total population of about 200,000. These communities are not connected to the North American electrical grid or piped natural gas network and are permanent or long-term (at least 5-year) settlements with at least ten permanent residences. Many of them are very dependent on imported oil and pay energy costs that can be up to 10 times higher than in the rest of Canada.

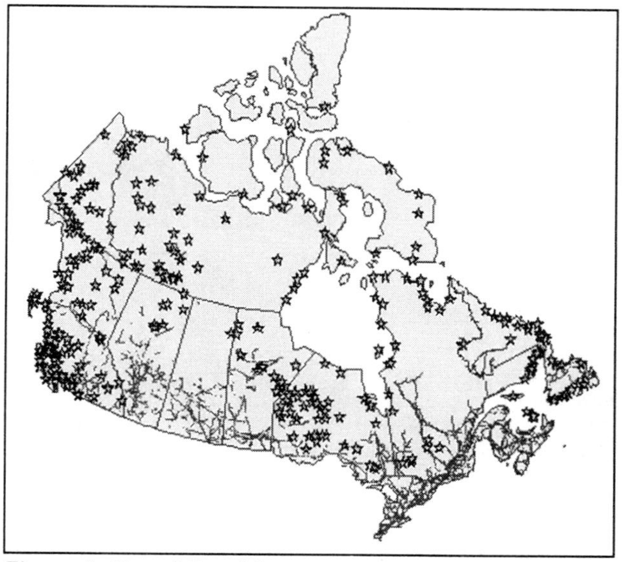

Figure 1: Map of Canada's remote communities.

Details of these communities' energy consumption, energy costs, fuel prices and other aspects of their socio-economic life have been described at length in other publications. Most produce electricity on isolated diesel grids by generators operated mainly by organizations such as electric utilities, Indian and Northern Affairs Canada (INAC) and aboriginal bands.

Though renewable energy resources such as small hydro, wind, solar and biomass are readily available, few of these communities have taken full advantage of them to meet their electricity and heating needs in a sustainable and cost-effective way. The main hindrances to the growth of renewable energy projects are a lack of local technical expertise, the presence of subsidies that distort market decisions and a lack of tools for evaluating projects.

A number of communities have nevertheless initiated renewable energy projects, after considering factors such as: (i) the size of their electrical or heating load, (ii) whether the community has renewable energy resources available near at hand, (iii) technical feasibility, (iv) economic viability, (v) community acceptance and (vi) environmental impact. This article will provide a brief overview of the renewable energy projects, by technology, in Canada's remote communities. Note that many renewable energy projects in remote areas, such as outpost exploration camps, will not be covered in this article since these areas are not considered permanent remote communities.

Small Hydro

Small hydro projects that have been built in Canada's remote communities are listed in Table 1 and account for about 173 MW of installed capacity, indicating that small hydro has been playing an important role in these communities' development. This is especially true in Yukon, where small hydro

A 700-kW small hydro plant installed in rural Ontario, (Photo courtesy of Ottawa Engineering).

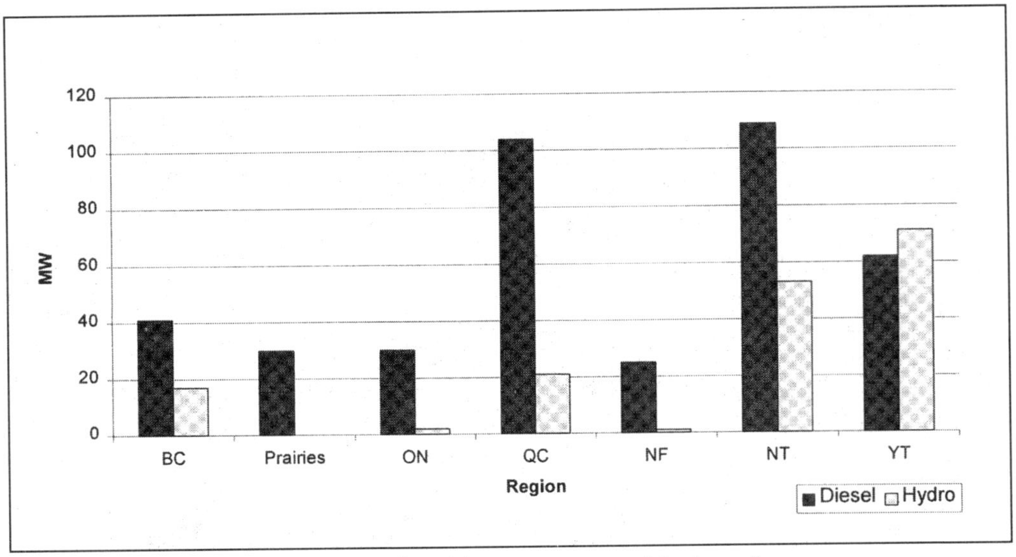

Figure 2: Installed electric capacity in Canada's remote communities by region.

accounts for a large portion of the installed electric capacity (see Figure 2) and produces electricity more cheaply than diesel generators. Small hydro plants have a useful life of more than fifty years and are quite cost-effective and suited to remote communities which have a good hydro resource. In the past few years, small hydro sites in remote communities have increasingly been developed by independent power producers who have been able to negotiate a long-term contract to sell power to the local utility.

3.5 kWp PV array incorporated into the façade of the Nunatta campus on Nunavut Arctic College, Northwest Territories.

Wind Energy

In recent years, wind energy projects in Canada's remote communities have been spurred by the improved reliability and affordability of wind energy systems and by the experience derived from earlier demonstration projects. In fact, these projects currently account for about 630 kW of installed electric capacity, as shown in Table 2. This is quite significant given that 15 years ago few projects existed. Wind energy projects in Canada's remote communities are generally tied to isolated diesel grids to form wind-diesel hybrid systems in which wind essentially is used to save on fuel. When there is no wind, all energy is supplied by the diesel generator, thus eliminating the supply risks associated with the variable nature of wind.

Two 80-kW wind turbines connected to an isolated diesel grid in Kugluktuk, Northwest Territories.
(Photo courtesy of Elliot Rodger).

Solar Energy

Electricity generation by photovoltaics (PV) is still not cost-effective in Canada's remote communities. Thus, deployment of such systems is limited to demonstration projects, like the diesel-grid-connected 3.5-kWp photovoltaic array incorporated into the facade of the Nunatta campus of Nunavut Arctic College in Iqaluit, Northwest Territories. However, PV is often the most cost-effective option for off-grid applications, like telecommunication sites, outpost and exploration camps, remote cottages/homes, coast guard buoys, lighthouses.

The use of solar energy for heating is currently cost-effective in Canada's remote communities. In fact, passive solar space heating designs and solar water heaters have been installed in a number of remote communities. Recent studies in several communities have also shown the benefits of installing solar ventilation air heating systems. A system of this type is now under construction in Fort Smith, Northwest Territories.

Ground-source heat pumps which use the earth as both a source for heating and a "sink" for cooling also show good potential.

Biomass

Since most remote communities are located below the tree line, they have access to good supplies of wood. As long as the wood is harvested in a sustainable way, it can be considered a renewable energy resource. Many homes in these communities burn wood in fireplaces, stoves and airtight furnaces to provide space heating and hot water. In a typical year they will burn four to five cords, with an energy content equivalent to about 1,600 to 2,000 litres of heating oil.

As indicated in Table 3, some communities have also been installing wood chip boilers that can provide space heating for schools or clusters of buildings. Electricity generation from biomass has not been as prominent, but the construction of a 9 MW cogeneration plant in Masset, British Columbia, scheduled to start operation in 1999, will likely demonstrate the benefits that can be derived from such plants.

Keys to Success

A concerted effort by all major stakeholders, such as electric utilities, community leaders, independent power producers and government, is key to the further deployment of renewable energy projects in Canada's remote communities. The recent launch of Natural Resources Canada's Renewable Energy for Remote Communities Program, the availability of reliable technology and the concrete actions of a number of electric utilities and community leaders are signs of a bright future for renewable energy in Canada's remote communities.

Passive solar design in a residential application.

Acknowledgments

The authors would like to acknowledge all the people who contributed to this article. Special thanks are due to the following people from whom more information on the listed projects can be obtained: Ken Boyd, BC Hydro, (604) 597-4928; Tim Farrell, NWT Power Corp, (867) 874-5204; John Maissan, Yukon Energy Corporation, (867) 667-8119; Bob Collins, Yukon Economic Development, (867) 667 5015; John Mann, Ontario Hydro, (807) 474-2812; Georges Fournier, Hydro-Quebec, (418) 368-4456; Keith Boone, Newfoundland & Labrador Hydro, (709) 737-1249.

Bibliography

Sigma Engineering, "Energy Overview Study of Remote Communities in Canada", prepared for the Remote Community Demonstration Program, Energy Mines and Resources Canada, March 1985.

Leng G. et al., RETScreen™ - "Renewable Energy Technologies Project Assessment Tool, Version 98," Natural Resources Canada, May 1998.

The CANMET Energy Diversification Research Laboratory web site: Web:www.cedrl.mets.nrcan.gc.ca.

The RETScreen™ website at: Web:www.retscreen.gc.ca

Table 1: List of small hydro projects in Canada's remote communities.

Location of Project	Year Started Operation	Project Owner	Installed Capacity
Taltson River, NT	1965	NWT Power Corp.	22 MW
Snare Forks, NT	1976	NWT Power Corp.	9.2 MW
Snare Rapids, NT	1948	NWT Power Corp.	8.2 MW
Snare Falls, NT	1960	NWT Power Corp.	7.4 MW
Snare Cascades, NT	1996	Dogrib Power Corp.	4.3 MW
Whitehorse Rapids, YT	1958	Yukon Energy Corp.	6 MW
Whitehorse Rapids, YT	1958	Yukon Energy Corp.	6 MW
Whitehorse Rapids, YT	1969	Yukon Energy Corp.	8 MW
Whitehorse Rapids, YT	1984	Yukon Energy Corp.	20 MW
Aishihik, YT	1972	Yukon Energy Corp.	30 MW
Fish Lake, YT	1950s	Yukon Electric Corp.	1.3 MW
Mayo, YT	1950s	Yukon Energy Corp.	5 MW
Clayton Falls, BC	1992	Synex Energy Resources Limited	2 MW
Ocean Falls, BC	1980	Central Coast Power Corp.	12 MW
Queen Charlotte, BC	1992	Queen Charlotte Power Corp.	6.5 MW
Dease Lake, BC	1997	Stikine Nation Power Corp.	3 MW
Kingcome Inlet, BC	1982	DINAC	75 kW
Deer Lake, ON	1998	Ontario Hydro	490 kW
Sultan, ON	1983	Ontario Hydro	150 kW
Lac Robertson, QC	1995	Hydro-Québec	21 MW
Mary's Harbour, NF	1987	Mary's Harbour Hydro (IPP)	175 kW
	Total	**172.8 MW**	

Table 2: List of wind energy projects in Canada's remote communities.

Location of Project	Year Started Operation	Project Owner	Number of Turbines	Installed Capacity	Turbine Model
Kuglugtuk, NT	1996	NWT Power Corp.	2	160 kW	Lagerwey 18/80
Sachs Harbour, NT	1998	NWT Power Corp.	1	50 kW	AOC 15/50
Rankin Inlet, NT	1998	NWT Power Corp.	1	50 kW	AOC 15/50
Cambridge Bay, NT	1994	Dutch Industries	1	80 kW	Lagerwey 18/80
Igloolik, NT	1988	Dept. of Public Works, NWT	2	20 kW	Aerowatt LM10
Haeckel Hill, YT	1993	Yukon Energy Corp.	1	150 kW	Bonus 150
Big Trout Lake, ON	1997	Ontario Hydro	1	50 kW	AOC 15/50
Kasabonika, ON	1997	Ontario Hydro	3	10 kW	Bergey Excel
Kuujjuaq, QC	1989	Hydro-Québec	1	60 kW	Bonus 60
		Total	**630 kW**		

Table 3: List of wood chip boilers in Canada's remote communities.

Community	Location	Year Started Operation	Capacity
Pelly Crossing, YT	Eliza Van Bibber School	1982	300 kW
Whitehorse, YT	Elijah Smith School	1992	1.2 MW
Whitehorse, YT	Klondike Central Heating	1987	1.8 MW
Haines Junction, YT	District system heating serving	1988	1.2 MW
	5 band-owned buildings		
Carmacks, YT	Administration building occupied by the	1992	0.6 MW
	First Nations of Little Salmon / Carmacks		
Burwash Landing, YT	Administration building occupied	1998	0.6 MW
	by the Kluane First Nation		
Masset, BC	Cogeneration plant in Masset	1999	9 MW
Fort Smith, NT	Water Treatment Plant	1983	1.2 MW
		Total	**15.9 MW**

CASE STUDY:

Lighthouse Automation Systems
Canada

POWER SYSTEMS have been supplied to 20 coast guard sites as part of a countrywide program to convert lighthouse stations from generators to solar power. The power systems operate weather monitoring equipment, and provide power to beacons and fog horns. The systems have been installed on both the West and East coasts of Canada. To date, there have been no site failures, and all sites are operating with almost zero maintenance. The cost savings over the existing generator sites is quite substantial.

System Specifications

Individual customized systems incorporating:
Siemens SP-75 PV modules with superior low light performance
Custom designed PV-3 charge control centres featuring temperature compensation, digital metering and alarm relays.
GNB Absolyte 2 volt sealed lead-acid batteries

Other Applications

All types of marine navigation and obstruction lighting systems.

The Big Picture

A solar photovoltaic system can be installed anywhere there is suitable exposure to the sun and a demand for electricity. In many cases, PV modules are the only way to keep batteries charged reliably in critical remote applications.

Client: **Canadian Coast Guard**

Contact: **SOLTEK Solar Energy Ltd.**
#2-745 Vanalman Ave.
Victoria, British Columbia V8Z 3B6
Tel: (250) 727-7720
E-mail: soltek@soltek.ca
Web:www.soltek.ca

Cost-Effectiveness of Renewable Energy in Remote, Northern & Rural Communities

In many off-grid locations in Canada's remote, northern and rural communities, renewable energy is an affordable solution when a customer is looking for a safe, small and reliable supply of electricity, i.e. a system under 25 kW. If the option exists for a customer to connect to a local grid, the consumer should be aware that they will have to pay high capital costs, upfront, and ever-increasing operating and maintenance. For example, Ontario Hydro charged between $18,000 to $60,000 per kilometre for grid extension into properties in 1996.

When comparing the costs to installation a renewable energy option and a traditional grid connection for remote or northern locations, it is essential that the maintenance costs are taken into consideration. A renewable energy system is capital intensive, yet low maintenance. A traditional system in a remote location is capital intensive and high maintenance. The remoteness of the location will raise maintenance and installation costs for a grid connection in the following ways:

- Additional fuel costs to access the remote location, either by helicopter, plane, or all-terrain vehicle
- Higher repair costs as diagnosis and repair may require multiple visits
- Loss of revenue by the utility as the load may lack power for an extended period
- Higher installation costs for the transportation of people and materials to the remote location
- Higher wages for installation and repair for employee time spent in transit
- Increased capital costs for grid extension

In examining the overall costs of a wind or photovoltaics system and to effectively compare it to the grid, it is important that the consumer understand the cost breakdown of each part of the system. The greatest expense is often the wind turbine itself or the photovoltaic modules (up to 33%), followed closely by the design and installation of the system (up to 20%). In both types of systems, batteries and inverters can also add another 15 to 20% onto the costs, with another 10 to 15% spent on controls and, in the case of photovoltaics, a tracker. In a stand-alone system, maintenance costs are slightly higher as the batteries require maintenance at least once a year and the batteries have a restricted life span of around 10 years, if they are maintained properly. Alternately, grid-connected renewable energy systems require little or no maintenance over their lifetime of at least 25 years.

When a system owner is comparing direct costs of a renewable energy system against the costs of a traditional electricity system, they often do not assign the societal benefits any real costs or savings. When installing a system in a rural, northern or remote community, the overall benefits to the local society include:

- A non-polluting energy source
- The elimination of the risk of an oil, gas or gasoline spill
- The environmental and direct savings associated with not having to build or maintain access roads to the lines
- An energy payback of one to five years in remote locations

CASE STUDY:

The MandoKey Project: Electrifying Canadian Native Communities with Solar Energy

La Verendrye Park, Quebec

THE MANDOKEY PROJECT was set up to test the EnerKey™ technology in a real-life situation, in a First Nation community in Canada. At the same time, it would encourage economic development of the community. The EnerKey technology is a prepayment controller which shuts off access to electricity if the user has not purchased energy units. The project has now been in operation for two summers in the community of Kitcisakik, an Algonquin Indian band living in the heart of the La Verendrye Park, north of Montreal.

Twenty 150 Wp PV systems were installed at the summer ground of the Kitcisakik Band, in July 1997, to meet the electricity demand of 18 families and power the community centre and the water pump. The summer ground is used only five months of the year when sunshine is relatively abundant. The systems were purchased by the Mando Co-operative formed by band leaders of the Kitcisakik community.

At their summer camp, most families used to produce their own electricity via individual one- to five-kVA gasoline generators. Each family had to carry its gasoline over water and land, for a distance of 65 kilometres. The average cost was C$80/month. Thus, the cost of electricity varied between C$3 to $5/kWh.

Upon installation of a PV system, each family had to pay an initial fee of C$500. In exchange, a PV system (owned by the Mando Co-operative) was installed on their house, plus a memory button programmed with 400 units (10 days of electricity, worth C$20). They are expected to use three of these buttons per month, providing them with an average of 14.4 kWh/month of electricity. This electricity is used to power one or two lamps, a television and a video set, a toaster, a radio-cassette and, in some dwellings, a small portable cool box. The community did express the desire to have a communal freezer, but this request has not yet been addressed.

The initial cost of installing the PV systems and the recurring maintenance costs which are paid by the Mando Co-operative are to be reimbursed by each family as they purchase key-buttons. Assuming that each user will purchase an average of three buttons/month, the co-operative will receive C$60/system/month. Including the cost of operation and maintenance and changing the batteries every three years, each system will be repaid in about eight years.

System information

System owner	Mando Energy Co-operative
System name	MandoSol-Pak-1
Date of starting operation	June 1997
System type	Domestic PV System
System location	47.5 °N, 77 °W Great Lake Victoria, La Verendrye Park, Quebec, Canada
Rationale and reasons for installation	Power the electric needs of Indian households (20) during summer These installations were done to demonstrate the EnerKey™ technology
System description	3-50 Wp LBK-50 Lenbrook modules + MandoSol-Pak-1 consisting of: 1-PSI PVCC 2012 Battery regulator 2-6 Volt, 220 Ah deep-cycle GNB batteries 1-StatPower ProWatt 800 W inverter 1-EnerKey™ 1280dc prepayment meter
Typical load description	Each system can typically power the following equipment: 1-2 lamps: 40-100 W 1 colour TV + video: 80-120 W 1 radio-cassette player: 10-50 W 1 toaster: 800 W The maximum average demand was estimated at 750 Wh/day
Installer	Environergie Quebec Supervised by Solener inc.
Funding of installation	The EnerKey™ technology and the Mando-Key pilot project have been supported by the Quebec government through its PADTE program of the Ministère des ressources naturelles Québec (25%) and the Canadian government through CANMET's division of Renewable Energy Technologies of Natural Resources Canada (25%). The Kitcisakik band (5% up front and 20% by prepayment), and Enersite International Inc. (25%) have also contributed to this project. CANMET's Energy Diversification Research Laboratory has kindly tested the system in its environmental chamber.

Other Applications

The EnerKey™ technology is a pre-payment technology which was developed to permit the commercialization of electricity in remote communities not served by grid power. In remote regions of the Canadian North and in rural areas of developing countries, communities are usually sparsely populated and the demand for electricity by individual families is small. This makes it relatively unattractive for utilities to install grid power. By using a pre-payment technology, an independent energy supplier can become a mini-utility using individual solar power systems or village-sized centralized power systems without facing the problems of actually doing the collection of small monthly fees, which would normally be too costly to collect. By using a pre-payment meter, it is the user who comes and purchases the energy units he needs at a service outlet or at his local merchant, just as he would purchase any other consumer good.

Comparing the Cost of PV and Gasoline Generator

Comparing the cost of purchasing and operating a PV system against the cost of purchasing and operating a gasoline generator, the PV system's initial investment is recovered after only three years. After 10 years, a PV system owner will have spent C$3,400 whereas a gasoline generator user will have spent C$6,600.

Figure 1: Cumulative cost of a gasoline generator and a PV system (utilization of five months/year)

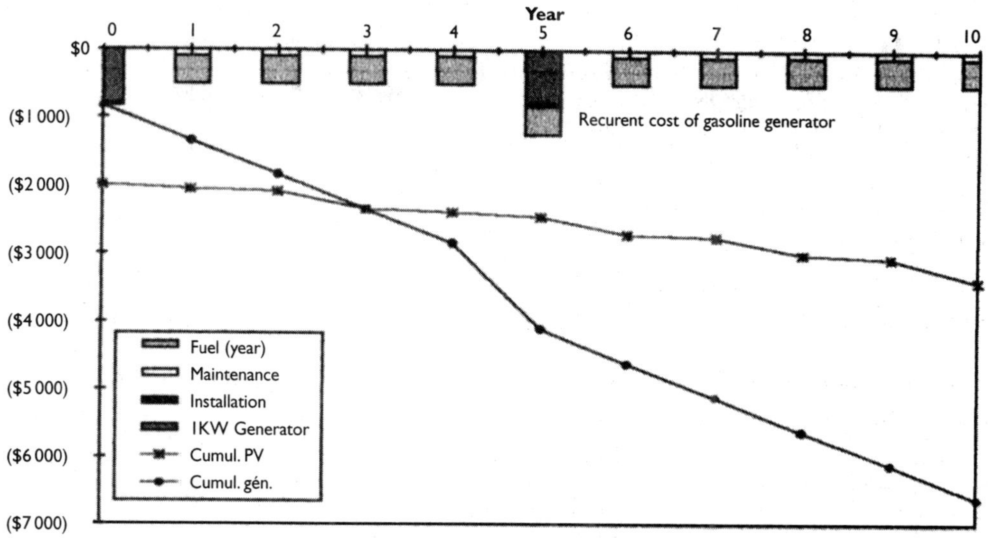

The price per kWh is based on a total cost for the PV system of C$2,000 (quantity purchase), a simple payback time of eight years and an initial investment of C$500 by the user. On average, each PV system owner will use 14.4 kWh/month for five month/year and will thus pay about C$300 /year. Maintenance cost is estimated at C$50, and battery replacement is expected to occur every three years.

Contact Information:

Enersite International Inc.
74 Auriga Dr., Unit 1
Nepean, Ontario
K2E 7X7
Tel: (613) 723-5714
Fax: (613) 723-5980
E-Mail: rthomas@newsun.ca

Quebec Office: Jimmy Royer
Solener inc.
442, rue Lavigueur
Quebec, Quebec
G1R 1B5
Tel: (418) 640-7444 Fax: (418) 640-7445
E-mail: Solener@qbc.clic.net

Effects of Climate on Renewable Energy Systems

In the Canadian climate, the installers of wind, micro hydro or photovoltaic systems must take into account the effect that weather will have on the operating efficiencies of the equipment. Parts of southern Canada have temperature differences of over 70°C from the dead of winter to the heat of summer, while in the north, temperature variations may be even greater from season to season. The higher latitudes must also deal with great variations in the amount of available sun, water and wind. During the summer months, parts of the North have 24 hours of direct sunlight, hence the name "Land of the Midnight Sun," and during the winter months, there is very little available sun. There are also special considerations for using PV in the Canadian climate as snow and ice accumulation may affect the structure directly and the efficiency of the systems. The storage batteries in the renewable energy systems are the most affected by the temperature extremes in the Canadian climate. The extreme cold in northern climates greatly reduces the storage capacity, efficiency and lifetime of the batteries.

In photovoltaics systems, the arrays, mounts and the electronic components are all affected by the cold and heat extremes. In colder temperatures, an array's output may be affected by snow and ice coverage and low light levels, but the cell's overall efficiency will actually increase. As the sunlight strikes a PV array, some energy is turned into electricity, but the majority of the energy, over 70%, heats up the module. In the colder temperatures, the PV effect operates more efficiently as the electrons in each cell are less agitated. Colder climate systems must also contend with snow, ice and freezing rain accumulation on arrays, thereby blocking the sunlight from reaching the array and reducing electrical power. Rarely does snow and ice accumulation damage the PV arrays as the angle of the modules in the mounts often ensures that heavy layers of precipitation cannot build up. Typically, in colder climates, the winter days are shorter and there is often more cloud coverage. Also, at locations in the higher latitudes, the angle at which the sun strikes the array is higher, thereby decreasing the overall amount of energy that strikes the array.

Stand-alone renewable energy systems rely on batteries for electricity storage. Special attention must be given to the design and operation of the batteries in cold climates. It is suggested that the battery bank should be protected from the outside environment by an enclosure, such as an insulated box, a room or shelter. These rooms or boxes must be properly vented, as open batteries produce hydrogen and oxygen gases. If the concentration of hydrogen in the air exceeds 4%, there is an explosion hazard. Closed lead-acid batteries also produce some small amounts of hydrogen, but venting is only required to ensure cooling and temperature regulation.

In many renewable energy systems, lead-acid batteries are used and they can be affected by cold climate conditions in the following ways:

1. Effect of temperature on the capacity of the battery. The capacity of batteries decreases significantly with a decrease in temperature.
2. Effect of temperature on the end of charge. The colder a battery gets, the less charge it will accept at a given voltage. For a battery to receive a full charge, the cut-off voltage will have to be set higher in colder temperatures.
3. Freezing of electrolyte. As a lead-acid battery is discharged, the water content of the electrolyte increases and the acid concentration decreases. As the concentration of acid in the battery falls, the freezing point rises from between –44ºC and –66ºC in a fully charged battery to between –6ºC and –13ºC in a fully discharged battery. In colder temperatures, there is the possibility that the electrolytes in a fully discharged lead-acid battery may freeze. Frozen electrolyte cannot be further discharged and recharge is difficult.
4. Effect of temperature on measuring state of charge. The specific gravity of an electrolyte solution is usually given at a temperature of 15ºC, but at lower temperatures, the reading will be higher, suggesting a state of charge that is higher than the actual reading.
5. Effects of deep discharge on lead-acid batteries. It is suggested that lead-acid batteries not be discharged beyond 80%, even if the battery is designed for deep discharge. Lead acid batteries that are deeply discharged may suffer from a permanent loss of capacity due to sulphation, acid dilution and greater tendency for freezing.

For further information on photovoltaic applications in Canada's north, refer to *Photovoltaics in Cold Climates*, 1998; details listed in Chapter 9, Books Section.

Photovoltaics in Canada's North

Contributed by: Michael Ross, Sylvain Martel & Eric Usher, CANMET Energy Diversification Research Laboratory, Varennes, Quebec.

THE TANQUARY FJORD warden station is nestled at the end of a long Fjord that penetrates the heart of the Ellesmere Island National Park Reserve, Northwest Territories. Though located at 81ºN, the surroundings are, for the High Arctic, lush and full of life. On a typical day, arctic hares dot the plain behind the warden station, muskox graze on the nearby mountain slopes, bright arctic flowers track the sun around the sky, and the wake of a duck creases the reflection of an iceberg in the waters of the fjord. In a harsh northern way, it is peaceful and beautiful. Peaceful, except for the drone of the gasoline generator that supplies the station with electric power. Beautiful, as long as one averts one's gaze from the hundreds of empty gasoline barrels forming a rusty echelon at the perimeter of the wardens' residences.

This, at least, was the situation until 1994. In that year, the gasoline generator was replaced by a hybrid power system. While a generator was kept as a backup, most of the power is now supplied by a small photovoltaic (PV) array and wind turbine. The turbine and the photovoltaic array, which converts sunlight directly into electricity, charge a battery bank. The batteries store electricity for those periods when it is cloudy and calm and consequently the electrical loads—lights, water pumps, washing machine, television, VCR, stereos, microwave, and other standard appliances—exceed the combined output of the PV array and the wind turbine. Now the generator runs only infrequently, reducing maintenance costs, fuel consumption, local pollution, and noise.

The hybrid power system was installed after a study by the CANMET Energy Diversification Research Laboratory (CEDRL) showed that photovoltaics were not only feasible, but also cost effective for this application. The study was conducted under the aegis of Photovoltaics for the North, a five year, C$1 million joint research, development and technology transfer program of the CEDRL in Varennes, Quebec, the Aurora Research Institute in Inuvik, Northwest Territories, and the Nunavut Research Institute in Iqaluit, Northwest Territories. The objectives of the program were to develop and implement clean and reliable energy supplies for northern climates and to improve the knowledge base and skills of the residents of the Northwest Territories. In this way, the program has strengthened Canada's commitment to sustainable energy.

Electricity Generation in the Northwest Territories

In southern Canada, most consumers requiring electricity are connected to the electric grid. In the Northwest Territories, on the other hand, *no* sites are connected to the North American grid. Rather, each community or cluster of communities has its own isolated grid and associated generation equipment. This situation reflects the region's low population density and enormous scale. Though more than one-third of Canada's land is contained in its borders, the Northwest Territories has only 68,000 residents, dispersed among 65 far-flung communities.

Presently, diesel fuel and water are the resources used to generate electricity in the North. The Northwest Territories Power Corporation sells approximately 450 GWh annually. Hydropower developments in the vicinity of Great Slave Lake supply about half of this; diesel generators scattered throughout the territory are responsible for the remainder. Private diesel-electric installations, located at mines, telecommunication stations, and remote hunting and fishing camps, generate an additional 120 GWh each year. Thus, about 60% of the region's electricity—or about 350 GWh annually—is generated from diesel fuel.

While hydropower is relatively inexpensive and reliable, there are significant drawbacks associated with diesel-electric generation, as the Tanquary Fjord example demonstrated. The most obvious of these is cost. Transporting the fuel to the communities is expensive; delivered fuel costs range from $0.55 to $1.30 per litre and at remote sites like Tanquary Fjord, the price is even higher. As a result, electricity costs between $0.25 and $1.00 per kWh, though residential and commercial consumers are subsidized and generally pay less than the full cost. Further, the heavy reliance on diesel ties electricity costs to fluctuations in the price of oil.

In many northern communities, local pollution is a problem. On cold, still days, emissions shroud the community like a thick fog. Noise makes diesel generators unpopular as well. Many generating stations were built on the outskirts of expanding communities and now that the town has been built around the generators, residents complain about their continuous hum. Moving the stations outside of the community is expensive, and precludes the use of waste heat from the generators. At remote sites, which often must operate unattended, the reliability and maintenance requirements of generators inflate their cost. Service calls can be very expensive, and the cost of a failed generator can be the loss of an essential telecommunications link, for example. In addition, there is the risk of fire and leakage associated with the storage and transport of fuel.

Why PV?

The problems associated with diesel-electric generation have created a willingness to consider alternatives, such as photovoltaic power. Initially, PV may not be an obvious choice, since the long, dark, and very cold northern winter seems at odds with the goal of using sunlight to generate electricity. In reality, PV operates more efficiently at cold temperatures, and the Arctic summer provides long days of plentiful sunshine. On a summer's day, a PV system in the Arctic can generate far more electricity than a similar system at the equator.

PV has a number of strengths. Importantly, it is a proven technology. PV systems are already installed throughout the Northwest Territories at installations ranging from critical military telecommunications repeater stations to houseboats on Great Slave Lake. Other advantages emerge when PV is compared with fossil fuel-powered generation. First, the system requires no fuel and can function for long periods without maintenance. Second, it has a reputation for high reliability because it contains no moving parts—a PV panel will generate electricity for 20 years or more without any intervention. Third, PV systems have no emissions—either noise or pollution. The combination of these advantages has made photovoltaics the least-cost option at many remote sites requiring a small, reliable power supply.

With such impressive characteristics, why has PV not supplanted generators? The principal reason is cost. A PV array costs about $5 per watt of peak generating capacity (Wp); the not insignificant cost of batteries, inverters, controllers and installation must be added to this. This cost can not be compared with the purchase price of diesel generating capacity. With a PV panel one is essentially paying up front for a 20-year fuel contract.

PV system costs have dropped significantly in the last 20 years, and will continue to do so in the future, as technological advances in PV cells and manufacturing techniques appear in the marketplace, and economies of scale become established. The cost of PV per Wp has fallen from about $100 in 1970 to $5 today, and is projected to decrease further to about $2 by 2010. As prices drop, PV systems will become increasingly attractive in the Canadian North.

The PV for the North Program

The need for clean and reliable energy sources combined with the high cost of diesel fuel in the Northwest Territories has created opportunities for photovoltaics. PV for the North helps Canadian companies and northern residents exploit these opportunities. The program consists of three phases. Phase A laid the groundwork for the further implementation of photovoltaics, identified promising applications and barriers to the use of PV, and studies demonstrated the feasibility of the technology to potential users. Phase B focuses on developing new PV technologies tailored for the northern climate, building awareness and transferring expertise to the North, and encouraging the use of PV in the most promising applications identified in Phase A. Phase C was a three day conference on renewable energy in cold climates, held in Montreal, Quebec, in May 1998. This international conference brought experts and users together, disseminated information, and showcased Canadian companies and products.

Cost-Effective Applications for PV

Phase A identified many applications for which PV is cost-effective right now. These applications tend to be remote and generally require a modest but reliable power supply. For example, in a CEDRL study it was shown that NorthwesTel could save $750,000 over 20 years by replacing the primary (non-rechargeable) batteries used to power its 12 VHF repeater sites with PV systems. NorthwesTel is now embarking on this transition. UHF and microwave repeater stations and navigational aids are also candidates.

The large seasonal variation in the amount of sunshine available complicates the use of PV in the North. While solar energy may be plentiful in the summer, it is scarce or even nonexistent in the winter. Battery storage can be used to transfer some energy from the summer to the winter, but this tends to be expensive. Phase A pointed out two other approaches to this problem: summer-only applications and hybrid systems.

The seasonal variation in solar energy affects not just northern PV systems, but northern residents as well. Outdoor activities that the hardiest soul would not contemplate during the winter are common during the summer. People visit cottages, groups of families set up remote hunting and fishing camps far from their communities, tourists come to explore the area's parks, firewatchers install themselves in firetowers, and researchers head off across the tundra to collect data. Often the electrical loads at these sites are fairly small—a radio, a few lights, a laptop computer— and the power system must be simple and reliable. For these summer-only applications, PV can be an excellent choice. In addition, the warmer summer temperatures can create other applications, such as exhaust fans for attics. These are commonly installed in houses with electric service, due to the simplicity of installation and control. One doesn't need to remember to turn these on and off, since the sun does it automatically.

When larger loads must be powered year-round, the benefits of PV can be combined with some other power source—a fossil-fuel powered generator, wind turbines, thermoelectric generators, or primary batteries—in a hybrid system. This system combines PV's low operating and maintenance costs with the power on-demand available from generators. Hybrid systems are often cost-effective for medium size loads, ranging from 2 kWh to 60 kWh per day. Hybrid systems can be ideal for remote residential, telecommunications, and cathodic protection applications.

Sometimes existing prime power generators for variable loads can be upgraded to hybrid systems by adding batteries to form a cycle charging system, then adding PV panels as capital becomes available. This has the advantage of spreading the high capital cost of the photovoltaic system over a longer period.

While these applications create a market for PV right now, the medium-term market has even more potential. As prices drop, it will become economical to use PV to reduce fuel consumption and maintenance costs on remote community grids. This suggests that PV may be widely implemented in northern communities before it appears at a significant scale in southern Canada and the USA. On the North American grid, the cost of building additional capacity tends to be high, but the unit cost of generating electricity is low. Unfortunately, since PV depends on an intermittent resource, it is hard to convince utilities that it has capacity benefits. In northern communities with diesel generation, however, the principal cost of electricity is fuel, and each litre of fuel that is not consumed because of PV generation represents a significant savings. In a CEDRL study, it was shown that these avoided cost benefits—based on fuel and maintenance savings only—could be valued at about \$2.50 to \$5.50 per Wp of PV installed. In other words, when PV's installed price in the North reaches this level, it will become economical to supplement diesel community grids with PV power; it is expected that this will happen within 10 to 15 years. It is interesting to note that at the best sites in California, PV has a similar avoided-cost benefit, but only when environmental credits for pollution reduction and transmission and distribution costs have been factored in. That is, PV seems to make as much sense in the Arctic as it does in sunny California.

While PV cannot replace all the diesel generation on the remote community grids, this market still has enormous potential. In a PV for the North study, it was shown that PV could be used to provide up to half of a community's power without having to dump much electricity during the summer. Even if PV represented just 5% of the remote grid power generation in the Northwest Territories, this would result in the installation of 15 MWp of PV—six times the total capacity of PV installed in Canada at the end of 1995.

Knocking Down Barriers to Market Penetration

While the high capital costs of PV and subsidized electricity prices are economic deterrents to large-scale implementation of PV in the North, Phase A also identified various cultural, social, technological and organizational barriers. Through various initiatives, Phase B has focused on removing these barriers.

Northerners tend to be familiar with the conventional energy sources, such as generators, with which PV competes. People are not aware of PV's capabilities and often assume that solar energy doesn't function when it is cold. In addition, some opinions of renewable energy have been formed on the basis of early experimental or demonstration projects, which lacked the reliability of today's technology. Phase B has attempted to address these problems by making PV more visible within communities and by disseminating information about PV.

On Canada Day, 1995, a 3.5 kWp PV array, discreetly incorporated into the facade of the Nunatta campus of the Arctic College, in Iqaluit, was switched on. Financed by the PV for the North program, designed by Dirk & Price Engineering Ltd., and installed by a local electrical subcontractor, this is the first diesel-grid connected PV system in the Northwest Territories. As a demonstration, it raises awareness of PV. Equally importantly, the system will provide valuable operating experience for the future, when grid-tied PV systems become cost-effective and provide hard data essential to fostering support for major investment.

Workshops and numerous meetings have heightened PV's profile in the North and helped southerners better understand the opportunities in the North. In 1995, workshops were held in Iqaluit and Inuvik, and the CEDRL met with over 30 northern companies, government departments, and nongovernmental organizations. In November of that year, seven groups from the Northwest Territories were funded to send representatives to the annual conference of the Solar Energy Society of Canada, in Toronto.

Phase A also noted the limited number of companies working in the field and the lack of northern technical expertise related to PV. Phase B has tackled this through training programs, training systems, and close co-operation with northern companies. In September 1996, a hybrid training system was inaugurated at the Inuvik Research Centre. This system, which consists of a 1.4 kWp PV array, an 850 Wp wind turbine, a 5 kW generator, batteries, and a 360º tracker for the PV array, will be used to learn about hybrid systems; it will also provide a test bench for investigations into the optimal operation of hybrid systems in the North. The PV panels were manufactured by CANROM Photovoltaics Inc. of Hamilton, the wind turbine was supplied by Prometheus Energy of Concord, Ontario, and the system incorporates an inverter from Statpower Technologies, of Burnaby, British Columbia. In the spring and summer of 1996, a

course covering PV, small wind turbines, and diesel generators in hybrid systems was taught in Inuvik, Yellowknife, Iqaluit, and Rankin Inlet, four of the main communities in the Northwest Territories. In addition, five northern residents have spent one to two months in PV training sessions at the CEDRL. One of these northerners has set up a PV company in the eastern Arctic. Since the inception of the PV for the North program, Midnight Sun Energy, has grown from one part-time to three full-time employees.

In the autumn of 1997, a third demonstration project was inaugurated One of NorthwesTel's UHF radio stations in the remote and mountainous Nahanni National Park was converted from a genset system to a hybrid system with photovoltaics. The experience and confidence gained through this demonstration has convinced NorthwesTel to install photovoltaics at other UHF stations.

In the last years of the PV for the North program, a book and a video were prepared so the expertise and momentum of the program would not be lost after its conclusion. *Photovoltaics in Cold Climates*, a book aimed at decision-makers, engineers and installers, was published in November 1998 by James & James. As a collaboration between PV for the North and the International Energy Agency's Photovoltaic Power Systems Program, the book was written by experts from Canada and around the world, and will be distributed internationally. For more general audiences, there is a video introducing photovoltaics and explaining their operation and use in Northern Canada from the perspective of a Northern Canadian.

Developing New Technologies

The harsh northern climate creates special difficulties for PV technologies which often are not fully appreciated in the southern climates where photovoltaic components are generally developed. While the PV panel itself performs extremely well in the cold, batteries and electronic components do not fare so well. This creates a niche for Canadian technologies and Canadian companies.

While they do not fail, batteries do perform poorly when cold. In order to increase the useable capacity of the batteries used in cold climates, CEDRL, working in conjunction with SOLTEK Solar Energy of Victoria, B.C., has developed an insulated battery enclosure with a thermal reservoir of phase change materials. This enclosure maintains the batteries at a reasonable temperature throughout the winter, ensuring that useful battery capacity does not drop below a certain level. This means that smaller battery banks can be used, reducing cost.

When snow and ice accumulate on PV panels, less light reaches the solar cells and they produce less electricity. In general, such accumulation is rare, but at certain sites this can be a problem. The Montreal consultancy of TN Conseil has developed a passive device that melts off snow and ice. CEDRL has performed modelling and testing to establish the range of conditions for which this technology functions and to optimize the design.

The mathematical models used to estimate the amount of solar radiation available for a PV system are not particularly accurate for sites at high latitudes. The University of Waterloo has conducted a study for the PV for the North program to determine how these models can be improved. The result is an improved version of the WATSUN-PV simulation software; this makes WATSUN-PV one of the most appropriate tools for modelling PV systems in northern Canada, Antarctica, Scandinavia, and northern Russia.

In order to assist Canadian installers and system houses, CEDRL has initiated various component testing programs. The ongoing battery test program focuses on the operation of lead-acid batteries at cold temperatures. Commercially available charge controllers are currently being tested in the CEDRL climate chamber. In the upcoming year, testing of photovoltaic modules in variable field conditions, as opposed to the standard test conditions, will begin at the CEDRL. The results from these testing programs will provide the information necessary for the design of more reliable and less costly PV systems.

Land of the Midnight Sun, Land of Opportunity

PV's high reliability, low maintenance and operating costs make it cost-effective for a wide range of small loads at remotely situated locations. The use of photovoltaics will continue to grow as the price of systems drops. Programs such as PV for the North, working in conjunction with Canadian companies and northern residents, are knocking down the barriers to the widespread implementation of this technology. The opportunities are considerable, not only for Canadian business but for ordinary people who wish to live in a sustainable environment.

For more information about photovoltaics or the Photovoltaics for the North program,

Contact: **Sylvain Martel**
CANMET Energy Diversification Research Laboratory
1615 Lionel-Boulet
PO Box 4800
Varrenes, Quebec J3X 1S6
Tel: (450) 652-6747
Fax: (450) 652-5177
E-mail: sylvain.martel@.nrcan.gc.ca.

CASE STUDY:

Remote Hybrid Power Systems: A Mountaintop Radio Repeater Station
Nahanni Range Mountains, Northwest Territories

MOST OF NORTHWESTEL'S large radio repeater stations across the North are currently powered by diesel fuel generators. In 1998, however, the situation changed for the station located in the Nahanni Range Mountains, Northwest Territories. At 61.5° N and 2,600 m above sea level, the repeater station now gets most of its power from the sun since the power supply was turned into a PV-diesel hybrid system in the fall of 1997. From 100% of its energy supplied with diesel, the telecommunication system can now be powered at about 75% by the sun, leaving the generators off for most of the summer.

Such conversion from a diesel to a hybrid configuration presents both environmental and economic advantages, mainly due to the reduction of fuel consumption. This reduction in diesel fuel translates into lower fuel bill; and less frequent expensive helicopter trips to the site. It also means less emission from the generators and the helicopter. While the PV upgrade at the Nahanni station cost about $50,000, the annual savings are about $10, 000 (of which about $8,000 is related to diesel fuel). This project should pay for itself within five years, leaving 15 more years of savings for the owners.

System Specification

The PV array was simply added to the existing generator and battery bank. The power generation and storage equipment of this station now includes:

- A 1.5 kWp PV array
- Two 10 kW diesel generator
- A 1400 Ah battery bank (@ 24 V)

Other Applications

With most of its territory sparsely populated, Canada has many applications needing autonomous power supply. The traditional power solution, diesel generators and sometimes batteries, is slowly being complemented by photovoltaics and converted to hybrid systems. While telecommunication applications hold the largest potential for PV in Canada's north, many other applications, such as parks and remote residences, can also use hybrid systems. In these latter cases where the users live on site, significant side-benefits include reduced run-time (less noise) and increased autonomy.

The Big Picture

NorthwesTel had been considering alternatives to diesel for some time when support by the Photovoltaics for the North program was offered to them in 1997. The Nahanni project, a collaboration between NorthwesTel, CANMET Energy Diversification Research Laboratory and SOLTEK Solar Energy Ltd., wanted to demonstrate that photovoltaics could meet reliably the need of such application even under northern conditions. After one year of operation, the system has proven successful by meeting reliability expectations.

Client: **NorthwesTel Inc.**

Contact: **Sylvain Martel, Project Officer**
CANMET Energy Diversification
Research Laboratory
1615 Lionel-Boulet, P.O. Box 4800
Varennes, Quebec J3X 1S6
Tel: (450) 652-6747 Fax: (450) 652 5177
E-mail: sylvain.martel@nrcan.gc.ca

CASE STUDY:

Microhydro Electric System
Purcell Lodge, British Columbia

PURCELL LODGE is high up in the Canadian Rockies, near the town of Golden, British Columbia, four miles beyond the nearest road. All supplies and guests are brought in by helicopter. Over 99% of the electrical requirements are met by the 12kW microhydro system. All lights, appliances, pumps, motors, tools, and electronics needs are met by the 24-hour-a-day, year-round hydro output. Backup electric power comes from a diesel-powered alternator which operates only during maintenance shut-down or extremely low creek flows. Storage batteries are not required. Excess hydropower is directed to low priority loads, such as water and space heating by the load control governor, which significantly reduces flown-in fuel costs.

The microhydro system is a classic high head (150 psi), low-flow system using a pelton-type hydraulic turbine. The intake weir is a small concrete wall on a bedrock location with screened inlet pipe. The system is at 7,000′ elevation but a deep snow cover provides insulation and prevents the stream and surface-laid pipe from freezing. There are 800 m of 150 mm PVC pipe carrying up to 300 gpm to the powerhouse. Electric power is transmitted to the lodge at 600 volts via a buried wire. The load control governor has computer-controlled electronic switches which turn auxiliary loads on and off as necessary to maintain balance with the turbine rpm output based only on frequency.

System Specifications

10 kW pelton turbine by IPD, Montana
Brushless alternator by Fidelity rated at 2 kW, 120/240 VAC, 1800 rpm
Electronic Load Control Governor by Thomson and Howe
Isuzu diesel generator 18.4 kW

The Big Picture

A stand-alone microhydro system can be installed anywhere there is suitable head and water flow, such as at remote lodges. With a good site, microhydro is a renewable energy technology with life-cycle costs lower than grid-connected electricity.

Client: **ABC Wilderness Adventures Ltd**
PO Box 1829, Golden, British Columbia
V0A 1H0
Tel: 604-344-2639

Contact: **Bob Mathews**
Energy Alternatives Ltd.
825 Laval Crescent
Kamloops, British Columbia, V0E 1M0
Tel: (800) 265-8898
Fax: (250) 374-3943
E-mail:
energyalternatives@bc.sympatico.ca

CHAPTER 5

Power Supply in North America

Restructuring the Canadian Electric Industry

IN ORDER TO CREATE a more sustainable and efficient energy industry, Canada must grapple with problems inherent within the existing utilities structure, including massive debt, bureaucratic inefficiency and, in the nuclear industry, concerns over safety and waste disposal and concerns over high cost. The Canadian electricity industry is moving rapidly towards restructuring as a result of technological change, increased customer pressures for lower electricity prices and examples of successful restructuring in other industries, including natural gas and telecommunications.[1]

Many Canadian provinces have already begun to implement utility restructuring initiatives. Many large Canadian utilities are grappling with complicated restructuring issues as they move towards open and competitive markets. The incumbent utility companies in each province have a distinct advantage in the expanding, competitive markets as they already have market share, while the competition must work to establish market recognition. However, the incumbent provincial utilities must still adapt to the modern corporate culture, develop new business strategies and develop new business partnerships and alliances.

The United States is moving ahead with restructuring at both the federal and state levels. The US Federal Energy Regulatory Commission (FERC) began the process in 1992, with the passage of the *Energy Policy Act*, mandating wholesale competition. In 1996, they passed Order 888, making transmission a common carrier service, open to all suppliers. It is predicted that the introduction of a competitive market may allow prices to drop between 8 and 15%,[2] and may result in increased service and a more consistent supply to the customer. Other countries, including the United Kingdom, Australia, Norway, New Zealand and some countries in Latin America, have also begun the move towards restructuring. Observers of the electricity industry predict that in the next ten years, all monopolistic franchises will have been replaced by competitive supply. The large encumbered companies will have given way to more efficient companies more accountable to their customers. Energy observers predict electricity will be traded openly, like other commodities. Electricity, gas and other businesses will converge and a new organisational structure will have emerged. The restructuring will also allow small, independent power producers (IPPs) to serve the local market directly, rather than having to feed into the provincial grid.[3]

Even as the utility companies are facing restructuring, they are also experiencing another significant change, called demand-side management or DSM. The utilities have now spent billions of dollars a year to encourage customers to become more energy efficient, through the reduction of the costs of lighting, cooling and refrigeration. This ensures the demand side, the customer, is efficiently managing its electrical needs. Some studies have shown that if customers improve energy efficiency, the utilities are able to cut costs and not invest in expensive new power plants. The

---

[1] Ontario Ministry of Energy, Science and Technology, Direction for Change: Charting a Course for Competitive Electricity and Jobs in Ontario, (Toronto: 1997), p. 1.
[2] Ontario Ministry of Energy, Science and Technology, p. 4.
[3] Ontario Ministry of Energy, Science & Technology, p. 4.

movement towards DSM began in California in the early 1980s. At the beginning, the changes were modest. Companies eliminated discounts for large power users and offered customers a free energy audit. They also offered cash rebates for purchasing low-energy appliances, energy efficient lights, high efficiency electric motors and solar water heaters. Some offered loans for home or industrial retrofits.[4] British Columbia and Ontario were the most active provinces in Canada in offering customers DSM solutions, while most provinces maintain DSM programs to some degree.

As the restructuring of the electricity industry is in transition, it is best that the interested parties contact either the local or provincial utility or provincial Ministry of Energy to obtain information on the status of restructuring. As of spring 1998, there is a province by province restructuring update.[5]

Atlantic Provinces

Nova Scotia and New Brunswick utilities are working on developing strategies that will allow them to compete in the restructured electricity market. These utilities are in a perfect position to take advantage of a large US Eastern Seaboard market, but they must reduce production costs and to remain competitive on both sides of the border.

Quebec

The Government of Quebec has instituted a new energy policy that sets up an independent energy board to regulate the new competitive provincial energy sector. Hydro-Quebec has been preparing for competition and is positioning itself as one of the larger electricity providers in North America.

Ontario

The Ontario Government recently restructured the electricity market and aims to provide Ontario energy customers with low-cost energy through competition in the retail, distribution and generation sectors. The restructuring plan, if implemented, would:

- Create a competitive market in the year 2000 for wholesale & retail customers
- Establish an independent market operator, and provide for an interim supply market for replacement power
- Gradually separate monopoly operations from competitive businesses, over the next 10 years
- Provide the Ontario Energy Board with an expanded mandate to protect electricity customers
- Take steps to ensure environmental protection
- Encourage cost savings in the local distribution sector
- Establish a level playing field on taxes and regulation
- Restructure Ontario Hydro into new companies with clear business mandates
- Take action to ensure that the new electricity companies are on a sound economic and financial footing.

The Ontario government also proposed that two new companies be formed to succeed Ontario Hydro after April 1, 1999: the Ontario Electricity Generation Corporation, which will take over the generating capacity, and the Ontario Hydro Services Corporation, which will be responsible for other electricity businesses, including transmission, distribution, retail and operating contracts.

Manitoba

Manitoba has had an incumbent utility, Manitoba Hydro, and customers have enjoyed some of the lowest electricity rates in North America. Recently, the government regulated the opening up of its transmission system for wholesale wheeling purposes. The utility is also getting ready to expand its market into the United States.

Saskatchewan

A recent review of SaskPower concluded they could be operating in a competitive environment by the year 2000. The review also concluded SaskPower is going to be vulnerable to open competition due to the utility's higher generation costs and existing cross-subsidies. The report suggested that SaskPower needs to balance its rates, reduce it costs, provide new services and consider partnerships with its neighbouring utilities.

Alberta

In January 1996, the Alberta government passed the *Electric Utilities Act* that created a power

[4] Christopher Flavin and Nicholas Lenssen, *Power Surge: Guide to the Coming Energy Revolution.* (W.W. Norton: New York, 1994), p 252.

[5] Canadian Electricity Association, World Wide Web Site, www.canelect.ca, "Restructuring the Electric Industry"

pool, competitive future generation and opened up the transmission system. The government also committed to the further liberalisation and deregulation by the introduction of competitive options for all classes of customers. One Alberta utility, TransAlta, quickly moved into the United States' market, obtained its FERC marketer's license and currently is the thirteenth largest power marketer in the US. Another Alberta utility, Canadian Utilities, has significant international independent power producers' investments, while Edmonton Power, a municipal utility, has expanded its operations to include water and gas.

British Columbia

BC Hydro has opened up its transmission system to the wholesale wheeling of power. It, along with its wholy owned power marketing subsidiary, Powerex, has obtained the necessary license to market electricity within the United States. Altough an extensive study has been completed on various degrees of deregulation in the province there are no plans to offer retail competition.

Green Power in the Post-Kyoto Period

Contributed by Jason Edworthy, Vision Quest Windelectric Inc., and Leon Burn, ENMAX

What is Green Power?

Green power is electricity from renewable energy sources, typically low or zero emissions sources such as solar photovoltaic, wind, microhydro or biofuel technologies. The electricity is produced at commercial-scale facilities and put on the electricity grid for customers, who choose this product as an alternate to the conventionally available power supply.

Where are GP Programs?

Green power programs have been developed in many locations, including northeastern United States, California, Colorado, British Columbia, Alberta and Ontario. Each program is slightly different, but the basic idea is the same. A customer can choose the source of electricity, and may elect to pay a premium over the 'regular' supply in order to get the desired product. Some programs ask customers to make a donation or contribution to a fund that will, when large enough, install renewable energy facilities. Others tie the supply to specific renewable energy facilities, which are set up and commissioned in order to meet demand. Some programs offer donations to food banks or wildlife funds as part of the product package, while others provide emissions reductions credits with the electricity.

Why do Customers want GP?

There are many sorts of customers interested in green power for various reasons, including private residential customers, commercial businesses, industrial customers or government customers.

Private customers are interested in green power as a way to indicate their choice in how their power is supplied. This choice is in part an empowerment of the consumer, a method to send a message to the suppliers of electricity about the sources that are used. It is a tangible way for people to reduce emissions created on their behalf.

Commercial businesses are interested in green power in order to differentiate their business from others, showing they are 'greener' and more environmentally correct, and drawing in more customers as a result. This approach may bolster an existing image, create brand recognition, or help improve the image of the business.

Industrial customers are interested in green power as a way to hedge or reduce risks associated with emissions. They may choose an emissions-free source of power in order to reduce the emissions they are responsible for, or to demonstrate to customers and shareholders that they are responding to environmental issues in a positive manner. They may see green power as one way to minimize increased regulation or possible carbon taxes.

Government customers want green power to meet their own requirements, demonstrate leadership, and foster policies that support new industries, diversification, and increased employment.

Canadian Examples

Alberta

Enmax

Greenmax™ — This program offers residential customers two optional purchase plans. One block of 125 kWh of wind energy costs $7.50 per month, two blocks cost $15 per month. A commercial program is also available.

Vision Quest Windelectric Inc.

Green Energy™ — Vision Quest offers residential, commercial and industrial blocks of green energy, and also offers emissions reductions credits. The residential program is 100 kWh for $9.50 per month (includes the certified emissions reductions), while the commercial program is 1000 kWh per month for $68. Industrial supply is available at quoted volumes and terms.

Ontario

Ontario Hydro

Clear/Green Choice Program — Ontario Hydro have offered green power from small hydro facilities to commercial customers. The program was initially called ClearChoice, but the name is reportedly being changed to Green Choice.

Others

West Kootenay Power

Green Power Tariff — West Kootenay Power in British Columbia has announced a green power rate option through a flyer included with bills and on its website. The new rate tariff has been submitted to provincial regulatory authorities. According to the announcement, green power will be procured from renewable sources such as solar or wind energy.

Standards, Certification, Registration

New products and new markets sometimes provide opportunities for unscrupulous suppliers to take advantage of consumers. In order to prevent this from happening, the Canadian green power industry has begun to lead by example, using standards, certifications and licensing set by others, and supporting registration processes.

The Environmental Choice Program, which licenses facilities, products and suppliers to carry the EcoLogo™, is one of these. Two criteria for alternative sources of electricity have been developed, one for renewable energy and one for biomass. An audit and investigation program is involved in qualifying for a licence and facilities are checked annually for compliance.

Some companies, such as Vision Quest, have published protocol standards for their products, involving multi-level independent third party monitoring, reporting, certification and verification. The ability to show that transactions and production can be tracked, and audited, provide a strong case for high quality product.

Green power products are now being vetted by independent panels, particularly their emissions reductions aspects. One panel, the Greenhouse Gas Emissions Reduction Trading (GERT) pilot, initiated by the Government of British Columbia, negotiated a trade between Enmax and Environment Canada for emissions reduced by wind energy in Alberta. The GERT pilot will be registered at VCR Inc., the private agency which publishes emissions reductions plans and progress reports of most Canadian organizations.

There is a need for further development of definitions and universal standards for green power. Some of this work has been initiated by Environment Canada, some by the EcoLogo™ program, while other work has been started by industry participants.

Future

There is a positive relationship between electricity industry deregulation and the availability of choice in power supply, including green power. The somewhat limited choices now in Canada are likely to expand, particularly in regions where the industry is deregulating and competition is allowed. The association between green power and emissions reduction is likely to become stronger, meaning customers anywhere in Canada can potentially purchase from suppliers located in other jurisdictions.

Green power can be brought on-line relatively quickly, and in modest-sized increments. This gives a potential advantage over conventional power supplies, which have long lead times. There is also a risk of increased environmental regulation and compliance requirements for conventional power plants, particularly those which are highly carbon intensive.

CASE STUDY:

Grid-Connected Wind Farm
Pincher Creek, Alberta

THE PE-KUN-NEE PROJECT, located west of Pincher Creek, Alberta, is a 9.9 megawatt (MW) plant fueled by wind power that started generating electricity commercially on August 12, 1994.

The Chinook Project Inc. and the Peigan Nation Indian Band of Brocket, Alberta, have been working together since 1991 to plan, locate, build and finance the Pe-Kun-Nee Project. Initially, the co-developers hoped to locate the project on the Peigan Reserve, 15 km east of Pincher Creek. However, difficulties in obtaining financing at that location led them to discuss options with Kenetech Corporation of San Francisco, which was developing a slightly smaller project near Cowley, Alberta. By using the same wind turbines, and locating adjacent to each other, both projects stood to achieve some economies of scale and financing advantages

Under the new arrangements, Kenetech would provide construction financing, locate permanent third-party project financing, and construct the wind farm using turbines manufactured by its subsidiary, US Windpower. The Chinook Project and the Peigan Band continued to negotiate the power purchase contract with TransAlta Utilities and satisfy the municipal district, Energy Resources Conservation Board (ERCB), and environmental permit requirements. Construction began on foundations before freeze-up in 1993 and continued in spring 1994. Towers with turbines began to stand at the site and electrical connection to the utility grid was completed before the end of July. All machines were tested and commissioned in early August and on August 12, the twenty-year power purchase contract with TransAlta began.

Wind energy co-exists with cattle grazing, has little environmental impact, and produces no air emissions. With more than 90% of Alberta's electricity produced by coal fired steam turbines, the introduction of clean wind energy provides opportunities for consumers to obtain electricity from a more sustainable, environmentally appropriate source. As well, this new industry creates jobs in manufacturing, construction, and operations—often in areas where jobs are scarce.

System Specifications

27 - US Windpower 33m rotor diameter wind turbines, 24m high—each rated at 360 kW
Total Capacity: 9.9 MW

Developer Contact:

Mike Bourns
The Chinook Project Inc.
Calgary, Alberta
Tel: (403) 289-4399

Client/Customer: TransAlta Utilities

CASE STUDY:

Le Nordais—Canadian Wind Energy Development

Gaspé, Quebec

Contributed by Drew McKibben and Diane Bailey. Photos provided by NEG Micon

ON THE GASPÉ PENINSULA in Eastern Canada, a crane operator watches carefully as he raises a giant three-bladed wind turbine rotor to its resting place on top of a 55m steel tower. The construction stands against the green hills and cold, gray sky of late September, and is the fifth of 133 machines to rise on the south bank of the St. Lawrence River. When all are complete they will supply enough power to serve 16,000 households in the province of Quebec and form Canada's largest wind farm—Le Nordais.

Reacting to the potential of this undertaking, an editor at the *Globe and Mail* headlines a story with, "The winds of change are blowing." Although accurate, the author had no idea how slowly the change would come. The story ran in October 1994. The Gaspé wind farm's fifth rotor was lifted into place September 1998.

Hydro-Québec kick-started the project in 1992 to bring new generation on-line and broaden its resource portfolio. The government-owned utility signed a 25-year power purchase agreement with an independent developer in 1994, guaranteeing a premium price, averaging 5.8 cents/kilowatt hour (kWh), for the capacity of a 100-megawatt (MW) wind-energy facility. The plant was scheduled for completion in 1996. With, however, 76 turbines operating at the end of the first phase in 1998, and the remaining 57 on-line by December 1999, Le Nordais' construction is three years behind schedule.

While preparation for the construction is formidable, the 750-kilowatt (kW) NEG Micon turbines can be erected at a rate of roughly one per day during the spring, summer or fall. And while there is no easy explanation for why the venture has taken five years to achieve full production, the ongoing story of its struggle reflects the challenge of trying to develop wind energy in Canada. Before wind turbines started going up in the Gaspé in the fall of 1998, only a handful of utility-connected machines had been installed in the previous four years.

"I see Le Nordais becoming the reference and the proof that it can be done in Canada, and done well," says Axor Engineering spokesperson Marie-Josée Gagnon. The Quebec-based construction and engineering company has steadfastly promoted the $160 million wind farm, and is one of two partners in its ownership.

Le Nordais, as a success story, is now reaching Canadians. The image of whirling blades and steel towers appear in the media as renewable energy sentinels, safeguarding the promise of a cleaner environment. The whole story, however, has been a lengthy saga of both triumph and failure. Although

the project achieved success, Axor's original partner, Kenetech Canada, has utterly collapsed. Its demise began in the spring of 1996 when Kenetech's parent company, the American turbine manufacturer Kenetech Windpower, filed for bankruptcy protection.

Suddenly without turbines or adequate financing, and already behind schedule, the Gaspé wind farm was in danger of going down with the titanic descent of Kenetech. Worse still, while Kenetech attempted to find a buyer for its Canadian investments, Hydro-Québec's final, in-service deadline was less than a year away, and a utility executive stated it would not provide an extension. As the months passed, members of the Canadian wind energy industry began to surmise Kenetech had been too hungry for the Gaspé mega-project, and that plans for the wind farm were suffering under tight profit margins.

Then, through a frantic process of corporate matchmaking, a new believer stepped forward in the fall of 1996. Axor partnered with California-based M&N Windpower, which is an association between Danish wind turbine manufacturer NEG Micon and the Japanese investment firm Nichimen Corporation. The promise of a Gaspé wind farm was reborn as Le Nordais.

"It's not only with profitability in mind that we are doing this project," says Gagnon. "We have to make a profit because we are in business, but this is more than just a project. It is the only project in Canada. We had to make it viable, so we did."

The viability of wind energy in Canada is, of course, dependent on environmental and economic variables. There are two other wind farms in the country, and one of them is a 2.25 MW plant owned by Hydro-Québec, demonstrating technology used by Le Nordais. The other is in western Canada, where, in a fashion similar to the Government of Quebec, the Province of Alberta deemed it necessary to develop alternative energy resources. As a result, in 1994, a 20 MW wind farm was commissioned on a windswept ridge near the western prairie hamlet of Cowley. Power from the 52 turbines on Cowley Ridge costs more, so the government instructed an Alberta utility company to sign a long-term contract with the independent wind-power producer, guaranteeing a premium price.

In 1997, after a number of small renewable energy projects were completed, a spokesperson for the province's energy minister stated, "decisions to buy or build new generation would be guided by market forces." Since that time, Alberta-based Vision Quest Windelectric has, in fact, prospered in the market, individually siting four 600 kW wind turbines. From these machines, consumers can now choose to buy wind power in 125 kWh blocks at a premium green rate. The additional 3.5 cents/kWh is 27% higher than the standard residential rate, but, by the end of 1998, 1,000 consumers were taking some portion of their power from the wind.

As the only other model for utility-scale wind power development in Canada, the Alberta green rate sounds promising. The province, however, is more than 90% dependent on coal for electricity. Roughly 2,000 MW of new generation is planned for completion before 2005, and if all these plans succeed, 97% of the new power will come from another low-cost, non-renewable resource, natural gas. Less than 0.2% of the plans announced are wind.

Canadian Wind Energy Association president Jim Salmon believes market forces will not develop meaningful quantities of sustainable energy, or create the diverse energy resource base that will be needed to achieve national objectives to curb greenhouse gas emissions, established under the Kyoto Protocol on global climate change. Salmon hopes the completion of Le Nordais will "have some political persuasion," and lead the way towards real change.

"I think the positive thing about Le Nordais is it's a demonstration that the

technology works," says Salmon, "and that it can produce utility-grade and utility-size electricity. Europeans have known this for years and years, but I don't think Canadians believe. I don't think they get it."

After the creation of the Le Nordais consortium, the Gaspé wind farm traveled a steady incline to success. Hydro-Québec decided to relax its deadlines, and the project was divided between two sites and two phases of development. The first phase was at Cap-Chat and the second 70 km away near the town of Matane.

Gagnon says there are two vital forces that insured the completion of the Gaspé wind farm. The first is the effort invested in the project by Axor and its partners, and while development has been slow and arduous, she says politics is the second factor driving the completion of Le Nordais. The people of the Gaspé region suffer the highest unemployment in Quebec, and Le Nordais has been designed to spur regional economic development.

"I think that's one of the reasons why there is considerable political interest in wind power," says Hydro-Québec spokesperson Renée Arsenault, speculating on the "considerable headway" Quebec's Minister of Natural Resources, Guy Chevrette, has made with wind development.

"The minister wants to favour local industry," says Arsenault. "He wants the people of Quebec to make this equipment. He doesn't just want the wind power, but all the infrastructure that goes behind it."

In time to supply 77 of Le Nordais' 133 turbines, NEG Micon established a Quebec assembly plant near Montreal in Boucherville. The president of NEG Micon operations in North America, Andris Cukurs, says the factory is an important step in the building of a Canadian wind energy industry, and that turbine foundations, cabling, transformers and towers have all been sourced in Quebec.

"Our future profitability is heavily driven towards finding local supplies," says Cukurs. "We are really interested in that. The problem we have is we need to have more contracts."

After the construction of Le Nordais is complete, Quebec is the only place in Canada with plans for the installation of more wind farms. A government-appointed energy board has recommended 50 MW of wind capacity be installed per year for nine years, beginning in 2002, and legislation is expected to enact the proposal.

Again, it sounds promising. It is, however, still a drop in Quebec's hydro bucket when compared to the province's plans to add new generating capacity, which will be almost entirely derived from hydraulic resources. The reason for this, says Arsenault, is hydro power can be produced for $0.03/kWh, almost half the price of power from Le Nordais. Undaunted, Gagnon has a job in the Canadian wind energy industry, and the success of Le Nordais, she believes, is evidence that the winds of change are, indeed, blowing.

"You know," she says, "wind energy made environmental sense. It now makes economic sense. In Canada, we have to go a little bit further. It has to make political sense, and I am sure it is going to in the near future."

CASE STUDY:

City PV High-Value Generation for the Grid
Edmonton, Alberta

EPCOR Utilities, Inc., Edmonton's electricity, water, and electrical services utility, is evaluating the technical, economic, and regulatory aspects of electricity generated from solar energy as one alternative for managing electrical demand. The immediate potential is to offset the very high-cost peak electrical demands caused by summer air conditioning.

In January 1995, EPCOR contracted Howell-Mayhew Engineering, Inc., (HME), to design, install, and operate a grid-connected photovoltaic (PV) system on HME's Edmonton research home. HME is monitoring the system's performance to evaluate its potential use by EPCOR in other high-value applications.

The PV system, which produces two kilowatts of alternating current (AC) electricity in bright sunshine, consists of 36 PV modules located on the roof at an angle of 27°. They power an inverter that converts the electricity to the standard 240-volt AC electricity which is synchronised with EPCOR's system and fully compatible with all household equipment.

On bright sunny days throughout the year, the system produces too much electricity for the house and so turns the meter backwards and feeds into the province's electricity system. At night this system supplies any additional electricity needed by the house. During an average year, 70% of the house's energy is solar generated. There are no batteries, and for safety reasons, the system automatically shuts down if EPCOR's system is not available.

Other Applications

There are over 20 such grid-connected PV systems in Canada and thousands in Japan, Holland, Germany, Italy, and America. The biggest market for such systems is as a component integrated into society's infrastructure: building facades, windows, roofs, awnings, car ports, and sound barriers.

The Big Picture

This project was so successful in reaching its goals that EPCOR installed another system on the roof of its downtown office building.

PV is in commonly used to generate power for communication systems, monitoring equipment and warning devices, and is increasingly being used for water pumping, vaccine refrigeration, and remote lighting. Small systems are found all over our cities on highway emergency telephones, traffic counters, remote lights, and ventilation fans for "rustic restroom facilities" in parks.

System Specifications

Number of modules: 12 single-crystal Siemens 75 W and 24 poly-crystalline Solarex 60 W.
Rated PV Power Output: 2.3 kW DC
PV Operating Voltage: 205 V DC
Inverter: StatPower ProSine 5000 GT
Inverter Output Rating: 3 kW AC
Inverter Output Voltage: 240 V AC
Inverter Characteristics: Automatic silent operation
Maximum power point tracking
Automatic synchronising with the utility grid
Shuts down when the utility grid is not available
Total Cost of Equipment: $26,000 installed

Utility Contact

Joel R. Nodelman, P.Eng.
EPCOR Utilities, Inc.
Environment Affairs and Sustainable Development
16th Floor, Capitol Square
10065 Jasper Avenue
Edmonton, Alberta T5J 3B1
Ph: (403) 412 3487
Fax: (403) 412 3346
E-mail: jnodelma@epcor-group.com
Web: www.epcor-group.com

Engineering Contact

D. Gordon Howell, P.Eng.
Howell-Mayhew Engineering, Inc.
15006-103 Avenue
Edmonton, Alberta T5P 0N8
Ph: (403) 484 0476
Fax: (403) 484 3956
E-mail: ghowell@compusmart.ab.ca

PV Pioneer I Program – SMUD customers volunteer to host utility-owned solar arrays on their roofs. The power generated goes directly to the SMUD electrical grid. Three Sacramento homes located at 57th Street and Broadway.

Sacramento Municipal Utility District

THE SACRAMENTO Municipal Utility District (SMUD), located in Sacramento, California, is a world leader in providing energy alternatives to its utility customers. SMUD's power supply includes energy from wind, hydroelectric, cogeneration, photovoltaic, geothermal, and other advanced and renewable resources. SMUD is committed to developing clean energy sources, and is a charter member of the United States Department of Energy's nationwide effort to reduce industrial carbon dioxide emissions by up to 30%. This is the equivalent of taking 200,000 cars off the road by the end of the decade. SMUD provides services to 1,123,400 customers in a 900 square mile service area.

In the late 1960s, SMUD built a 900-MW nuclear reactor, called Rancho Seco. For over 20 years, Rancho Seco operated inefficiently due to frequent shutdowns for repairs. Costs to the utility customer increased to the point where SMUD faced insolvency. For many years, the managers at SMUD and its customers argued over the economics and safety of the nuclear plant. Finally, in 1989, the customers voted to shut down Rancho Seco. That year, SMUD began to search for alternative sources of energy and encouraged customers to become more energy efficient. Eight per cent of their gross revenues were spent on improving the efficiency of customers lights, windows and other energy efficient equipment and planting 500,000 shade trees.

In 1993, SMUD began the PV Pioneer Program, developing partnerships with customers who were willing to adopt photovoltaics into their lifestyles. That first year, SMUD purchased, installed and operated 100, three-to four-kilowatt, utility-owned, residential rooftop systems, with plans to install an additional 100 systems per year. The customers who volunteered to participate in the PV Pioneers Program pay an additional $4 per month on their utility bills and provide the utility with a minimum 400 square feet of south, southwest or west facing unshaded roof area for the PV modules. SMUD anticipates the $4 premium will decrease as non-PV rates increase, due to the increasing costs associated with traditional energy fuels, such as oil, gas, natural gas and nuclear fuels. The PV Pioneer Program also allows SMUD to gain experience in the installation, operation, maintenance, pricing strategies and other aspects of residential PV systems.

Since 1993, SMUD has sited utility-owned solar units on the rooftops of over 450 homes and 35 businesses in the Sacramento area as part of the PV Pioneer I Program.

SMUD has calculated that over the 30-year life of the PV system, each 4 kW PV Pioneer system will produce about 263,000 kWh of electricity, help defer the need for new power plants and power sites, provide dramatic reductions to air pollutants and contribute significantly to the conservation of natural resources.

Based on the United States' national power generation resource mix, each PV Pioneer system provides the following savings:

Atmospheric Pollutants

Global Warming (CO_2)	433,000 lbs
Acid Rain (SO x)	2,990 lbs
Smog (NO x)	1,660 lbs

Reduced Natural Resource Consumption

Coal, or	253,000 lbs
Oil, or	21,000 Gal
Natural Gas	27,970,000 cu. ft.

Since 1993, SMUD has initiated many commercial PV projects, including the Sacramento International Airport. Completed in May 1997, the utility installed a 130 kW system in the unused air space above the airport's parking lots. Locating the PV panels above the cars also provides shade for cars in the open space of the large parking lots. The panels are located west of the terminals and keep approximately 70 parking spaces in the short-term parking lot shaded. SMUD estimates there is enough space on rooftops and in parking lots in the county to provide room for PV systems capable of producing 400 MW. SMUD has also installed 140 kW onto the roof tops of churches and other SMUD buildings. Seventeen 6.6 kW systems have been installed at nine churches, and four 6.6 kW systems are scheduled for installation at SMUD's Freshpond maintenance building. In addition to the obvious clean-energy advantages of PVs installed on rooftops and over parking lots, there are other advantages, including the provision of shade in previously exposed areas and the use of "free" space with a minimal environmental and community impact.

SMUD PV Projects through 1996

Project	Project Name Installation Year(s)	Size	System Description	Year(s)
1980s	Rancho Seco PV1	1000 kW (1230 kW, EPF)	Arco System	1984
1980s	Rancho Seco PV2	1000 kW (1230 kW, EPF)	Arco/Solarex/Mobil	1986
1993	PV Pioneers 93	400 kW	108 3-4 kW Siemens Systems	1993-1994
1993	Hedge PV1	210 kW (258 kW, EPF)	UPG/Siemens Systems	1993-1994
1993	SMUD Warehouse PV	30 kW (37 kW, EPF)	SEA Systems	1993-1996
1993	PVEV Charging Station	11 kW	Arco System	1992
1993	PV Res Demo Systems	10 kW	Various	1992-1993
1994	PV Pioneers 94	400 kW	119 3-4 kW Solec Systems	1994-1995
1994	PV Pioneers Commercial 94	144 kW	8 10-30 kW Solec Systems	1994-1995
1994	Hedge PV2	108 kW	APS System	1994-1995
1994	Hedge PV3	102 kW	RMI/Solarex System	1994-1995
1994	Hedge PV4	107 kW (132 kW, EPF)	UPG/Siemens System	1994-1995
1994	WAPA BI-Roof Demo	3 kW		1994
1995	PV Pioneers 94/95	87 kW	25 3-4 kW RMI/ Solarex Systems	1994-1995
1995	PV Pioneers 95	329 kW	80 3-4 kW Placer/ Solarex Systems	1995-1996
1995	Rancho Seco PV3	214 kW (263 kW, EPF)	UPG/Siemens	1995-1996
1995	PVEV Airport	8 kW	Arco System	1995-1996
1995	Solar Carport	128 kW (158 kW, EPF)	UPG/Siemens	1995-1996
1996	PV Pioneers 1996	300 kW 80 3-4 kW	Solarex Systems	1996-1997
1996	PV Pioneers Commercial 96	140 kW 7 20 kW	Solec Systems	1996-1997
1996	PV Substation 96	129 kW (160 kW, EPF)	UPG/Siemens	1996-1996
1996	WAPA BIPV 96	40 kW roof integrated,	Powerlight	1996
1996	BIPV 96	10 kW	Demonstration Projects	1996-1997
Total PV kW Installed to 1996			**5120 kW**	

In June 1997, SMUD introduced the Greenergy program, a clean-air program that allows its customers, both residential and commercial, to contribute directly to green energy. Nearly half of the electricity supplied by SMUD to its customers is generated from one of SMUD's own green resources— water, wind, natural steam from the earth and solar. The Greenergy program offers customers two options. First, the Community Solar Program, which allows customers to contribute directly to the replacement costs of fossil fuel driven power plants, with clean, renewable electricity, generated by photovoltaic panels located on community rooftops. For an extra one cent per kWh, the SMUD customer helps purchase large PV systems for the roofs of non-profit community agencies, including schools, churches and other community buildings. The first site chosen for this program was the Sacramento Zoo. Second, the Renewable Energy Option lets commercial and residential customers buy electricity from clean, renewable resources, such as wind, solar, biomass and photovoltaics. There are two different levels of participation in the Renewable Energy Option. The All Renewable Option guarantees that 100% of the customers electricity comes from a non-polluting renewable resource. This option costs the customer an additional one cent per kilowatt hour on its electricity bill. The Renewable Energy Advocate ensures that the customer receives more energy from clean and green energy sources and costs the customer an additional half cent per kilowatt hour.

SMUD's PV 1, 2, 3 and 4 – Located adjacent to the now closed Rancho Seco nuclear plant. PV 1 was in operation five years prior to Rancho Seco closing, PV 2 was completed three years prior to plant closure.

In October 1998, SMUD introduced the PV Pioneer II Program which, like the original program, offers homeowners the opportunity to purchase rooftop solar units. But this PV Pioneer II Program is net metered, meaning the power generated by the system goes directly to serve the customer's needs with any excess power produced by the panels being sold back by the customer to the utility. On days the unit cannot meet the household electrical needs, the customer buys the additional energy from SMUD. With the new program, SMUD will "buy down" over half the cost of the solar system for participating SMUD customers. A typical two-kilowatt rooftop system will generate enough energy to offset about half of the average yearly energy needs of a SMUD customer and cost under $4,500. The full cost of the system is over $10,000 and SMUD has arranged a financing program to provide a loan for the customer's portion of the solar unit.

SMUD's PV/EV – SMUD charging station harnesses solar energy that is converted and used to provide electricity for electric vehicle charging.

PV inverter – converts solar energy into usable electrical energy.

Under the new PV Pioneer II Program, two types of PV systems are available for purchase through SMUD. Solar shingles, best suited for a new construction or a home in need of a new roof, covers the whole roof in PV panels instead of the traditional asphalt or terra-cotta tiles used on roofs. For existing roofs in good condition, a modular unit is available. Both types of systems have a 30-year operating life and SMUD will provide or arrange for any needed service for both warranted and out-of-warranty items.

SMUD, its Greenergy subscribers and PV Pioneers are sending a serious message to the energy industry, policy makers and other utilities. They want and will pay slightly more for technologies and resources that are not as harmful to the environment as traditional fossil fuels and nuclear energy. They are promoting a cleaner environment because the energy comes from projects using clean, renewable energy technologies and not from fossil fuels, which emit smog forming chemicals and contribute to global warming. By promoting sustainable energy resources through resource diversity, SMUD is also reducing its compliance on finite fuels.

SMUD obtains its Greenergy renewable energy for its customers from four sources. A solar thermal electricity plant, the Solar Tower project, in conjunction with California Edison, generates up to 10MW at its plant in Barstow, Southern California. During daylight hours, 2,000 mirrors at Solar Two track the sun and store its energy as heat in molten salt. This energy can then be used to generate electricity when needed, such as during periods of peak demand for power. The sun heats salt in the receivers and the salt is stored in hot storage tanks. Hot salt pumped through a steam generator drives a turbine/generator to produce electricity. Through photovoltaics, the combination of commercial systems and individual residential systems, SMUD generates over 3.5MW of PV-produced solar power. Phosphoric acid fuel cells, which convert chemical energy directly into electricity, are used in two locations in SMUD, including a hospital and the SMUD head office. The only emissions are pure water and a minimal amount of heat. Also, in the Spring of 1994, SMUD installed a commercial size wind farm of 17 wind turbines in the Montezuma Hills of Solano County. The SMUD Wind Farm runs at its peak during the hot summer months, when the cooling sea breezes run through the Sacramento River Delta into the hot Sacramento Valley.

Since 1993, SMUD has installed a generous amount of PV in its district and will continue through its partnerships with commercial and residential customers. It has installed 1.5 MW through the PV Pioneer Program and another 4.2 MW from the rest of the ground-mounted substations and parking lots. SMUD has continued to commercialise PV for widespread use and developed the experience needed to successfully integrate PV as distributed generation in the utility system. SMUD is part of a nationwide program intended to accelerate the cost-reductions necessary for PV to be successful in the competitive utility market. SMUD plans to install another 10 MW of PV from 1998 to 2002 and bring PV manufacturing to Sacramento. Through the efforts of SMUD and other like-minded utilities, it is expected that PV systems will drop in price to the point PV will be cost competitive in the retail, grid-connected market as well as continue strong growth in off-grid markets and overseas.

For further information on the Sacramento Municipal Utility District, visit their World Wide Web site at www.smud.org

Chapter 6

International Applications of Renewable Energy

Renewable Energy in the Developing World

ENERGY IS A KEY COMPONENT to fostering economic growth in the developing world, where more than two billion people still rely upon kerosene, wood and batteries to provide them with light and power. It is expected that fossil fuels will continue to play a significant role in the rapidly growing economies of the developing world. However, in the rural economies, renewable sources of energy, including wind, photovoltaics and microhydro, are slowly replacing the traditional forms of energy and fuel, such as wood products, animal dung, kerosene and batteries, as these items become more costly and scarce. In the rural environment, renewables are increasingly attractive as they are environmentally friendly, quiet, stand-alone, modular systems and, in many instances, cost-competitive with fossil fuels.

The growing energy demands of the developing world have put a great strain on the limited resources of the centrally or state-owned energy companies. Many of these companies are attempting to plan and, in some cases expand, their energy supplies but they are restricted by ongoing financing difficulties, government bureaucracies and corruption. The United Nations is interested in promoting "a reliable, efficient and environmentally sound supply of energy services"[1] with the full support and co-operation of the private sector and non-government organisations.

There has been a resurgence of interest in renewables, due to the environmental and sustainability concerns associated with fossil fuels. This interest is illustrated by the increased use of photovoltaics and small wind systems for communications and home systems in regions off the grid, solar heating and biogas for cooking, and microhydro systems in remote areas. The United Nations has identified that in the current energy environment, renewable energy is placed at a disadvantage. While traditional forms of energy, including fossil fuels and large hydro projects, continue to receive large subsidies and tax considerations, many of these same benefits do not exist for renewables.[2] According to the UN, it is imperative that research and development in renewable energy again be fully funded; a reversal from the last half decade when R&D funding was dramatically reduced.

Renewable Energy Policies in the Rural Setting

In the rural areas, the provision of energy services is spartan owing to the dispersed nature of the population and the traditionally low income levels of the customers. There are an estimated 3.1 billion people in rural areas, with approximately 2 billion who do not have access to electricity and who must rely on wood, charcoal, animal and plant wastes to support their energy demands.[3]

In developing countries, as in the developed world, electrical generation and distribution are being privatised, with subsidies for rural power being phased out. This has the potential to drive up the costs of commercially traded energy, opening

[1]United Nations Committee on New and Renewable Sources of Energy for Development, Report on the Second Session, 12-23 February, 1996. Chapter III, Section A2, No. 22.
[2]Ibid. Chapter III, Section B5, No. 61.
[3]United Nations Department of Economic and Social Affairs, Energy and Sustainable Development: Development and Implementation of Rural Energy Policies, Report of the Secretary General, Summary.

the door for renewable energy to become cost-competitive. The United Nations is concerned with relying on the private sector to supply rural economies with energy. They feel that as the potential for large returns on investment is lower or possibly nonexistent, the private sector may not develop the rural sector of the national grid without stringent contracts between the state and private sector energy companies.[4]

The UN has examined the potential of setting up local co-operatives to increase the supply of energy in rural areas. In Kenya, renewable energy technologies can be cost effective in the rural areas, as transportation and transmission costs are not a significant portion of total costs.[5] The adoption of renewable energy technologies in rural areas will provide modern energy services to rural customers with none of the environmental problems associated with conventional fuels.

In the rural areas of developing countries, energy is used for cooking and for water pumping, irrigation, operating machinery, and processing and preserving food in the agricultural sector. Many households use non commercial sources of energy, such as firewood, crop residues and animal dung, to cook the food, with women spending up to 24 hours each week collecting the fuel.[6] But the burning of those materials presents health risks as particulates and carbon monoxide, are released, causing the women, children and elderly to be stricken with chronic respiratory diseases, including bronchitis and emphysema. The movement away from traditional fuels for cooking would significantly improve the quality of life in the rural setting, improve women's productivity in the household, and dramatically improve their health.

Renewable Energy Technologies in the Developing World

Biomass, such as wood, crops and animal wastes, will continue to be an important renewable energy in the developing world because of widespread availability, ease of use for differing applications and improving efficiencies. If biomass is used in an efficient and sustainable manner, it has many environmental and societal benefits, including job creation, the provision of modern energy carriers to rural communities, the implementation of better land management practices, and the reduction of carbon dioxide and sulfur emissions into the atmosphere.[7]

In many developing countries, biomass is used mostly for cooking in open hearths, an inefficient use of energy. Many developing countries are encouraging the use of more efficient stoves or solar cookers.

Research is being done into the use of biogas, which is gas generated from the anaerobic digestion of animal dung, as a convenient and sustainable fuel for cooking in rural areas, instead of stripping the local forested areas. In many developing and developed countries, biomass has been turned into

---

[3]United Nations Department of Economic and Social Affairs, Energy and Sustainable Development: Development and Implementation of Rural Energy Policies, Report of the Secretary General, Summary.

[4] Ibid. Chapter II, Section B3, No. 25.

[5]Ibid. Summary.

6Ibid. Chapter II, Section A, No. 9.

[7]United Nations Committee on New and Renewable Sources of Energy for Development, Report on the Second Session, 12-23 February, 1996. Chapter III, Section B1, No.44.

liquid or gaseous fuels, such as ethanol, for uses in the agricultural, transportation and other commercial sectors.

The use of solar energy in developing countries is growing, as the applications of active solar and photovoltaics continue to become more commonplace and cost-effective. As the tourism industry expands, more hotels and resorts are utilising active solar systems for hot water heating. Hospitals in rural areas are using active solar systems to heat and distill water. Active solar energy is being used to dry crops, such as tea and grain. Solar cooking is another application that is gaining favour among the United Nations and private aid agencies. The use of photovoltaics in developing countries is growing rapidly with the expectation there will be over 900, 000 MW in new energy needs over the next 15-20 years.[8] Photovoltaics will be able to provide modular, affordable electricity for lighting and electricity in homes and institutions, communications, water pumping, battery charging and railway signaling. In regions with small health clinics, photovoltaics provide electricity to the refrigerators that store life-saving vaccines.

In developing countries, the costs of similar-sized wind energy projects tend to be higher than in the developed world, mainly because they are more expensive to install and after-sales services cost more. If the wind conditions are favourable, the investment will be paid back faster than investments in traditional energy sources. Wind turbines are more cost-efficient to operate and once the wind systems are installed, there can be substantial savings in energy imports, such as coal, gas and oil. The diversification of the country's energy sector helps ensure a consistent supply of energy. Wind components can be manufactured locally, stimulating the local economy by providing jobs in the region. Wind energy in the developing world is providing energy for water pumping, individual residences or institutions. By offering to install wind energy, local and national governments are able to provide electricity in rural areas quickly and affordably. Wind systems can also be used as back up systems for stand-alone operations.

Microhydro has great potential in the developing world but needs to be encouraged and developed further. Currently, there are many small streams and rivers around the world that have not been utilised for local hydro generation due to the lack of head or drop. But many of these streams and rivers can be utilised through the use of "run-of-the-river" turbines or "water current turbine technology." The turbines sit submerged in a river or stream and turn, with sufficient current, extracting the energy from the flowing water, providing a stand-alone power source of 240V. In the developing world, water current technology can replace diesel water pumps for irrigation, pump

[8]"Solar Facts: Solar Going International" Factsheet taken from Solar Energy Industries Association World Wide Website, www.seia.org, P.1.

water for villages, generate electricity for villages, schools and hospitals, and charge storage batteries.

When sizing a water current system, the installer will take into account seasonal variations of water flow and water levels. It is possible to manufacture water current turbine technology locally, providing jobs in the local economy and ensuring replacement parts can be obtained quickly. Unlike other microhydro systems, water current turbines do not require dams, thereby, minimising costs and environmental impacts. But water current systems are vulnerable to floating or submerged debris, which tend to collect at the part of the machine which breaks the surface, not damaging the actual turbine.

International Applications for Solar Cooking

Solar cooking has applications worldwide. There are many solar ovens in use around the world that have been distributed through non-governmental organisations and people working at the grassroots level. Factors related to the continued development of solar cooking include access to simple materials, such as cardboard, aluminum foil and glass. Other factors influencing the use of solar cooking include the availability of traditional cooking fuels, climate, food preferences, cultural factors and technical capabilities.

Applications of solar cookers include sterilizing of medical equipment, pasteurizing water, desalinating sea water, distilling or boiling water for drinking. They can also be used to dye wool, melt candle wax, separate honey from comb wax, make soap and dry foods.

The benefits of using solar cookers in the developing world are well established. They can eliminate the health problems caused by cooking with dung or wood stoves, including unsanitary conditions and the inhalation of burned particulates. In a related benefit, they help reduce the deforestation in the developing world. They allow dung to be reallocated as a fertiliser, which contributes to better crops and more food.

CASE STUDY:

Portable Solar Cookers for Families, Worldwide

Introduction

IN MOST DEVELOPING AREAS of the world, cooking is the most significant use of energy. With the increase in the population of the world and the increasing use of local biomass for commercial and industrial purposes, deforestation is taking place at an alarming rate. The development of programs, encouraging the use of solar cookers, would help reduce the populations' dependency on local biomass fuels. Not only would local air quality would be improved, but it also would free women and households from the arduous task of walking many kilometres to collect scarce firewood. Solar cookers should be utilized in locations where the instantaneous solar radiation intensity is often in excess of 700 watts per square metre and where the conventional cooking fuels are either costly, unavailable or the direct cause of pollution to the local environment.

System Specifications

Solar cookers were originally designed for semi-arid developing areas of the world, under a Quebec government grant. They are made entirely of plastic, insulation, aluminum, reflective film, sealants and accessories. A solar cooker weighs less than six kilograms, is easily transportable and is made of a minimum number of components. Food is cooked by boiling, steaming or baking. Cooking is done in two flat shallow cooking pans, covered with a specially adapted thin, rigid, transparent polycarbonate cover. These pans cover the entire exposed area of the cooking chamber. The incident solar radiation is directly absorbed by the ingredients in the cooking pans. A rear adjustable reflector increases the radiation available into the pans.

Solar cookers have been developed and tested for many years by the Brace Research Institute, located in Montreal, Quebec and associated organizations in different parts of the world. The cookers have undergone extensive testing in Canada, Argentina, Senegal, India and Haiti, to name a few. Questionnaires are available for potential users to determine local traditional cooking practices as well as the availability of fuels, artisans, materials and financial resources. A second questionnaire specifically permits the user to determine whether solar cookers are appropriate in any given location. The cooker is provided with operating instructions in English, French or Spanish, depending on its intended area of use.

The cooker is exposed to the sun on a flat surface and is oriented several times a day in order to benefit from optimum use of the solar reflector. Generally, a wide variety of different dishes can be cooked within a period of one to two hours under favourable solar conditions.

Other Applications

When not used for cooking, the solar cooker can be utilized for heating water, disinfecting bacterially contaminated water, and some drying operations. Solar cookers have a major advantage over other systems: it is relatively impossible to burn or destroy food in the cooker in the first few hours of use.

The cooker has been specifically designed as a valise to meet the needs of the increasing number of Canadians with family connections in developing areas of the world. It is particularly useful for people visiting development agencies overseas who utilize the cooker as a model for local craftsmen to build and use in their areas.

The Big Picture

The cooker is easy to operate and maintain. The instructions must be followed carefully in order to achieve success in the cooking process. There has been very little problem introducing solar cookers to those areas where the cost of available fuels is high and delivery unreliable. The cost savings are very much a local function. In high fuel cost areas blessed with a good solar regime, savings can be substantial. The potential client base is of the order of one billion people in developing areas at least and this number is much more significant if one considers use by campers, in backyards or in suburbia.

Client: To date, most clients have been people connected with religious or development agencies working in developing areas of the world as well as for overseas families of Canadian residents.

Contact Information:

T.A. Lawand
Brace Research Institute
Faculty of Engineering
21111 Lakeshore Road
3 Stewart Park, P.O. Box 900
Macdonald Campus of McGill University
Ste. Anne de Bellevue, Quebec H9X 3V9
Tel: (514) 3898-7833
Fax: (514) 398-7767
E-mail: ae12000@musica.mcgill.ca

CASE STUDY:

Solar Drying for Food Colouring
Coimbatore, India

ONE OF THE LARGEST solar dryers has recently been commissioned at the Synthite Industrial Chemicals Ltd. factory near Coimbatore, India. The 1300 m^2 installation is actually a two-stage solar dryer combining SOLARWALL® transpired solar collectors as the first stage with a smaller second stage glazed solar booster heater.

The wall and roof mounted solar heating system dries marigold flowers, which are used for animal feed stock. With increasing awareness of carcinogenic properties of artificial pigments, those derived from natural sources are gaining importance. Synthite's solar dried marigold flowers are all natural and are used by farmers as a food supplement.

System Specification

1300 m^2 black galvanized steel, mounted on existing roof with an air gap to allow for heated air to be drawn in the back of the panel. Air flow of 40,000 cfm into the drying chamber; savings of US$70,00 per year in avoided fuel cost; payback of two years.

The Big Picture

India, like many tropical countries, is ideal for SOLARWALL® due to its more than 250 sunny days each year. Crops are grown continually or have a long period over which they are harvested and dried, making SOLARWALL® drying economic in most cases. Other drying methods use oil, which is imported and expensive, and wood, which is also expensive and becoming more scarce.

Client: **Synthite Industries Inc.,** manufacturer of agricultural extracts

Supplier: **Conserval Engineering**
200 Wildcat Rd
Toronto, Ontario M3J 2N5
Tel: (416) 661-7057
Fax: (416) 661-7146
E-mail: Conserval@globalserve.net
Web: www.solarwall.com

CASE STUDY:

Solar Distillation in the Southern Chaco Region of Argentina

THE WATER in the area is salty and contaminated with arsenic, therefore, it is not suitable for drinking. In addition, there is no electricity or fuel to operate small-scale water purifiers. Currently, the local expertise does not exist to adapt the technologies necessary to resolve either the water quality issue or the energy issue. Brace Research Institute collaborated with local institutions to develop capabilities in water distillation using renewable energy sources.

System Specifications

4-2 square metre solar stills
Insulated fibreglass boxes, glass covers and plastic piping
Cost: about $1000
Capacity: 24 to 40 litres of water per day

Contact:
T.A. Lawand
Brace Research Institute
Faculty of Engineering
21111 Lakeshore Road
3 Stewart Park, P.O. Box 900
Macdonald Campus of McGill University
Ste. Anne de Bellevue,
Quebec H9X 3V9
Tel: (514) 389-7833
Fax: (514) 398-7767
E-mail: ae12000@musica.mcgill.ca

CASE STUDY:

Solar Dryer for Tea
Malabar, Indonesia

CONSERVAL DEVELOPED the SOLARWALL® for solar air heating in northern climates; however, this technology is also being used in tropical climates for solar crop drying. One of its most interesting applications is drying tea leaves in Indonesia. The leaves are dried at the wilting state.

System Specifications

SOLARWALL® area: 600 square metres
Ventilation rate: 75,000 cubic metres/hour
Hours of operation: 70 hours/week
Capacity: 4500 kilograms/day

The Big Picture

The SOLARWALL® can be installed in most applications that use conventional fuels and open air drying.

Future Plans

The success of the SOLARWALL® has led to a number of other applications around the world. The newest design is a photovoltaic SOLARWALL® system to provide both heat and electricity.

Other Projects

Installations throughout North America include the Ford plants in Oakville and Talbotville, Ontario; Toledo, Ohio; Chicago, Illinois, and Buffalo, New York.

Contact: **Conserval Engineering Inc.**
200 Wildcat Road
Downsview, ON M3J 2N5
Tel: (416) 661-7057
Fax: (416) 661-7146

CHAPTER 7
Renewable Energy and Climate Change

Introduction

ANY MEANINGFUL DISCUSSION about renewable energy and energy efficiency in the 1990s cannot be undertaken without emphasizing their significance in the context of climate change. Renewables and energy efficiency are two of the most important mitigation tools available, and the climate change issue will provide significant opportunities for increasing their market penetration.

With few exceptions, climate change is about energy consumption. The fuels we currently consume are among the greatest emitters of the greenhouse gases that promote climate change. Indications that we have been paying a high environmental and socio-economic cost for these fuels have been around for many years. In the 1970s, the energy crisis raised fears over resource scarcity. In the 1980s and early 1990s, environmental concerns were raised over the impact associated with the production and consumption of energy. Each time there have been calls for increased energy efficiency and renewables as a means of mitigating the impacts associated with traditional sources of energy.

Climate change is a harbinger that our current energy consumption patterns are not sustainable. This chapter offers a brief introduction to the climate change issue, alternatives for its management, and the mitigative role of renewable energy and energy efficiency.

For further information on climate change, the role of renewables in mitigating climate change and Canada's position, please see the references in Chapter 9, Books section.

The Enhanced Greenhouse Effect and Climate Change

The greenhouse effect is a natural and necessary process for life on earth. The earth's climate is maintained by a continuous flow of energy from the sun, some of which is reflected back into space while some is locked in by the earth's atmosphere (Figure 1). Our environment is a result of an equilibrium between the energy radiated back into space and that which is retained. The characteristics of different gases within the atmosphere controls the amount of energy trapped. As such, increasing the concentrations of these greenhouse gases (GHG) can change this equilibrium. The enhanced greenhouse effect is the term given to increases in the intensity of the greenhouse effect as a result of human activities.

The primary gases contributing to the enhanced greenhouse effect are carbon dioxide (CO_2), water vapour, methane (CH_4), nitrous oxide and chloroflourocarbons (CFCs). Not all GHGs are

Figure 1: The Greenhouse Effect[1]

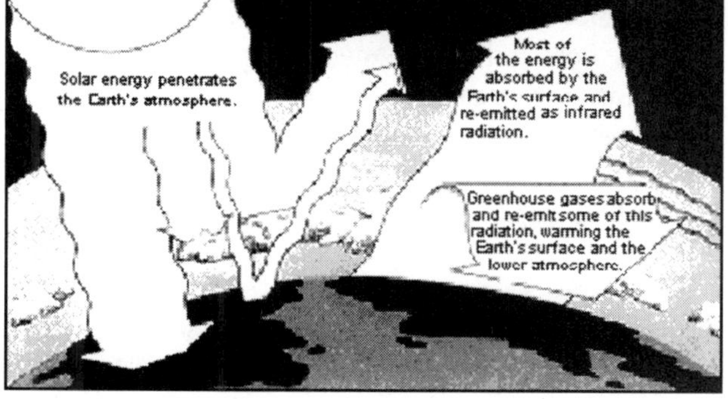

[1] Canada's National Action Program on Climate Change from website, http://www2.ec.gc.ca/climate/resource/cnapcc/c3part01.html

in terms of their ability to trap the sun's energy or in terms of the time the gases are retained in the atmosphere. For example, if over a 100 year span, carbon dioxide is considered to have a global warming potential of one, then methane is 21, nitrous oxide is 310 and some hydroflurocarbons are in the order of 7,000 to 9,000. Despite this wide range in warming potentials, globally, CO_2 is responsible for 60% of the enhanced greenhouse effect and, as a result, it receives much of the attention.

The Energy Link

The substantial contribution of CO_2 and other energy related emissions to the enhanced greenhouse effect makes the link between the energy and climate change undeniable. The consumption of fossil fuels, such as coal, oil and natural gas, generates CO_2 and evidence shows that since the Industrial Revolution, concentrations of CO_2 in the atmosphere have increased at unprecedented rates. This combustion of fossil fuels has essentially taken the earth's natural carbon cycle out of balance by removing carbon from the earth (stored as fossil fuels) and putting it into the atmosphere faster than it can be absorbed back into the earth's biomass.

Impacts of Climate Change

Changes to the earth's climate resulting from GHG have the potential to impose substantial impacts on society with significant environmental, economic and social dimensions. The International Panel on Climate Change (IPCC) predicts that, if strong policy actions are not taken, the mean global temperature will increase 1ºC to 3.5ºC between 1990 and 2100 with the best estimate being 2ºC. The magnitude and direction of these impacts will vary dramatically from region to region around the world. In general, however, climate change during the next century is projected to bring about warming trends, changes in precipitation, a greater frequency of storms and unusual weather patterns, and increased sea levels in the range of 50 cm by 2100.[2]

International Responses

Climate change has been recognized as a serious problem for some time within international scientific circles. It was not, however, until the Earth Summit, held in Rio de Janeiro in June 1992, that the issue started to penetrate into the mainstream consciousness. The Earth Summit saw the first signing of the "Framework Convention on Climate Change" (FCCC), an international mechanism to initiate the discussion of management options between countries. The FCCC seeks to stabilize atmospheric concentrations of GHG "at a level that would prevent dangerous anthropogenic (human-induced) interference with the climate system."[3] A preliminary target of stabilizing GHG emissions at 1990 levels by the year 2000 was agreed upon.

Since 1995, the signatories to the FCCC have held annual meetings (referred to as Conference of the Parties(COP)) to discuss the issues relating to its implementation. The most recent of these meetings were held in Buenos Aires, Argentina, in November, 1998. In Kyoto, Japan, at the December 1997 meeting Canada and 160 other countries agreed to a protocol that called for further reductions in greenhouse gas emissions over the next 15 years. Canada's target was set at 6 percent below 1990 levels for the period spanning 2008 to 2012.

Responding to Climate Change

Developing a sound response to the threat of climate change is a two-stage process. First, appropriate measures, which are technically capable of dealing with the problem, must be identified. Second, governments must develop policy mechanisms that will ensure that the measures will be adopted by society.

Measures for dealing with climate change generally fall into two categories—adaptation or mitigation. To adapt, society must adjust to the ensuing climatic changes, through initiatives such as moving populations from low lying coastal regions or adapting crops. On the mitigation side, attempts are made to stop or reduce the GHG emissions that are contributing to global warming. Although adaptive measures might ultimately be necessary to reduce the impact of climate change that are already underway, they are considered, by most, to be too expensive and risky to form the core of a response strategy.

The ability of energy efficiency and renewable energy to reduce energy intensity and dependency on fossil fuels makes them essential tools for the mitigation of climate change. As a result, this chapter will focus on the mitigative side and where renewable energy and energy efficiency fit into the scheme.

Mitigation

Mitigating climate change means decreasing the release of GHG into the atmosphere. Although there are many options available for reducing emission levels, managing directly the sources of

[2] Rolfe, Chris, *Turning Down the Heat.* West Coast Environmental Law Association: Vancouver, 1998.
[3] Canadian Global Change Program, *Cross Canada Briefings Before Kyoto*, Participants' Briefing Book, unpublished workshop briefing notes, 1997.

GHG emissions is one of the simplest to address. From a CO_2 perspective, this means bringing the earth's carbon cycle back in balance, by either reducing the amount of CO_2 released into the atmosphere or increasing the amount that is absorbed back into the earth's biomass. Measures available for reducing emissions of CO_2 can be grouped into four general categories:

- Reduction—energy conservation in the plants and facilities that generate emissions
- Avoidance—introducing energy efficiency, renewable energy or switching to less CO_2 intensive fuels
- Sequestration—absorbing CO_2 before it has a chance to enter the upper atmosphere, through such actions as planting trees
- Storage—mechanisms that will take CO_2 out of circulation, such as capturing it and storing it in the ocean or underground reservoirs

Other options for mitigating climate change include addressing the more fundamental questions about the impact of population growth, wealth and the economic structure of our economies on GHG emissions.

Evaluation of Measures

Decisions about what emission reduction measures are undertaken can be based on a number of factors, including the cost of putting the measures in place (dollars per tonne of CO_2 reduced) and additional non-GHG benefits. The evaluation of these costs is a complex process and the focus of contentious debates about the underlying economic and technical assumptions. These costs will also vary from region to region within a country and among countries, depending on factors such as energy prices and the existing stock of energy consuming technology.

Measures can be prioritized with those that have the lowest cost implemented first. There are, in fact, many measures that have negative costs (have an economic benefit regardless of whether CO_2 emissions are reduced) and these are referred to as "no-regrets." Many energy efficiency measures are no regrets since they are cost effective, based on their energy savings alone. Many other measures may also be considered no regrets depending on how additional non-GHG related benefits are valued.

Non-GHG related benefits can have many social, economic and environmental dimensions. They could, for example, include reductions in other energy related emissions and pollutants that contribute to acid rain, smog and a variety of health risks. Measures that reduce transportation related GHG emissions by reducing the demand for transportation could see benefits in reduced insurance premiums (resulting from fewer accidents), reduced traffic noise, and investments in infrastructure. Increased energy efficiency can also improve the competitiveness of Canadian industry, and reforestation initiatives can help to ensure the long-term stability in forest industries.

Another important non-GHG benefit of adopting energy efficiency and renewable energy is the potential for job creation. Not only are there more jobs associated with these types of initiatives compared to traditional approaches of meeting energy needs, but additional spin-offs may also be realized. Most significantly these include local economic development through indirect job creation and respending effects.

Policy Mechanisms

The second stage in developing a response to climate change, after establishing measures, is for governments to establish policy mechanisms to ensure that the measures are adopted. These policies may focus on overcoming the barriers (social, economic, informational) that would normally prevent the measures from being accepted. The basic policy tools available for governments to encourage emission reductions include:

- Economic instruments
- Education
- Regulation
- Voluntary programs

Extensive experience by governments and utilities has been gained on the use of education and regulation to promote energy efficiency and renewable energy. Because of the high correlation between energy consumption and CO_2 emissions, much of this experience can be directly applied to mitigation policies for climate change.

Voluntary programs, such as Canada's Voluntary Challenge and Registry (VCR), have been used by many governments as a policy tool to encourage emission reductions. These programs are attractive because they provide corporations with flexibility to set and obtain their own emission reduction objectives. They have, however, been met with considerable criticism because they provide little incentive to actually reduce emissions.

Economic instruments are considered by some to be the most efficient means of attaining emission reductions. Two widely considered economic instruments are a carbon tax and emission trading. In the case of a carbon tax, the government levies a tax on CO_2 emissions, thus creating a direct economic incentive to reduce emissions. Although a carbon tax is simple to put in place and administer, and could provide revenue to help promote economic expansion in desirable sectors, its net impact on the economy is often questioned. The primary reason cited against the use of a carbon tax is that it would be politically unpopular.

Emissions trading is becoming more widely accepted, by some, as an efficient mechanism to promote reductions in GHG emissions. Under a trading scenario, those individuals, corporations or governments who can reduce emissions easily, ie cost-effectively, below a level set by regulation, are permitted to sell emission credits or allowances to an emitter whose cost of reducing emissions is prohibitively high. A number of different trading programs have evolved; however, the basic theory remains the same. That is, setting appropriate regulations will create a demand for emission reductions, the price of which will be established by market forces and will be dependent on the supply of and demand for emission reductions.

The introduction of economic instruments as government policy creates significant opportunities for energy efficiency and renewable energy since it puts a monetary value on environmental benefits. As a result, cost effectiveness increases compared to technologies that have higher GHG emissions, thus helping to level the playing field.

Despite the fact that Canada has not yet adopted a carbon tax or emission trading, there have been some voluntary initiatives taken by some corporations to pursue emission offsets and trades. These initiatives are speculative in nature due to the uncertainty of upcoming regulations and, as a result, the market value of emissions reductions is low. They are particularly attractive for corporations who may be vulnerable to future restrictions on GHG emissions. One example is the Alberta Schools Program.

Alberta Schools Benefit from Energy Efficiency Emission Offsets

In October 1998, Suncor Energy announced that it would contribute more than $300,000 towards energy efficiency programs in Alberta schools. As part of the agreement, Suncor, which operates an oil sands plant in Ft McMurray, Alberta, will receive credit for the emission reductions resulting from the energy savings. These credits would be applied against the emissions generated at Suncor's facilities.[4]

To test the feasibility of emission trading for GHGs, an emissions reduction trading pilot, Greenhouse Gas Emission Reduction Trading, know as GERT, has been established in British Columbia. The project involves the federal government and representatives from industry and several environmental groups. The pilot provides participants with practical experience related to GHG emission reduction trading. It will help create the foundation for a possible full-scale emission trading system in the future.

Climate Change in Canada

Climate change is an important issue for Canada for two reasons. First, in terms of its GHG contributions and, second, Canada has a lot to lose from adverse climate change.

Canada contributes only about 2% of total global GHG emissions. However, on a per capita basis, it is the second highest emitter in the world. This position can be attributed to some of Canada's identifying characteristics including our northern climate, sparse population, high standard of living and resource based economy. Each one of these contributes to Canada's large appetite for energy which, under our current fuel mix, translates directly into GHG emissions, most particularly CO_2 and methane.

In Canada, energy production, transformation, and consumption are the major producers of carbon dioxide (CO_2) and methane (CH_4), which, in 1995, accounted for 93% of Canada's total greenhouse gas emissions. Fossil fuel combustion and production are the dominant sources, accounting for about 89% of greenhouse gas emissions in the same year. Figure 2 outlines Canada's total the GHG by emissions in 1995 by gas and by source.

Forecasts of GHG emissions indicate that there will be a substantial gap between Canada's stabilization commitment and actual GHG emissions. Figure 3 outlines the projections of GHG emissions.

The magnitude, timing, and regional impacts of climate change in Canada are highly uncertain. But current predictions suggest they will be far-reaching and damaging. Most significantly for Canada, climate change will hurt coastal communities, the vulnerable ecosystems of the far north, and the health of some individuals. Droughts, hot spells, and insect infestations may become more frequent on the Prairies. Water

[4] Suncor Inc. web site, http://www.suncor.com

Figure 2: GHG Emissions in Canada, 1995[5] (Total 618 Megatonnes)

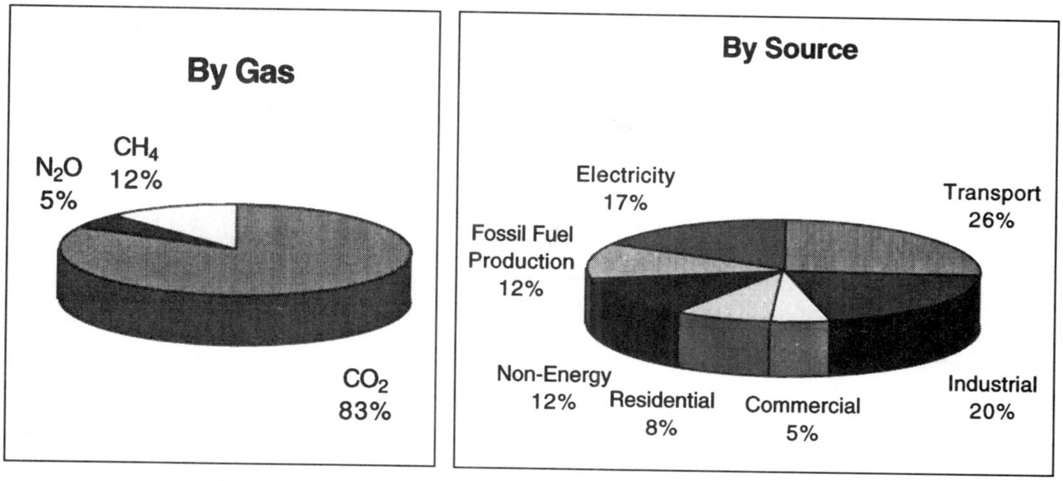

Figure 3: Canada's GHG Emissions Forecast[6]

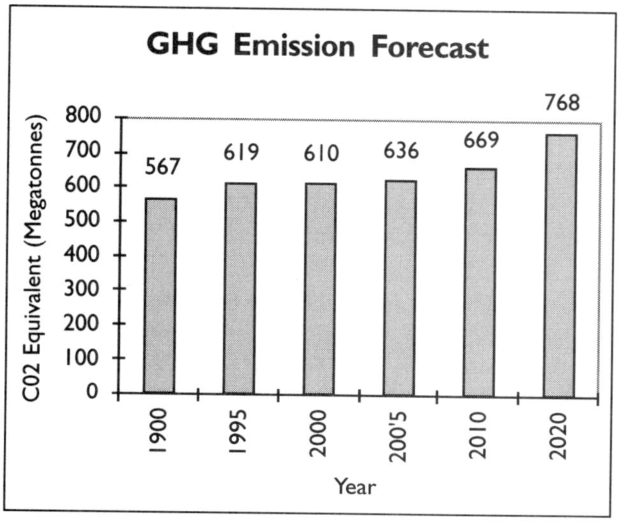

resources in southern regions may come under stress. The boreal forest may be unable to adapt to the relatively rapid changes in climate. Stock in coastal fisheries may be seriously diminished.

Canada's Response to Climate Change

In response to Canada's initial commitment under the FCCC to stabilizing its anthropogenic GHG emissions at 1990 levels by 2000, the National Action Program on Climate Change (NAPCC) was developed. The NAPCC is the first stage of Canada's plan to address climate change. It sets directions for achieving the country's commitment. Despite this program, Canadian emissions could be 13% above 1990 levels by 2000 (Canada's National Action Program on Climate Change). In terms of how renewable energy fits into Canada's response, the NAPCC states Canada will try to increase the use of renewable resources in the nation's energy supply. In support of this goal, the Renewable Energy Strategy was released by Natural Resources Canada in 1996.

Renewable Energy, Energy Efficiency and Emission Reductions

Energy efficiency and renewable energy have an obvious role to play in the mitigation of climate change. The barriers, which have prevented their adoption in the past, still exist. However, the climate change agenda is helping to lower the 'bar' somewhat.

Opportunities exist for increased levels of energy efficiency and renewable energy in all sectors of the Canadian economy. Canada's low energy prices and weak commitment to policies promoting renewable energy and energy efficiency have meant that only the most cost effective applications have been adopted in the past. As a result, there remains a huge untapped potential for their application and the resulting reduction in GHG emissions. Several recent studies have identified a set of "made in Canada" measures and policies that can be used to meet our international commitments. These include the *Rational Energy Program* by the Sierra Club of Canada and *Canadian Solutions* by the Pembina Institute of Appropriate Development and the David Suzuki Foundation. Both of these studies emphasize the importance of establishing more aggressive policies for promoting energy efficiency and renewable energy in all sectors of the economy.

The degree to which renewable energy and energy efficiency can contribute to emission reductions depends on the fuel they are displacing. As an example, if a natural gas hot water heater were replaced by a solar hot water heater, the emission reductions associated with the installation would be equal to the amount of gas saved (GJ)

[5] Natural Resources Canada, *Canada's Energy Outlook: 1996-2020, 1996 Update*, Ottawa: Supply and Services Canada, 1997.
[6] Canada's National Action Program on Climate Change from website, http://www2.ec.gc.ca/climate/resource/cnapcc/c3part01.html

times the emission factor of the fuel. Table 1 outlines the CO_2 emission factors for a variety of fossil fuels.

Table 1: CO2 Emissions from Fossil Fuels

Combustion Source	Emission Factor (tonnes CO_2/TJ)
Coal(Bituminous)	94.3 - 83.0
Natural gas	49.68
Propane	59.84
Light Fuel Oil	73.11
Diesel Oil	71.69

(Environment Canada, 1992)

The CO_2 emission factors associated with the consumption of grid connected electricity is more complex. Several factors contribute, including the interconnected nature of the North American electrical systems and the changing mix of fuels and generation technology that feed the system. These factors make it difficult to determine which source of electricity is actually being displaced. This is the subject of considerable debate. In the absence of any conclusive methodology for dealing with this, the average CO_2 emission factors for a region can be used as a rough proxy. Table 2 outlines the average emission factors for electricity across Canada.

Table 2: Average Electrical CO2 Emission Factors by Province (1996)

Province/Territory	EmissionFactor (kg CO_2 /MWh) *
Newfoundland	30.9
Nova Scotia	752.3
New Brunswick	420.7
Quebec	3.1
Ontario	82.4
Manitoba	8.3
Saskatchewan	838.2
Alberta	1,019.3
British Columbia	16.0
Yukon	208.0
Northwest Territories	453.2

* Calculated from NRCan publication *Electric Power in Canada 1996* Table 6.2 (Electrical Energy Production by Fuel Type 1996) and Table 6.7 (Emissions from Electricity Generation 1996).

Emissions Reductions In Canada

As discussed above, Canada has a large potential for emission reductions through the application of renewable energy and energy efficiency. These are, however, awaiting the appropriate policy and market signals to see them realized. The following sections discuss some specific climate change issues and emission reduction potentials for renewable energy resources and their role in different sectors of the economy.

Solar Energy

The adoption of any type of solar energy—passive or active, thermal or electric—has the potential to reduce CO_2 emissions if it displaces emissions from another energy source—fossil fuels. As with any renewable resource, the cost effectiveness of solar technologies depends on the specific application. The competitiveness of solar technologies, however, increases in a society concerned about climate change.

A study commissioned by the Efficiency and Alternative Energy Technology Branch/CANMET, entitled *Active Solar Heating in Canada to the Year 2010,* investigated the penetration of specific solar heating technologies through to 2010. Although the study is dated in terms of current market conditions, it gives a general idea about the potential for emissions reduction.

The study looked at solar applications for domestic, commercial and industrial hot water heating, pool heating, and commercial/industrial ventilation air heating. By making assumptions about the technical potential and penetration rates for each technology under different energy prices and fuel mixes (electricity, gas and oil) across the country, the potential for CO_2 and NO_x emission reduction in 2010 could be estimated. Table 3 summarizes some of the results of the study.

The expected annual solar savings in 2010 (14.38 PJ or 11% of the technically feasible market), does not account for any of the environmental benefits that would increase its value under certain policy conditions. The study further assumed that, if some of the environmental benefits were accounted for (and monetized into the calculations), savings would increase to 21.12 PJ in 2010 (17% of the technically feasible market). Although the study has not been reevaluated recently, it might easily be imagined that these savings could be increased considerably with the added motivation of the climate change agenda.

In terms of CO_2 emissions reductions, the study evaluated two different scenarios. The first, a

Table 3: Impact of Active Solar Heating in 2010

SECTOR	Total Mkt. Energy Use (PJ/yr)	Technical Feasibility (PJ/yr)	Solar Savings (PJ/yr)	CO_2 Emissions Reductions (Tonnes/yr) Ave. Fuel Mix	Fossil Fuel Only
Residential					
DHW	233.45	81.71	3.63	110,380	245,106
Outdoor Pools	14.58	11.66	6.66	186,220	342,268
Industrial					
Low Temp. Process Water	13.09	2.61	0.03	1,463	6,132
Ventilation Air	106.4	4.04	3.67	87,147	277,221
Commercial					
Service Hot Water	121.76	24.35	0.17	11,290	35,464
Ventilation Air	228.31	8.68	0.22	10,458	33,267
TOTALS	**719.59**	**133.06**	**14.38**	**406,958**	**939,458**

low emission scenario, assumes that solar technologies displace the average provincial generating mixes. The second assumes that solar displaces the fossil fuel portion of the mix only. This high emission scenario represents about 1% of the expected 2010 emissions gap shown in Figure 2. With appropriate policies implemented under a climate change agenda, emission reductions from the application of active solar heating in Canada could be increased.

Wind Energy

From a climate change perspective, wind energy provides an outstanding opportunity to reduce CO_2 emissions at a reasonable cost. Wind energy is already in the cost competitiveness range of existing fossil fuel generating technologies and, since it is virtually emission free, the cost of avoiding emissions is correspondingly low.

A study completed in 1996 for Environment Canada and the Canadian Wind Energy Association, entitled *The Environmental and Economic Benefits of Wind Energy Development in Canada: Lost Opportunity and Future Promise,* investigated alternative scenarios for wind energy development in Canada. Although the report was not directly focused on emission reduction potentials, an illustrative scenario was developed. This scenario called for 1000 MW of installed capacity by the year 2000 and 6400 MW by 2010. Table 4 outlines the potential CO_2 emission reductions from this scenario.

Table 4: An Emissions Reduction Scenario for Wind Generated Electricity

Year	Assumed Installed Capacity	CO_2 Reductions (Megatonnes)	Reductions in Electricity Sector Emissions Gap*
2000	1000	1.9	13.1 % of 14.5 Mt Gap
2010	6400	9.4	12.5 % of 75 Mt Gap

* The electricity sector gap refers to the electricity sector's share of Canada's anticipated emissions gap, as was anticipated at the time the study was completed.
Although these growth rates seem optimistic at this time, they were based on wind development rates experienced in Germany, which saw the installation of 3,647 wind turbines totaling more than 1100 MW in four years.

Emission Trading: Credits Blowing in the Wind[7]

The following emission trading agreement was established as part of the BC led Greenhouse Gas Emission Reduction Trading Pilot (GERT). The trade involves the purchase of emission reduction credits created through the displacement of fossil fuel generated electricity by wind generated electricity. The details of the project are as follows:

Seller: Enmax Power Corporation (a subsidiary of Enmax Corporation, the City of Calgary's reorganized municipal electrical utility).

Buyer: Environment Canada and Natural Resources Canada facilities in Alberta.

Source of Emission Reductions: Two 600 kW Vestas wind turbines owned and operated by Vision Quest Windelectric Inc. and connected to Alberta Interconnected Electric System supplying "Green Power" under contract with Enmax Power Corporation. Enmax subsequently sells the Green Power to Environment Canada and Natural Resources Canada with Environment Canada retaining the emission credits under the contract.

Emissions Reduced: 31,104 tonnes of CO_2 equivalent over 10 year life of project. This was based on an emission factor of 0.9081595 tonnes/MWh (an average for each of all Alberta utility-owned coal-fired plants; natural gas-fired plants; and oil-fired units) and an average annual energy demand of approximately 3,400 MWh

Biomass

Biomass resources are important for the mitigation of climate change because they play a role on both the source and sink sides of the carbon cycle (as an energy resource and as a tool to take CO_2 out of the atmosphere).

Key to the role of biomass as an energy resource, and a mitigative tool for climate change, is the assumption that biomass resources are consumed sustainably. That is, the same amount of carbon released in the energy utilization of biomass, is used in new biomass growth. From a climate change perspective, this puts biomass energy resources at a considerable advantage over fossil fuels.

There are many forms of biomass energy that can play a role in emission reductions. Some of the most economically promising include wood residues, pulping liquor, the direct combustion of municipal solid waste (MSW), the combustion of landfill gas and the combustion of biogas from anaerobic digesters.[8]

Energy from biomass can be either obtained from biomass grown and harvested explicitly for energy, or from the consumption of biomass wastes (wood waste or MSW). The use of waste, and particularly MSW, to generate energy is often criticized from a sustainability perspective since, as soon as waste becomes valuable as an energy resource, there is less incentive to reduce its production. With this said, the incineration of waste can help to mitigate a number of other problems associated with the landfill issue.

Metropolitan Toronto Landfill Gas Recovery Project[9]

The Metropolitan Toronto landfill gas recovery project has proposed to capture and generate a total of approximately 60 MW of electricity from the recovery of methane at three landfill sites. These facilities are expected to displace coal fired electricity and in doing so eliminate the equivalent of annual CO_2 emissions of more than one million cars while generating royalties exceeding $2.5 million from electricity sales

Land filling MSW creates a particular problem for climate change because of the methane released from decaying garbage. Since methane has 21 times the global warming potential of carbon dioxide there is considerable incentive to reduce the amount released. Two solutions are generally considered, burning the MSW: as discussed above, or collecting and burning the landfill methane. Flaring the methane to reduce it to CO_2 will have a net climate change benefit. However, using this energy to displace other more carbon intensive fuels could give even greater GHG reductions benefits in many Canadian jurisdictions.

International Applications

The global nature of climate change makes GHG emission reduction projects effective regardless of their location. This opens up additional opportunities for governments and corporations when looking for cost-effective emissions reductions.

[7] GERT website - http://www.gert.org

[8] Thermoshare, *Biomass Energy in Canada: With Emphasis on Electricity*. (Natural Resources Canada: Ottawa, 1998).

[9] Federation of Canadian Municipalities website: www.fcm.ca

The FCCC recognizes that developed countries have traditionally been the major source of increased GHG emissions and, as a result, they should shoulder the initial responsibility to reduce emissions. It also recognizes that future economic growth within developing nations may soon outweigh any mitigative measures adopted in developed countries. The FCCC provides a framework for international co-operation to provide financial and technical support to address climate change.[10]

Within Canada, the Canadian Joint Implementation Initiative (CJII) has been established to encourage Canadian companies to take part in international projects to limit GHG emissions. Since its inception, the CJII has worked with 10 companies, on 15 projects, in eight different countries, on projects in the energy sector dealing with energy efficiency, renewable energy, microhydro development and co-generation.[11]

Transportation

Figure 2 shows that the transportation sector is the single largest contributor of GHG emissions in Canada. Transportation and its associated social, economic and environmental impacts have kept transportation management on the agenda of policy makers for many years. Since the energy crisis, and more recently the environmental movement of the 1980s, gains have been made in increasing the energy efficiency and decreasing the pollution of vehicles. Despite these improvements, progress has been slow in reducing vehicle use because Canadian society has a high willingness to pay for personal mobility. Many variables impact personal transportation decisions other than economic efficiency, such as personal safety, style, preference and a lack of efficient public transportation systems. As a result, despite the significant potential for CO_2 emission reduction and the other benefits, the transportation sector continues to be problematic.

Alternatives for emission reduction in the transportation sector generally fall into three categories:

- Decrease the demand for transportation, by decreasing distances traveled
- Switch to less intensive modes of transport by switching from personal vehicle to public transport
- Switch to lower GHG intensive fuels

CO_2 Emissions and Your Car[12]

The average car in Canada emits more than its own weight in carbon dioxide each year. An improvement in the average fuel efficiency of these cars of just one litre per 100 kilometres would reduce Canada's carbon dioxide emissions by about 3.3 million tonnes per year. That would have roughly the same impact as closing down two modern 450-megawatt coal-fired power plants
(Federation of Canadian Municipalities, website: www.fcm.ca)

Alternatives one and two are the most the most complex to deal with as they often need to address behavioural issues of transportation demand. Item three focuses on technological fixes, which could include alternative transportation fuels such as ethanol gasoline blends, hybrid cars and zero emission vehicles. From a climate change perspective, electricity-based zero emission vehicles can be a misnomer if they displace the local emissions to emissions from electrical generation facilities.

Community Based Initiatives

As discussed, Canada is one of the highest energy consumers in the world on a per capita basis, largely due to the inherent characteristics of the country. Appropriately, Canadian communities have evolved in a manner that mirrors the characteristics of the country. These include inadequately insulated buildings, poor land-use planning, sprawling suburbs, and an inefficient municipal infrastructure to match. There are limits to what efficiency improvements can attain when the underlying community structure is not conducive to a lower energy and CO_2 existence.

Community energy management (CEM) is quickly evolving as a mechanism for addressing some of the fundamental issues governing energy use within society. CEM attempts to find synergies between urban planning objectives for livable communities, energy management objectives of minimizing use and associated environmental effects. In doing this for a given standard of living, CEM measures typically focus on:

- Land use planning—density and land use patterns.

10 Canada, *Canada's Second National Report on Climate Change.* (Supply and Services Canada: Ottawa, 1997).
11 Canada.
12 Federation of Canadian Municipalities website: www.fcm.ca
13 Jaccard, M., Failing, L. and Berry, T. "From Equipment to Infrastructure: Community Energy Management and Greenhouse Gas Emission Reduction" *Energy Policy*, Vol. 25, No. 13, 1997.

- Transportation networks—developing appropriate modes of public and private transport.
- Building and site design—addressing barriers influencing the ability of buildings to operate efficiently and make use of alternative and renewable energy supply technologies, including preserving access to solar radiation.
- Fostering efficiency and environmentally benign energy supply and delivery systems—supporting distributed generation, district heating and waste heat recovery.[13]

A study completed in British Columbia looked at four representative communities in the province and compared a business-as-usual scenario to one focusing on CEM over the period 1995 to 2010. It concluded that adopting CEM could result in a decrease of energy consumption by 15-30% and a decline in air emissions (CO_2 and NO_X) by 30-45%. Extrapolating these results to the provincial level, but focusing only on the land-use planning aspects of CEM, resulted in CO_2 emissions reductions of 17%.[14]

Municipalities and Emissions Reductions[15]

In accordance with Canada's pledge to achieve further reductions in greenhouse gas emissions, the municipal governments of Edmonton, Montreal, Ottawa, Toronto, Regina, Vancouver, Victoria and the regional government of Metro Toronto have already made commitments to reduce emissions 20% from 1988 baseline levels by 2005. The actions and leadership of these municipalities make an important contribution to Canada's National Action Program, since urban centres are significant sources of greenhouse gases, such as CO_2 and methane. It has been estimated that 72 Canadian municipalities with populations in excess of 20,000 (approximately 44% of the national population) were responsible for 169 Mt, or about 37%, of Canada's total CO_2 emissions in 1990.

In recognition of the leadership role these jurisdictions are taking, the Government of Canada has been working to establish a "20% Club" with those municipalities who have made a commitment to reduce their greenhouse gas emissions by 20% from 1988 levels by 2005.

Remote Communities

In Canada, remote communities have always represented a significant market opportunity for renewable energy and energy efficiency. These communities hold the same promise in terms of inexpensive CO_2 emission reductions. Traditionally supplying energy to these areas has meant importing fossil fuels to meet the electrical and much of the thermal demands. Because the fuels used in these communities generally have a higher CO_2 intensity (diesel fuel oil compared to natural gas) and because fuel transportation costs are relatively high, the costs of emission reductions can be correspondingly low. Adopting a CEM approach within remote communities with a focus on local renewable resources and energy efficiency will reduce emissions, energy costs and the flow of money out of the community to pay for imported fuels.

Even within the existing energy infrastructure, reductions in emissions can be made. In diesel based communities for example, district heating systems using the waste heat from the community generator can decrease heating costs and reduce emissions by displacing the use of heating oil. Pilot projects are currently in place in six communities in the Northwest Territories and are proving to be a practical, economical and an environmentally friendly alternative.

Summary

There is a strong and undeniable link between climate change and the world's energy production and consumption patterns. Canada's climate and socio-economic structure makes it one of the world's highest per-capita energy consumers and, as a result, also one of the highest per-capita emitters of the greenhouse gases. One of the leading mitigative measures available to combat climate change is to reduce society's dependency on carbon intensive fuels. Renewable energy and energy efficiency are key tools available to meet this end.

Despite the significant potential for both renewable energy and energy efficiency in Canada, little progress has been made to reduce the barriers to their adoption and realize the benefits. Climate change can play a role in reducing these barriers.

---

13 Jaccard, M., Failing, L. and Berry, T. "From Equipment to Infrastructure: Community Energy Management and Greenhouse Gas Emission Reduction" *Energy Policy*, Vol. 25, No. 13, 1997.

14 Ibid.

15 (Canada's National Action Program on Climate Change from website, http://www2.ec.gc.ca/climate/resource/cnapcc/c3part01.html)

With the appropriate policies mechanisms in place, the climate change agenda provides an opportunity to internalize one of the many benefits of renewable energy and energy efficiency (i.e., reduced CO_2 emissions). Should renewable energy and energy efficiency be adopted to combat climate change, additional environmental and socio-economic benefits can also be realized.

Economic instruments, such as emissions trading, provides an alternative to supplement the more traditional command and control approach to government policy. Although no formal mechanisms have been established to promote emission trading, some speculative trades are already being made between renewable energy and energy efficiency advocates and carbon intensive industries.

Any application of renewable energy or energy efficiency can realize GHG related benefits if it reduces or displaces the consumption of carbon intensive fuels. These can occur in the residential, commercial, industrial, transportation and electricity production sectors of the Canadian economy, through the adoption of energy efficiency and/or solar, wind, biomass, or other sustainable energy resources.

Climate change represents a substantial challenge to Canada and renewable energy or energy efficiency can be used to overcome this challenge. By adopting a more sustainable approach to energy management, Canada has an opportunity to capitalize on past research, meet its international commitments and obtain significant non-GHG related benefits. Although we have the tools, the changes we need will not happen without the appropriate signals being sent. Sending these signals means increased communication between the renewable energy community, government, industry and the public to identify creative alternatives and move the renewable energy and energy efficiency agenda forward as a mainstream alternative to current trends.

Chapter 8

Canadian Industry Overviews

THE SECTORS of the renewable energy industry described in this chapter serve consumer, industrial and commercial sectors. The brief profiles presented, by members of each respective industry, attempt to provide an up-to-date overview of the industries in Canada. In the last three decades, there has been considerable progress as customers can now obtain renewable energy products for lower costs, higher efficiencies, improved quality and increased reliability. Today, a knowledgeable, competitive Canadian industry is part of the multi-billion dollar, global renewables industry.

The Active Solar Industry in Canada

by Joe Thwaites, Taylor Munro Energy Systems Inc.

The solar thermal industry consists of companies engaged in the design, manufacture and installation of solar air and water heating systems.

Canada has a relatively small industry compared to other nations. This is reflective of the size of the domestic market more than the strength of the technology. Despite these conditions, Canadian solar heating technologies are world class as they have been developed in a country with both harsh climate and inexpensive energy.

Solar heating technology is well established in many countries around the world. Europe has a very strong market, with over 700,000 m2 of glazed collectors sold in 1996, mostly for domestic hot water. In fact, Europe and the Mediterranean countries make up approximately 40% of the world market for solar collectors. The Netherlands is a leader with installations approaching 10,000 systems per year. Much of the success of solar in Europe can be attributed to the strong government support in terms of both legislation and subsidies, as well as much higher energy prices it has compared to Canada.

Current Status

Most companies are small and medium sized enterprises with strong technical abilities. The largest area of application is residential pool heating. Companies servicing this market have the infrastructure of designers and installers that will be required as the market grows. Domestic hot water installations remain at a fairly steady, but relatively low level. Recent interest from utilities in more wide spread distribution is promising.

Niche commercial markets, such as aquaculture, municipal pool heating, and dairy farms are seeing increased activity. Industrial ventilation air heating continues to make serious market penetration on new construction. This increase in activity is assisted by support from Natural Resource Canada's Renewable Energy Deployment Incentive program.

Opportunities

The solar industry experienced tremendous growth in the 1970's and 80's as a result of the energy crisis. While great strides were made in the development of solar technology and in the reduction of system costs, the use of solar did not increase significantly because the price of conventional energy stabilized during this period. This left little incentive for adoption of solar heating or other applications since solar could not compete in costs with the status quo.

Now the spotlight is again on renewables due to a global desire to fight climate change, as demonstrated by the Kyoto commitment to reduce carbon dioxide emissions. While the exact form of Canada's actions is still to be determined, innovative mechanisms such as emission reduction trading could create opportunities, particularly for the solar thermal industry as it is one of the most cost effective renewable technologies.

Trends

One of the more interesting trends in solar heating equipment is the increasing use of the strategy of building integration. This approach has been popular with industrial ventilation air heating in Canada, and with solar water heaters in Europe. There are several advantages to this approach, not the least of which is the clean aesthetic look which is created. While this strategy lends itself best to situations of new construction, retrofits are also possible.

Sales of solar energy equipment will significantly expand when solar presents a more economical option than conventional energy sources. In several applications of solar technology, this is now the case. The interest and growth opportunities for this industry will continue as non-renewable energy sources diminish.

Challenges

There are two key challenges that the solar active market faces in Canada - financing and consumer awareness.

Traditionally customers have been asked to make a significant up front capital purchase which then pays for itself in monthly energy savings. In essence, the customer is comparing economic apples to oranges. Even when a system pays for itself in just a few short years, there is a natural reluctance to part with the capital. The industry needs to find ways to finance these systems through leasing, retail financing, and energy service contracts. These will enable the customer to more easily see the clear advantages of the technology.

Although solar was very high profile in the 1970s and 80s, due to the energy crisis and significant government support, it suffered when this support was withdrawn. Today solar heating is relatively unknown to consumers, so they are full of questions concerning operation, performance, longevity and maintenance. The credibility of the technology is naturally improving as more and more Canadians are exposed to the benefits of solar heating systems. Increasing awareness and understanding of solar thermal through public education will lead to increased use of the technologies.

Summary

The Canadian solar active industry has good reason for optimism as markets are expanding and it has a vital role to play in providing clean energy for Canada's future.

The Biomass Industry in Canada

by Roger Samson, Resource Efficient Agricultural Products Canada

BIOMASS or plant matter is the largest renewable energy source used in the world today. It accounts for about 15% of the global energy supply and 38% of that in developing countries. In Canada, biomass provides approximately 7% of our energy use or 540 petajoules (PJ).

Most of Canada's biomass energy at present comes from the forest, but most new growth in the biomass energy industry will likely come from the agricultural sector. Wood residues from forest industries is our main source of biomass energy in Canada. The forest industry uses 390 PJ of biomass energy to provide space heat as well as steam and electricity for industry processes. Approximately 150 PJ of energy are provided by cordwood for space heating in Canada mainly from unmerchantable timber.

Most of the biomass energy used in the forest industry sector is used internally in pulp and paper manufacturing. In total there are about 60-80 combustion system installations in paper mills across Canada. They have been made by large engineering groups such as ABB combustion, Babcock-Wilcox and Foster-Wheeler.

The major player in Canada in institutional and commercial installations has been KMW of London, Ontario, which has installed approximately 70 wood combustion systems. About 8-10 new installations are being made each year by KMW in recent years and export sales are starting in developing countries. Most installations are in the 5-15 MW range and burn bark or sawdust waste from the wood processing industry. Approximately 20 systems have been installed in institutions, such as schools and hospitals, although this sector is not growing as fast as the solid wood products forest industry sector. Some large district heating systems have been installed in Canada in the past 15 years. These systems work best where buildings are spaced close together (reducing piping costs and heat losses).

In smaller scale biomass heating systems there has also been progress. Efficient small commercial wood chip heating systems have been gaining in popularity. Grovewood Heat, manufacturers of the Bio-blast, of Little York, Prince Edward Island, have installed approximately 120 wood chip combustion systems in the 75-300 kW range for heating enterprises such as greenhouses, hog farms, fish farms, garages, motels as well as smaller wood industries.

Wood pellets have also entered the market place and have shown significant growth in the past five years. Over 100,000 tonnes of wood pellets are being manufactured in Canada at present. Quebec is the largest producer with 50,000 tonnes per year followed by Ontario (20,000 t/yr), the Maritimes (12-15000 t/yr), the Prairies (12-15,000 t/yr) and British Columbia (8-10,000 t/yr). There is good potential for expansion of the pellet market from both forest and agricultural sources. A new close-coupled gasification pellet stove developed by Del-Pointe Technologies of Blainville, Quebec, and Natural Resources Canada combustion laboratory entered the market place in 1997. It is more efficient than current stove designs and is capable of burning higher ash fuels which will stimulate the development of the pellet fuel market.

There is over 1000 MW of electrical capacity from biomass in Canada. Almost all of this is done by cogeneration in the pulp and paper industry. Facilities solely dedicated to electricity generation from wood have generally not been economical in Canada and only three systems exist. The industry in the United States has been characterized by low efficiency (<25%) and high capital costs. Many of the older installations in the United States are no longer operating after government subsidies were withdrawn. A promising medium term strategy for modernizing electricity generation from biomass is to produce electricity or cogenerate heat and electricity using gasified biomass with advanced conversion technologies.

For the transportation sector, ethanol blended gasoline is now being sold in all provinces except those in the Maritimes. Canadian production has now increased to 180 million litres per year with the completion of a new 150 million litre per year plant in 1998 in Chatham, Ontario. The barriers to a larger grain ethanol industry in Canada are high and relatively unstable feedstock costs and byproduct market saturation if a large industry develops. The competing technology with grain ethanol made from starch is cellulosic ethanol. Recent advances in enzymatic hydrolysis has made it possible to efficiently convert lower cost raw materials, such as straw, grass or wood waste, into ethanol. Feedstock costs are only approximately one quarter those of grain and the only byproduct created, lignin can be burned to create the steam and electricity for the conversion process. A fifty tonne per day cellulosic ethanol demonstration plant is being constructed in 1999 in Gloucester, Ontario. The partners in the project are PetroCanada and Iogen Inc., an enzyme manufacturer. Cellulosic ethanol appears poised to become an important renewable energy fuel for the transportation sector. If biomass is to have a greatly expanded energy future for Canada it will probably come from the agricultural sector. There are large quantities of agricultural residues in western Canada that could be used for energy, although some of these are beginning to be used for fibre industries. As well, fast growing perennial grasses, such as switchgrass can produce annual yields of biomass 10 times that of the natural forest on an annual basis and are well adapted to marginal agricultural lands. These systems may contribute not only to our future energy supply, but may help reduce the pressure on the nation's forests for our expanding fibre requirements.

The Earth Energy Industry in Canada

by Bill Eggertson, Executive Director, Earth Energy Society of Canada

SOLAR ENERGY is viewed as one of the most environmentally benign sources of energy available; more solar radiation falls on the earth each day than humans can feasibly consume. If these rays of the sun are not falling onto solar cells or onto flat plate collectors, they are absorbed into the surface of the earth.

Transferring this energy from the ground into space conditioning applications is the role of a ground-coupled heat pump, or earth energy system (the term 'geothermal' is also used, but is confused with deep boreholes that generate electricity from the venting steam). In the same way that a refrigerator transfers heat out of food and into the kitchen, an earth energy system takes solar heat from the ground and brings it into the house or office. It can also heat domestic water, and the entire process can be reversed in summer to transfer building heat into the nearby ground, thereby providing air conditioning and dehumidification.

The efficiency of an earth energy unit is measured by its co-efficient of performance (COP), which is the ratio of electricity required to handle the transfer, for every joule of useable solar heat that is delivered. By law, units in Canada must have a COP of at least 2.8 (2,800 watts of solar heat delivered for every kilowatt of motive power consumed); most units hover at 3.5, and some boast a COP of more than 4.0 (four units of space heating or cooling for every unit of electricity that is consumed).

Just below the frost line, the ground temperature in most of Canada remains above 7°C throughout the year, and an earth energy unit uses a compressor to upgrade this level to 50°C, pushing warm air into the ductwork or heating water for use in radiant floor heating.

The first common use of heat pumps was to keep submarines warm during the Second World War but, in the 1970s, Sweden became a pioneer in market development, and units became popular there when they were promoted as a method to avoid the construction of nuclear plants. Currently, about 25% of Sweden's earth energy units are installed north of the Arctic Circle. The United States embraced earth energy as an effective method to provide cooling for buildings in southern states, and Canada combined these two early approaches to build its own domestic industry. More than 30,000 units have been installed across the country, displacing 100,000 tonnes of carbon emissions each year through the displacement of 'dirty' conventional power generation. Market niches include schools, institutions, hotels, prisons, strip malls and, of course, private homes.

When Ontario Hydro was involved in demand-side management in the early 1990's, the utility's strong promotion of earth energy technology resulted in the installation of more than 6,700 residential systems. Incentives now are offered in many US states, as Congress promotes earth energy as a plank in the fight against climate change.

The earth energy industry is dominated by US imports, and many of the big names in the HVAC industry have offered heat pumps as part of their product line. Some Canadian firms have carved viable niches for their product (one is the 'supplier of choice' for federal correctional centres on the East Coast) and others are involved in exporting units to Eastern Europe.

Earth energy, literally, is an invisible technology. The loop is buried underground and the compressor and other working components are hidden in the cellar. It is one of the four heating sources recognized under NRCan's Renewable Energy Deployment Incentive, (REDI) but the Canadian industry unanimously agreed to reject incentives in favour of heightened awareness. A comprehensive market study (due to be released early in 1999) will allow the industry to work with government to promote cost-effective space conditioning applications and explain how the technology can help meet Canada's commitment from the Kyoto protocol on emissions.

Of 7,314 PJ of energy consumed in Canada (1994), earth energy can supply 67% of total commercial demand (628 PJ for space heating, space cooling and water heating) and 84% in the residential sector (1,392 PJ for heating, cooling and DHW). This 1,800 PJ represents 25% of all secondary energy consumed in the country.

In a report on space conditioning by the US Environmental Protection Agency, electric resistance, oil, natural gas and air-source heat pumps were evaluated against ground- coupled heat pumps in six cities across the US. Earth energy provided the highest source heating performance and highest source cooling performance in all locations, and it tied with gas-fired heat pumps in annual operating costs. It had the lowest total annual cost for all electricity-driven options, except in the city of Portland. The only class in which it did not score highest was in water heating, which was tied between gas-fired and air-source heat pumps.

Suffering from the market downturn, the industry is working to build its infrastructure through a timely revision to national standards for unit performance and system installation, and to develop training for contractors that will ensure quality installations. Strong demand for earth energy products is expected as environmental concerns grow, and as sales continue to climb south of the border.

One of the greatest impediments for the Canadian industry is the perception that earth energy costs more than conventional heating options. In some installations, payback can be almost immediate, and promising commercial sites are as low as three years. The current depressed price for oil and natural gas makes it more difficult to compete on operating costs, but earth energy is almost always calculated as the lowest lifecycle cost for space conditioning options.

Another problem is that many of government-sponsored design tools do not include earth energy as an option. The RETScreen preliminary assessment tool will release an earth energy module in 1999, and the industry has asked to be included in the HVAC Advisor that is used by many heating contractors to compare heating options. Omission in these, and other important vehicles of awareness, has kept the industry's image low and impeded penetration in the domestic market.

With greater awareness by commercial and residential building owners, the industry can become more aggressive in its marketing approach, and promote its economic and environmental benefits to a wider audience of Canadians.

Earth energy systems provide solar heat at less than one-third of the cost of conventional resistance electricity, and reduce emissions and effluents by the same degree, making the technology a good hybrid between renewable energy and energy efficiency. It offers superior climate control, decentralization and long service life, all of which will combine to make earth energy one of the pre-eminent space conditioning options in Canada.

Microhydro Industry Status

by Robert Mathews, Energy Alternatives Ltd.

HYDROELECTRIC POWER is one of the oldest, most reliable and economical sources of electricity. Sixty per cent of Canada's electricity is hydro-generated by large provincial utilities, some of which are among the largest in the world.

Hydroelectric technology, in its current technological state, has been in use for over a century; has been well optimised (water-to-wire energy efficiencies of 60 to 90%); and is completely proven technology with plants operating 24 hours per day for 20 to 40 years, sometimes even longer, and often out-living their amortisation period. The operating costs for these plants are low because the energy source is free, clean and renewable. Current technologies allow micro and minihydro systems, from 5kW to 500kW on cost-effective sites for between $1500 to $2500 per kW of installed capacity. In comparison, new mega-scale hydro plants are in the range of $5000 per kW of installed capacity. In microhydro electric generation, the cost per kW is significantly lower than both wind energy and photovoltaics. Microhydro can convert from 50 to 100% of its energy into electricity, while wind and photovoltaics can only convert from between 12 to 30% of its energy into electricity.

Currently, large-scale hydro development has fallen out of favour with the general public, due to concerns with environmental impact. The environmental concerns associated with large-scale hydroelectric are mostly caused by dams and reservoirs, which alter the seasonal flow of streams, by storing water, for example, from spring run-off for release through the turbines months later. The dam itself is less of a problem for fish and fish habitat than the altering of the natural flow patterns. The reservoirs themselves are also a concern as they flood land which prevents other uses. The resulting "lake" is of marginal recreational use due to extreme fluctuations in water level.

"Run of the river" projects do not present serious environmental impacts and are generally viewed as benign, requiring only a small weir, as an intake, on a steep-gradient stream. In many cases the steepness of the river will have prevented fish migration before development.

Small hydro (less than 15 MW) has been a significant part of Canada's history, with many cities and towns in Eastern Canada settled around streams and rivers with hydro sites that could power mills and later, generate electricity. Many smaller sites fell into disuse during the 1950s with the advent and development of the megaprojects, both hydroelectric and nuclear, the centralisation of the power generation for the large provincial utilities, and establishment and economical expansion of the provincial grids. Small-scale generation could also be provided by portable, flexible and low-cost diesel generators, replacing the fluctuating nature of the small mills and power plants which were somewhat water-level dependent. Some 30 or 40 years after being shut down, many of the early small sites have been refurbished and are economically viable, due to the ingenuity of the developers and modern cost-effective methods of automation.

Diesel generators are still inexpensive to purchase, but increased fuel costs and maintenance costs have made them more expensive to operate. The national electrical grids are extensive, but remote locations may not be connected and the additional expense for individuals or communities to connect can be prohibitive. People in off-grid areas are investigating the installation of waterpower as it offers a stable, inflation-proof source of electricity.

Small and microhydro installations have historically been inexpensive to run but expensive to build. But recent advances in microhydro technology are helping to reduce those costs including; inexpensive plastic piping, smaller, lighter, higher-speed turbine equipment, and lower-cost electronic speed and load-control systems. Recent technological advances in battery storage and inverter systems have been made through photovoltaic development and experience, and the advances are transferable to microhydro systems. Capital investments are higher in microhydro than investing in diesel equipment of comparable capacity but the lifetime of a hydro system is longer and the operating costs are minimal, in comparison to a diesel system.

The substantial growth of a sustainable micro and minihydro industry would be assured if the federal, provincial and municipal governments and utilities would allow net metering. This simple interconnection between the renewable energy source and the existing utility would encourage customers to purchase renewable energy systems and they would be spared the significant expense of a bank of batteries and inverter. The net metering system is run through the single existing kilowatt-hour meter, with the meter running backwards, when generation exceeds the load and runs forwards when the load exceeds generation.

Canada has several world-class manufacturers and packagers of small and microhydro equipment exporting their products internationally. These products include:

- electronic load control governors and load managers,
- a cost-effective 4" Turgo turbine capable of up to 4500 watts,
- a specifically-designed permanent-magnet brushless 1 kW generator (for battery-based systems),
- Propellor and Kaplan turbines made at reduced cost by accurate casting techniques and computer-controlled machining,
- development of technology that uses mass-produced water pumps-as-turbines.

In the international hydroelectric market, there are many large companies competing for the opportunity to build the mega-scale projects, but in the under 250kW market, there are few companies with the necessary expertise in design, manufacturing and installation of the systems. In Canada, where there is an abundant source of rivers on which to develop small economical generation sites, there is also an opportunity for Canadian companies to train people to build micro and minihydro systems. Canada could then export its trained professionals and products to Central and South America, Asia and Africa where there is an increasing demand for inexpensive, reliable electricity.

Photovoltaic Industry Overview

by Mike Cannon, Powersource Energy Systems Inc.

SOLAR POWER, or photovoltaics, is the most reliable source of energy in the world today, silently converting the energy in sunlight directly into electricity. Early applications for solar electricity were limited to high value specialty markets, such as powering equipment on satellites. Today, increased efficiencies and lower costs are helping promote PV's acceptance, both as a power source for applications beyond the power lines and as a replacement or supplement for utility supplied electricity.

Current applications for photovoltaics range from consumer products, such as calculators, watches and radios, to stand-alone remote sites and grid-tied residences. In off-grid sites, such as remote homes, recreational vehicles, water pumps, lights, telecommunications sites, and industrial applications, solar is often the power supply of choice, providing electricity that is both cost effective and reliable. In Canada and around the world, thousands of such stand-alone systems produce millions of watts of electricity. PV offers real benefits over conventional sources of power, such as prime generators—it is clean, silent, requires virtually no maintenance and is available at any time of day or night. Even larger generator powered sites are now using solar and inverters to reduce their fuel consumption, only running the generators for short periods of time per day.

A recent ground swell of interest has been seen in the use of solar electricity for grid connected applications. Fueled by environmental concerns, government bodies around the world are promoting PV as an alternative to conventional utility networked fossil fuel and nuclear energy. In Europe and the US, these systems are often utility-interactive, allowing excess solar power to be sold back to the utility. A few demonstration systems have been installed in Canada and there is considerable interest in developing more. The use of solar as a back-up or emergency source of power is also increasing. Many residential and business owners are installing emergency power systems to protect themselves from the loss of power due to storms, accident or generation problems. Collectively, these systems have the potential to make a huge impact in lowering the amount of carbon dioxide we put into the atmosphere.

It is anticipated that solar energy will supply an increasing amount of the world's power requirements. As this effectively unending form of energy is one of the most environmentally friendly power sources, the current political and economic climate grant it a very high social acceptability. There is no production of toxic pollutants, and the need for extensive power corridors is reduced with on-site generation augmenting central power stations. Decentralized power delivers improved electrical quality to locations farthest from central generating stations and decreases the occurrence of transmission problems, such as downed lines, overloading and surges. Using today's technology, the energy required to produce a PV system has a two-year pay back period. With systems expected to last 30 years, PV offers over 25 years of net energy gain, no pollution and operational cost savings in many applications.

Technological improvements in solar module design are helping to fuel the increase in demand for solar products. The vast majority of commercially available solar modules use polycystalline or monocrystalline solar cells. The efficiency of these cells at converting light energy into electricity has been increasing and now ranges from 12 to 18%. Higher efficiencies increase the power generated per square metre of module, reducing the amount of silicon material required and the total surface area of the array. Laboratory cell efficiencies as high as 26% have been reached, but their commercial viability remains to be proven. Considerable research and development in thin film solar technologies over the past few years has yielded significant improvements in their operational characteristics as well. These modules use much less active material and have the potential to be manufactured at a considerably lower cost. PV products are now being developed to integrate into building materials as well, both as window and roofing products. It is anticipated that these improvements will reduce costs by a factor of two over the next five years, further fueling the growth in the solar market. It also brings us a step closer to the time when solar electricity will become competitive with electricity generated by conventional technologies, both in Canada and throughout the world.

The manufacture of solar cells and modules (the building block of PV systems) requires a large, capital intensive assembly plant to operate competitively in world markets. Canada currently has no such facilities, with a single module lamination firm in Ontario offering limited volume of product for the domestic market. Although our PV module manufacturing is minimal, Canadian made equipment used for module lamination, solar simulation, solar cell and module testing are in operation worldwide.

Most of the Canadian photovoltaic industry has developed by integrating systems using key imported components. In recent years, a number of manufactures of 'balance of systems' components have offered product that is both technologically advanced and competitively priced. There are several companies manufacturing charge controllers and Canadian made batteries are increasingly being used in PV systems, throughout North America. Inverter manufacturers export around the world, with one pioneering the design of high frequency inverters. A number of companies make specialized PV systems such as solar powered water pumps and lighting systems, which are sold for domestic and export use.

The vast majority of photovoltaic systems installed in Canada today still rely on imported components. The companies installing these systems are typically small, operating in specific geographic regions and are supported by a small number of larger distribution firms operating on a national level. Most specialize in off-grid residential systems, including seasonal cottages and remote homes. Canadian companies have developed considerable expertise in the design and installation of PV stand-alone and generator hybrid systems for use in harsh, remote environments. Cumulatively, these firms are capable of meeting the needs of customers in Canada and internationally for any size photovoltaic system for any application.

The photovoltaic industry will continue to grow into the new millennium, roughly matching an estimated international growth rate of 20% per year. These are clearly exciting times for the PV industry —*the future has never looked brighter!*

Windpower Energy in Canada

Drew McKibben, McKibben & Bailey

WIND POWER is the world's fastest growing source of energy, according to the Washington D.C.-based policy organization Worldwatch Institute. Its annual measurement of new generating capacity installed worldwide, for four years beginning in 1995, shows wind outpaces any other energy resource, growing 35 per cent in 1998 alone. At the start of 1999, enough electricity to service 3.5 million homes - 9,600 megawatts (MW) - was generated by wind.

Germany, Denmark, the United States and Spain are the current leaders in wind power market development, and the huge majority of installations are in Western Europe, India and the United States. Germany alone has 2,800 MW of capacity, more than any other country.

Canada more than tripled its wind energy production in 1998 to 82 MW, nearly enough to power 30,000 homes. Across the country hundreds of wind turbines are used for remote battery charging, water pumping and utility-scale generation. The bulk of Canadian wind-generated power is for utility distribution in southern Alberta, where there is one 20 MW wind farm and four individually-sited 600 kW turbines, and Québec's Gaspé Peninsula, where the 100 MW Le Nordais project will be completed before the end of 1999.

Off-Grid Applications

Individual consumers who own or consider buying a wind generator usually require it as a remote supply of electricity. Small wind-powered machines, however, also have agricultural applications, used for aerating and pumping water.

Wind electrical generators may be stand-alone systems, or operate as part of a hybrid system with other generators using solar photovoltaic (PV) or diesel power. Small wind turbines typically range in power from less than 50 watts to 10 kilowatts (kW), with most machines less than 1 kW.

Stand-alone wind turbines are usually made with alternators instead of generators and have simple, effective solid-state charge controller/rectifiers. They are available primarily in 12- or 24-volt DC; however, larger systems are available in up to 120-volt DC configurations. The trend is toward simpler machines that retain high reliability and produce more energy at lower cost.

Virtually all stand-alone wind turbines can be used as part of a hybrid system, with each generation source usually having its own controller. During periods of low wind or high energy demand the inverter automatically starts an engine generator to charge batteries. Solid-state and computer controls, along with high efficiency sine-wave inverters, make these systems dependable. The growing demand for electricity in developing countries make these hybrid systems attractive for small rural villages.

The agricultural use of wind turbines in Canada started in the early 1900s, when thousands of machines pumped water on prairie farms. Those familiar mechanical pumpers are still available, but Canadian companies have advanced designs using new materials and bearings and introducing wind electric systems. Canadians have also adopted technology from abroad to match high-powered stand-alone wind turbines with high-quality, high-power AC pumps to direct drive water pumping systems using special controller technology.

While most water pumping systems are used seasonally for stock watering, they're also designed to pump air into standing water, maintaining water quality in ponds and dugouts. And larger systems are now capable of transporting water for irrigation.

Grid-Connected Applications

The commercial generation of electricity from wind resources, supplying energy to provincial power grids and small isolated communities, comes from wind farms, power station-sized machines or wind-diesel hybrid systems.

Turbines suited to large-scale utility service are between 200 kW and 1.5 MW in size, and machines between 300 and 750 kW are most commonly used. Isolated community grids, like those found in northern Canada, can add wind capacity to supplement fossil fuel generation, usually diesel oil. These machines range from 10 kW to 200 kW.

The smallest of these machines are single phase, with three-phase units beginning at about 50 kW. All grid-connected wind turbines have safety controls to shut down in the event of grid failure, protecting both the grid and the turbine from voltage or frequency excursions. Smaller units may use line commutated inverters for interconnection, while larger units usually use induction generators. These systems tend to have simple solid-state controls and displays.

The commercial production of utility-scale power is driving the ongoing development of wind energy technology. Dozens of reliable turbines are available in the market for this use. Designed with and for utility companies, they have onboard computerized controls with telephone or wireless communications to central control systems.

With the increasing improvement, proliferation and reliability of turbine technology, the cost of wind power is dropping precipitously. In 1981 the installed cost per kW was $3,900. By 1999, however, larger machines, an improved understanding of siting and manufacturing efficiencies have brought costs down to $1,200/kW.

The Canadian Industry

Progressing from turn-of-the-century water pumpers to farm battery chargers, the Canadian wind industry now manufactures, distributes, installs and maintains all kinds of wind turbines. It also offers wind resource assessment, and has advanced the technological design of equipment operating in cold climates. As developers, Canadian firms have a great deal of experience in devising all sizes of systems, bidding on large domestic and international wind farm projects.

While a number of wholly-Canadian turbine designs exist in the agricultural market, strategic partnerships with foreign manufacturers are beginning to encourage the production of Canadian wind generators. Although this part of the industry is still in its infancy, utility-scale European turbines are assembled in Québec, rotor blades are manufactured in Ontario and 25 kW Canadianised turbines, remodeled from a French design, are now being built, also in Ontario.

The Future of Wind Energy in Canada

Visionaries in the Canadian wind industry, and the Worldwatch Institute, predict the development of wind energy will blossom in Canada, although it is currently enormously underutilized. Mounting environmental pressure, economic circumstance and transformations in the electricity industry are now encouraging this to change.

Coal, oil, large hydro and nuclear power plants, which now supply Canadians with energy, are, in some cases, no longer cheaper to build than wind farms, and all are associated with environmental problems. Because these facilities already exist, however, they are often cheaper to maintain or repower. As new generation plants are constructed in Canada, natural gas, wind, small hydro, landfill gas and biomass are all contenders in the market, with natural gas clearly in the lead.

People in the far north, representing a niche market of 68,000 residents in 65 remote locations, may be the first to see increased wind development in Canada. The need to transport diesel fuel makes their electricity five to 12 times more costly than the national average. These communities, understandably, are now evaluating wind energy.

Québec is the only province in Canada with substantial plans for the installation of more wind power. A government-appointed energy board has recommended 50 MW of wind capacity be installed per year for nine years, beginning in 2002, and legislation is expected to enact the proposal. The province believes development on this scale will create economic growth around a new industry.

One of the strongest levers of change in the whole energy business, which has yet to be pulled, is national regulation on greenhouse gas (GHG) emissions. Many Canadian companies attempting to curb GHGs anticipate voluntary efforts will be replaced by imposed limits on emissions. This will create a market shift towards the purchase of technology and services offering energy efficiency and greener resources. Wind power is now being purchased by the federal government as a demonstration of how emissions can be reduced.

Clean energy is also a commodity which all Canadian consumers, showing a preference for green products, may someday purchase. Utility companies in Alberta and Ontario, the first two provinces with plans to deregulate their electricity sectors to allow private marketing, immediately offered a green-rate option to customers. In Alberta the resource is identified, exclusively, as wind.

Chapter 9
For More Information

THIS CHAPTER has been developed to assist the reader in locating further information on renewable energy. It gives information on renewable energy associations, World Wide Web Sites and a detailed books list, sorted by renewable energy type and subject.

In the past, information was given on government departments and contacts, but the editor has found that it is difficult to keep these lists up to date, due to changing provincial and federal government departmental structures. To contact a specific branch of the government about a renewable energy related issue, first, look in the local "blue" or government pages of the phone book and look for departments relating to energy or the environment. The federal department that deals with energy issues is Natural Resources Canada, federal environmental issues are dealt with by Environment Canada. For specific contact names in each of these and other departments, contact the related renewable energy association or your local MP.

Associations

Canadian Solar Industries Association (CanSIA)

The Canadian Solar Industries Association (CanSIA) was formed by the unification of the Canadian Photovoltaic Industry Association and the Canadian Solar Industry Association in August of 1992. The two associations joined to promote the increased usage of solar technology in Canada. CanSIA is a registered not-for-profit organisation, whose membership is comprised of private individuals, companies and governments interested in solar technology.

Benefits of Membership

Recognition of Excellence—CanSIA members have agreed to a code of ethics outlining a professional high standard of quality and performance

CanSIA Newsletter—published a minimum of six times per year to keep members abreast of the latest developments in the solar industry worldwide

Website—open access for national and international business

Membership Directory—to identify solar companies specialising in thermal or photovoltaics, or both

Display Booth—available to Corporate and Regular members to promote the solar products and services they offer

Trade Shows—members are offered substantial discounts to jointly participate in various trade shows across Canada

Generic Brochures—highlighting the benefits of solar energy products

Media Links—promotion of successful solar applications to various media

Training Courses—accredited training programs ensures installers have the proper knowledge to install renewable energy systems, safely and efficiently

Standardisation—products and installation procedures to improve the quality, safety and reliability of clean energy sources

Professionalism-communicates the vision of the solar industry to federal and provincial governments

CanSIA Membership Fees

Sustaining $1,500 - $2,000
Corporate $250
Regular $125
Associate $75

Contact

Canadian Solar Industries Association
Suite 250-2415 Holly Lane
Ottawa, Ontario K1V 7P2
Tel: (613) 736-9077
Fax: (613) 736-8938
E-mail: cansia@magmacom.com
Web: www.cansia.ca

CanWEA Association

The Canadian Wind Energy Association (CanWEA) is a national, non-profit organisation that promotes wind energy to government, utilities and the public.

CanWEA's mission is to represent the wind energy community, comprised of organisations and individuals who are directly involved in the development and application of wind energy technology, products and services, and to support the appropriate development and application of all aspects of wind energy, including the promotion of a suitable policy environment.

CanWEA publishes a quarterly newsletter for its members called "Windsight" and holds an annual conference and trade show that brings industry players together. If more technical information is required, CanWEA offers an Industry Directory for $7.00 that lists products and services available throughout Canada. The CanWEA bookstore also holds a wealth of information. Please contact the office for further details.

CanWEA Membership Fees (includes GST)

Annual Dues

Sustaining	(2 votes)	$1,605.00
Corporate	(2 votes)	$ 401.25
Associate	(1 vote)	$ 160.50
Individual	(1 vote)	$ 80.25
Student	(no votes)	$ 32.10

Canadian Wind Energy Association
Suite 100, 3553 31 Street NW
Calgary, AB T2L 2K7 Canada
Tel: 1 800 9 CanWEA (in Canada)
or (403) 289 7713
Fax: (403) 282 1238
E-mail: canwea@canwea.ca
Website: www.canwea.ca

Earth Energy Society of Canada

A non-profit organisation that promotes the feasible application of ground-coupled heat pumps (GeoExchange or geothermal heat pumps) for space conditioning.

The Earth Energy Society provides information to Canadians on the energy and environmental benefits of ground-coupled heat pumps. It represents the interests of dealers, manufacturers, distributors, academics, consultants, utilities and environmental officials. It is aligned with the Geothermal Heat Pumps Consortium, Washington, and the International Ground Source Heat Pump Association, Oklahoma, and works closely with national and regional Heat Pump Associations around the World.

It co-sponsors regular international satellite teleconferences on earth energy technology; and supports seminars across Canada and the United States. It publishes a monthly fax newsletter, available by subscription, and administers an extensive email network for industry officials. It also has representation on the Buildings Table of the climate change discussion, and has developed an extensive market plan with government assistance.

Its predecessor association, the Canadian Earth Energy Association, was responsible for the installation of more than 6,500 residential systems.

Earth Energy Society of Canada
#1050-130 Slater Street
Ottawa, Ontario K1P 6E2
Tel: (613) 230-2332
Fax: (613) 237-1480
E-mail: info@earthenergy.ca
URL: www.earthenergy.ca

Solar Energy Society of Canada Inc. (SESCI)

Established in 1974, the Solar Energy Society of Canada Inc. (SESCI) operates as a non-profit, non-government association with registered charity status. Energy conservation, building a strong Canadian renewable energy industry, preserving a healthy, sustainable environment by encouraging the use of solar and other renewable energies in Canada are some of the issues that SESCI works on. SESCI is also a national chapter of the International Solar Energy Society (ISES), an international organisation promoting solar energy, with over 5,500 members worldwide.

SESCI's educational programs stimulate interest and increase awareness of sustainable energy among students at all levels. SESCI is a sponsor of the Regional Science Fair Awards of the Youth Science Foundation Canada. SESCI actively supports the exchange and application of scientific and technical information through its annual Conference, trade shows and exhibits, public seminars, workshops and tours. The Society has a quarterly publication *Sol* ,available with all levels of membership. Recently, SESCI has become an active participant in the Federal Government's climate change discussions, presenting information, and critiques of position papers to various government departments in the energy resource, industry, finance and environment sectors.

Annual Membership Rates (effective 1 June, 1998)

Sustaining Lifelong	$3000
Supporting Corporate	$1500
Corporate 6 or more employees	$ 350
Corporate under 6 employees	$ 175
Library	$ 100
Individual	$ 65
Senior Citizen	$ 35
Student	$ 20
ISES	$ available upon request

Solar Energy Society of Canada Inc.
116 Lisgar Street - Suite 702
Ottawa, Ontario K2P 0C2 Canada
Tel: (613) 234 4151
Fax: (613) 234 2988
E-mail: sesci@cyberus.ca
Website: www.solarenergysociety.ca

On-Line Information

We have provided 13 Uniform Resource Locator's (URL's) related to Renewable Energy that will help you start sourcing internet locations. Many of these sites also contain indexes of other related sites. This list is intended as a guide only.

When using a search engine to locate renewable energy sites, it is best to be as specific as possible. For example, a search for information on solar energy in Ontario, Canada needs to be broken down. If the search engine is given "solar energy" it may yield 5-10 million responses, but if the search engine is used correctly and it is given "solar+energy+ontario+canada", it should narrow the field down significantly.

Canadian Sites

Canada Mortgage and Housing Corporation
www.cmhc-schl.gc.ca/
This site is set-up to provide information on housing, markets trends and general CMHC information. Gives information on publications products and services offered by the Canadian Housing Information Centre (CHIC).

Canadian Solar Industries Association (CanSIA)
www.cansia.ca
The Canadian Solar Industries Association provides information on its solar industry members. It offers workshops and has books available on installation of solar energy systems.

Canadian Wind Energy Association (CanWEA)
www.canwea.ca
The Canadian Wind Energy Association site provides general information on wind energy applications in Canada, as well as information on its annual conference, its members and its bookstore.

Earth Energy Society of Canada
www.earthenergy.ca
The Earth Energy Society of Canada provides general information on earth energy applications in Canada. It has a detailed list of its industry members and installers of earth energy systems.

Environment Canada
www.ec.gc.ca
The Green Lane on the Information Highway features the latest Environment Canada news, including all the updates on Canada's commitment to Climate Change and Canadian Climate Change initiatives, press releases etc.

Local Climate Data
www.cmc.ec.gc.ca/climate
Local climate data, including solar insolation and wind data can be obtained by calling Environment Canada's General Climate Service at (416) 739-4328 or by accessing their web site.

Natural Resources Canada
www.nrcan.gc.ca
Natural Resources Canada has a mandate to promote the sustainable development and responsible use of Canada's mineral, energy and forestry resources, and to develop an understanding of Canada's landmass.

Solar Energy Society of Canada
www.solarenergysociety.ca
The Solar Energy Society of Canada site provides general information about solar and renewable energy applications in Canada. It also provides information on its membership, annual conferences and links to other sites.

International Sites

American Solar Energy Society (ASES)
www.sni.net/solar/index.html
The American Solar Energy Society has detailed information on Solar Energy and its applications in the United States. It has a comprehensive book store of renewable energy information from all around the world.

Centre for Renewable Energy and Sustainable Technology (CREST)
www.solstice.crest.org
Solstice is the Internet information service of the Center for Renewable Energy and Sustainable Technology (CREST), and is a premiere site for Sustainable Energy and Development Information.

Energy Efficiency and Renewable Energy Network (EREN)
www.eren.doe.gov/
The United States Department of Energy's (DOE's) Energy Efficiency and Renewable Energy Network (EREN), a site that locates and organizes information about renewable energy and energy efficient technologies.

EnviroLink Network
www.envirolink.org/
The Envirolink Network is a non-profit source for all types of environmental information.

European Wind Energy Association
www.ewea.org
The European Wind Energy Association site has information on wind energy applications in Europe and around the world. It also has useful links to other wind energy sites.

International Solar Energy Society
www.ises.org
The International Solar Energy Society's site provides information on international applications of renewable energy. It has many international web links to renewable energy sites and information on its many national chapters.

Publications

Listed below is a wide range of recently published books that would be useful for further information. The list is by no means complete and the inclusion of any title should not be taken as an endorsement, it is intended as a guideline only.

General

Ander, Gregg., *Daylighting Performance and Design*, VanNostrad Reinhold, 1995. ISBN 0-442-01921-1

Anderson, Bruce, *The Fuel Savers*. Morning Sun Press, 1991. ISBN 0-9629069-0-5

Brower, M.. *Cool Energy: Renewable Solutions to Environmental Problems*. Boston: MIT Press, 1992.

Cole, Nancy & P.J. Skerrett, *Renewables are Ready: People Creating Renewable Energy Solutions*. Chelsea Green Publishing, 1995. ISBN 0-930031-73-3

Cross, Bruce. (ed.) *European Directory of Renewable Energy Suppliers & Services 1998*. London: James & James (Science Publishers) Ltd., 1998.

——*World Directory of Renewable Energy Suppliers and Services 1997*. London: James & James (Science Publishers) Ltd., 1997.

EUREC Agency, *The Future for Renewable Energy*. London: James & James (Science Publishers) Ltd., 1996.

International Energy Agency, *Renewable Energy Policy in IEA Countries, Volume I: Overview*. 1997. ISBN 92-64-15495-7

——*Renewable Energy Policy in IEA Countries, Volume II: Country Reports*. 1998. ISBN 92-64-16186-4

——*Benign Energy? The Environmental Implications of Renewables*. 1998. ISBN 92-64-16183-X

——*Key Issues in Developing Renewables*. 1997. ISBN 92-64-16009-4

Kosloff, K.L. and Dower, R.C.. *A New Power Base – Renewable energy policies for the nineties and beyond.* WRI: Baltimore, 1995.

Kristoferson, L.A. and Bokalders, V. *Renewable Energy Technologies: Their Applications in Developing Countries*. Rugby: Intermediate Technology Publications, 1991.

LeClair, Kim and Rousseau, David , *Environmental by Design*, Vancouver: H&M, 1992

Scheer, Hermann, ed. *The Yearbook of Renewable Energies 1998/9*. London: James & James (Sciences Publishers) Ltd., 1998.

Steen, Nicola. ed *Sustainable Development and the Energy Industries: Implementation and Impacts of Environmental Legislation*.. London: Earthscan Publications Ltd., 1994.

Twidell, J. *Energy for Rural and Island Communities.* Oxford: Pergamon Press.

Twidell., J and Weir, T.. *Renewable Energy Resources.* London: E. & F.N. Spon, 1990.

World Energy Council. *Energy for Tomorrow's World: The Realities, the Real Options and the Agenda for Achievement.* Kogan Page: London, 1993. (ISBN # 0-749411-17-1)

—— *Renewable Energy Resources - A Guide to the Future.* London: Kogan Page, 1994.

Biomass

Bassam, N.E. ed. *Energy Plant Species: Their Use and Impact on Environment and Development.* London: James & James (Science Publishers) Ltd., 1998.

Grassi, G., Trebbi, G. and Pike. D.C. (eds). *Electricity from Biomass.* Newbury: CPL Press, 1992.

Hall, D., Grassi, G. and Scheer, H. (eds). *Biomass for Energy and Industry.* Bochum: Pont Press, 1994.

Climate Change

Bruce, James, et al (1996) *Climate Change 1995: Economic and Social Dimensions of Climate Change* Cambridge University Press, New York.

Canadian Global Change Program, 1997: *Cross Canada Briefings Before Kyoto* - Participants' Briefing Book, unpublished workshop briefing notes.

Canadian Global Change Program, no date, *Changes*, An Information Bulletin on Global Environmental Change, Issue No. 4, "Reducing Greenhouse Gas Emissions: The Additional Benefits", Canadian Global Change Program, Ottawa.

Canada, 1997 *Canada's Second National Report on Climate Change*, Supply and Services Canada, Ottawa.

Carpenter S., et al 1993 *Active Solar Heating in Canada to the Year 2010*, Efficiency and Alternative Energy Technology Branch, Energy, Mines and Resources Canada, Ottawa.

Environment Canada, 1992, *Canada's Greenhouse Gas Emissions: Estimates for 1990*, Report EPS 5/AP/4, Supply and Services Canada, Ottawa.

Greenhouse Gas Emission Reduction Trading Pilot (GERT), 1998, website, http://www.gert.org

Jaccard, M, Failing, L and Berry T. 1997, *Energy Policy*, "From Equipment to Infrastructure: Community Energy Management and Greenhouse Gas Emission Reduction", Vol. 25, No. 13.

Natural Resources Canada, 1997 *Electric Power in Canada 1996*, Supply and Services Canada, Ottawa.

Natural Resources Canada, 1997 *Canada's Energy Outlook: 1996-2020, 1996 Update*, Supply and Services Canada, Ottawa.

Passmore Associates International, 1996. *The Environmental and Economic Benefits of Wind energy Development in Canada: Lost Opportunity and Future Promise.*

Pembina Institute of Appropriate Development, David Suzuki Foundation, 1998, *Canadian Solutions*, from Pembina Institute of Appropriate Development web site, www.piad.ab.ca.

Rolfe, Chris (1998) *Turning Down the Heat*, West Coast Environmental Law Association, Vancouver.

Royal Society of Canada (1993) *Final Report of the COGGER Panel to the Canadian Global Change Program and the Canadian Climate Program Board*, Canadian Global Change Program (Report 93-1), Ottawa.

Sierra Club of Canada, 1998, *Rational Energy Program, Draft*, from Sierra Club of Canada website, www.sierraclub.ca.

ThermoShare, 1998, *Biomass Energy in Canada: with Emphasis on Electricity*, Report prepared for Natural Resources Canada.

TransAlta Utilities, 1994, *Pursuing Greenhouse Gas Offsets*, TransAlta Utilities, Calgary.

Homes

Bainbridge, David, Athena Steen and Bill Steen, *The Strawbale House.* White River Junction, VT: Chelsea Green Publishing, 1994. ISBN 930031-71-7

Bower, John. *The Healthy House*, Lyle Stuart Books, 1989

Carmody, John, Stephen Selkowitz and Lisa Herschong. *Residential Windows: A guide to New Technology and Energy Performance.* New York: W.W. Norton & Co., 1996.

Crosbie, Michael J.. *The Passive Solar Design & Construction Handbook*. New York: John Wiley, 1997. ISBN 0-471-18308-3

Easton, David. *The Rammed Earth House*. White River Junction, VT: Chelsea Green Publishing, 1996. ISBN 0-930031-79-2

Grady, Wayne. *Green Home: Planning and Building the Environmentally Advanced House*. Camden East: Camden House, 1993.

Johnson, Lorraine. *Green Future: How to Make a World of Difference*. Markham: Penguin, 1990.

Johnson, S.. *Greener Buildings: Environmental Impact of Property*. London: MacMillan Press, 1993.

Kachadorian, James. *The Passive Solar House – Using Solar Design to Heat and Cool Your Home*. White River Junction, VT: Chelsea Green Publishing, 1997.

Manning, Richard. *A Good House: Building a Life on the Land*. 1993.

Marshall, Brian and Argue, Robert *The Super-Insulated Retrofit Book*,, Toronto: REIC, 1981

Ministry of Natural Resources Canada. *Home Energy Retrofit in Canada: Overview & Opportunities*. Ottawa: Ministry of Supply & Services Canada, 1994. *Aussi disponible en francais.*

Myhrman, Matts and S.O. MacDonald. *Build it with Bales: Version Two*, Out on Bale, (un) Ltd., 1997. ISBN 0-9642821-1-9

Pearson, David. *The Natural House Book*, Toronto: Simon & Schuster, 1989.

Potts, M., *The Independent Home: Living Well with Power from the Sun, Wind and Water*. USA: Chelsea Green Publishing, 1993.

Roaf, S. and Hancock, M. *Energy Efficient Building*. Oxford: Blackwells, 1993.

Rocky Mountain Institute. *The Energy Efficient Home*. Boulder: Rocky Mountain Institute, 1994.

Stein, Richard G.. *Architecture and Energy* New York: Doubleday, 1978.

Thomas, Dirk., *Guide to Wood Heat*. Toronto: Camden House Publishing, 1992

Wells, Malcolm., *Gentle Architecture*. New York: McGraw-Hill, 1991.

Hydro and Tidal

Baker, A.C.. *Tidal Power*. Stevenage, U.K.: Institute of Electrical Engineers, 1991.

Harvey, A. and Brown, A. *Micro-Hydro Design Manual: A Guide to Small-Scale Water Power Schemes*. London: IT Publications, 1992.

Ontario Ministry of Energy. *Streams of Power: Developing Small Scale Hydro Systems*. Toronto: Renewable Energy in Canada, 1986.

Solar

Canada Mortgage and Housing Corporation. *Tap the Sun, passive solar techniques and home designs*, Ottawa: Ministry of Supply & Services, 1998. comes with CD-ROM.

Davidson, Joel. *The New Solar Electric Home: the photovoltaics how - to handbook.*, Ann Arbor: Aatec Publications, 1990.

Hestnes, Anne-Grete, S. Robert Hastings and Bjarne Saxhof, *Solar Energy Houses: Strategies, Technologies, Examples*. London: James & James (Science Publishers) Ltd.

Humm, O. and Toggweiler, P. (eds.) *Photovoltaics in Architecture*. Basel: Birkhauser, 1994.

Jeffery, Kevin. *Independent Energy Guide: Electrical Power for Home, Boat and RV*. White River Junction, VT: Chelsea Green Publishing, 1996. ISBN 0-9644112-0-2

Kirkby, Noel and Barbara *RVers' Guide to Solar Battery Charging*.. Ann Arbor: Aatec Publications, 1990.

Markvart, T. *Solar Electricity*. Chichester: John Wiley & Sons, 1994.

Norton, B. *Solar Energy Technology*. New York: Springer-Verlag, 1991.

Ross, Michael and Jimmy Royer, *Photovoltaics in Cold Climates*. London: James & James (Science Publishers) Ltd., 1998.

Rostvik, Harald N. *The Sunshine Revolution*. Stavanger: SUN-LAB Publishers, 1992.

Schaeffer, John and the Real Goods Staff. *Real Goods Solar Living Source Book* Ukiah, CA: Real Goods Trading Corp., 1994.

Scheer, Hermann. *A Solar Manifesto: The need for a total solar energy supply... and how to achieve it.* London: James & James (Science Publishers) Ltd., 1994.

Sick, Friedrich and Thomas Erge, *Photovoltaics in Buildings: A Design Handbook for Architects and Engineers.* London: James & James (Science Publishers) Ltd., 1996.

Sklar, Scott and Sheinkopf, Kenneth.. *Consumer's Guide to Solar Energy: Easy and Inexpensive Applications for Solar Energy.* 1991.

Strong, S.J.. *The Solar Electric House.* White River Junction, VT: Chelsea Green Publishing, 1993. ISBN 0-9637383-2-1

Treble, F.C. (ed.) *Generating Electricity from the Sun.* Oxford: Pergamon Press, 1992.

Zweibel, Ken. *Harnessing Solar Power: The Photovoltaics Challenge.* New York: Plenum Press, 1990.

Wind Power

Gipe, Paul. *Wind Power for Home & Business: Renewable Energy for the 1990s and Beyond.* White River Junction, VT: Chelsea Green Publishing 1993.

Hunter, R.. and Elliott, G. (eds.). *Wind-Diesel Systems.* Cambridge: Cambridge University Press, 1994.

Marier, Donald. *Wind Power for the Homeowner.* Emmaus, PA: Rodale Press, 1981.

Park, Jack. *The Wind Power Book.* Palo Alto, CA: Cheshire Books, 1981.

Swift-Hook, D.T. (ed.) *Wind Energy and the Environment, IEE Energy Series.* London: Institution of Electrical Engineers, 1989.

CHAPTER 10

Classified Listings

IN THIS CHAPTER you can look up the name of a Canadian Company active in renewable energy. Companies listed may be active in biomass, earth energy, microhydro, solar energy or wind energy. They include listings for dealers, distributors, manufacturers, consultants, designers, architects and installers.

Each company is listed according to their activity in applications, products and services.

Full company details are in the A-Z section immediately following this chapter.

Province codes used in this listing.

AB Alberta

BC British Columbia

MB Manitoba

NB New Brunswick

NF Newfoundland

NS Nova Scotia

NT Northwest Territories

ON Ontario

PE Prince Edward Island

QC Québec

SK Saskatchewan

YT Yukon Territory

Active Solar Applications

Air Conditioning

Jack's All Trades Inc. (ON)
Solar Utilites Ltd. (AB)
Thermomax Industries Ltd (BC)

Aquaculture

David Nixon & Sun (ON)
Enermodal Engineering Ltd. (ON)
Kelln Consulting Ltd. (SK)
March Solar (ON)
Sun Stream Solar (ON)
Taylor Munro Energy Systems (BC)
Thermo Dynamics Ltd (NS)
Thermomax Industries Ltd (BC)
Wenvor Technologies Inc. (ON)
Zerowatt Energy (BC)

Cooling Systems

Conserval Engineering (ON)
Jack's All Trades Inc. (ON)
Solar Utilites Ltd. (AB)
Solcan Ltd. (ON)
Taylor Munro Energy Systems (BC)
Thermomax Industries Ltd (BC)

Crop Drying

Conserval Engineering (ON)
Enermodal Engineering Ltd. (ON)
Kelln Consulting Ltd. (SK)
Matrix Energy Inc. (QC)
Solar Utilites Ltd. (AB)
Sun Stream Solar (ON)
Wenvor Technologies Inc. (ON)

Desalination

Solar Utilites Ltd. (AB)

District Heating

Jack's All Trades Inc. (ON)
Thermomax Industries Ltd (BC)

Domestic Hot Water

David Nixon & Sun (ON)
Echo Systeme (QC)
Enerconcept Inc. (QC)
Energy Alternatives Ltd. (BC)
Enermodal Engineering Ltd. (ON)
Jack's All Trades Inc. (ON)
Renewable Energy of Plum Hollow (ON)
Rocky Mountain Solar (CO)
Solar Systems (ON)
Solcan Ltd. (ON)
Sun Ross Solar Systems Ltd. (NS)
Sun Stream Solar (ON)
Sun Volts Unlimited Inc. (ON)
Sunpower Co. (ON)
Suntime Enterprises (NS)
Sustainable Home Solutions (QC)
Taylor Munro Energy Systems (BC)
The Electric Sun & SH Engineering (AB)
Thermo Dynamics Ltd (NS)
Thermomax Industries Ltd (BC)
Wenvor Technologies Inc. (ON)
Zerowatt Energy (BC)

Industrial Process Heat

Conserval Engineering (ON)
Enerconcept Inc. (QC)
Matrix Energy Inc. (QC)
Solar Utilites Ltd. (AB)
Solcan Ltd. (ON)
Taylor Munro Energy Systems (BC)
Thermo Dynamics Ltd (NS)
Thermomax Industries Ltd (BC)
Wenvor Technologies Inc. (ON)

Space Heating

Conserval Engineering (ON)
Echo Systeme (QC)
Enerconcept Inc. (QC)
Enermodal Engineering Ltd. (ON)
Jack's All Trades Inc. (ON)
Matrix Energy Inc. (QC)
Rocky Mountain Solar (CO)
Solcan Ltd. (ON)
Sun Ross Solar Systems Ltd. (NS)
Sustainable Home Solutions (QC)
Taylor Munro Energy Systems (BC)
The Electric Sun & SH Engineering (AB)
Thermo Dynamics Ltd (NS)
Thermomax Industries Ltd (BC)

Swimming Pool Heating

David Nixon & Sun (ON)
Echo Systeme (QC)
Enerconcept Inc. (QC)
Energy Alternatives Ltd. (BC)
Enermodal Engineering Ltd. (ON)
Jack's All Trades Inc. (ON)
Kelln Consulting Ltd. (SK)
March Solar (ON)
Matrix Energy Inc. (QC)
Renewable Energy of Plum Hollow (ON)
Rocky Mountain Solar (CO)
Solar Systems (ON)
Solcan Ltd. (ON)
Sun Ross Solar Systems Ltd. (NS)
Sun Stream Solar (ON)
Sun Volts Unlimited Inc. (ON)
Sunpower Co. (ON)
Suntime Enterprises (NS)
Sustainable Home Solutions (QC)
Taylor Munro Energy Systems (BC)
The Electric Sun & SH Engineering (AB)
Thermo Dynamics Ltd (NS)
Thermomax Industries Ltd (BC)
Zerowatt Energy (BC)

Active Solar Products

Absorber Surfaces

Allen Kani Associates (ON)
Conserval Engineering (ON)
Enerconcept Inc. (QC)
Northern Lights Energy Systems (ON)
Renewable Energy of Plum Hollow (ON)
Solar Utilites Ltd. (AB)
Solcan Ltd. (ON)
Suntime Enterprises (NS)
Taylor Munro Energy Systems (BC)
Thermo Dynamics Ltd (NS)

Antifreeze

David Nixon & Sun (ON)
Enerconcept Inc. (QC)
Jack's All Trades Inc. (ON)
Northern Lights Energy Systems (ON)
Rocky Mountain Solar (CO)
Solcan Ltd. (ON)
Sovran Energy Inc. (BC)
Sun Ross Solar Systems Ltd. (NS)
Sun Stream Solar (ON)
Sun Volts Unlimited Inc. (ON)
Suntime Enterprises (NS)
Taylor Munro Energy Systems (BC)
Thermo Dynamics Ltd (NS)
Enerfree Enterprises (MB)

Collectors - Air Heating

Conserval Engineering (ON)
Echo Systeme (QC)
Enerconcept Inc. (QC)
Jack's All Trades Inc. (ON)
Matrix Energy Inc. (QC)
Rocky Mountain Solar (CO)
Sovran Energy Inc. (BC)
Sun Ross Solar Systems Ltd. (NS)
Sun Stream Solar (ON)
The Electric Sun & SH Engineering (AB)
Wenvor Technologies Inc. (ON)

Collectors - Evacuated Tube

Echo Systeme (QC)
Energy Alternatives Ltd. (BC)
Frank's Alternate Energy (ON)
Rocky Mountain Solar (CO)
Solar Utilites Ltd. (AB)
Sovran Energy Inc. (BC)
Suntime Enterprises (NS)
Taylor Munro Energy Systems (BC)
Thermomax Industries Ltd (BC)

Collectors - Flat Plate

David Nixon & Sun (ON)
Echo Systeme (QC)
Enerconcept Inc. (QC)
Energy Alternatives Ltd. (BC)
Kelln Consulting Ltd. (SK)
Northern Lights Energy Systems (ON)
Renewable Energy of Plum Hollow (ON)
Rocky Mountain Solar (CO)
Solar Freedom International (SK)
Solar Utilites Ltd. (AB)
Solcan Ltd. (ON)
Sovran Energy Inc. (BC)
Sun Ross Solar Systems Ltd. (NS)
Sun Stream Solar (ON)
Sun Volts Unlimited Inc. (ON)
Suntime Enterprises (NS)
Taylor Munro Energy Systems (BC)
Thermo Dynamics Ltd (NS)
Wenvor Technologies Inc. (ON)

Collectors - Unglazed

Conserval Engineering (ON)
David Nixon & Sun (ON)
Enerconcept Inc. (QC)
Enerfree Enterprises (MB)
Energy Alternatives Ltd. (BC)
Kelln Consulting Ltd. (SK)
Matrix Energy Inc. (QC)
Renewable Energy of Plum Hollow (ON)
Solcan Ltd. (ON)
Sun Ross Solar Systems Ltd. (NS)
Sun Volts Unlimited Inc. (ON)
Taylor Munro Energy Systems (BC)
Thermo Dynamics Ltd (NS)

Controllers

David Nixon & Sun (ON)
Echo Systeme (QC)
Enerconcept Inc. (QC)
Energy Alternatives Ltd. (BC)
Jack's All Trades Inc. (ON)
Northern Lights Energy Systems (ON)
Rocky Mountain Solar (CO)
Solcan Ltd. (ON)
Sun Ross Solar Systems Ltd. (NS)
Sun Stream Solar (ON)
Sun Volts Unlimited Inc. (ON)
Suntime Enterprises (NS)
Taylor Munro Energy Systems (BC)
Thermo Dynamics Ltd (NS)
Thermomax Industries Ltd (BC)
Wenvor Technologies Inc. (ON)
Allen Kani Associates (ON)
Frank's Alternate Energy (ON)
Northern Lights Energy Systems (ON)
Solar Freedom International (SK)

Heat Exchangers

Echo Systeme (QC)
Enerconcept Inc. (QC)
Enerfree Enterprises (MB)
Energy Alternatives Ltd. (BC)
Rocky Mountain Solar (CO)
Solar Utilites Ltd. (AB)
Solcan Ltd. (ON)
Sun Ross Solar Systems Ltd. (NS)
Sun Stream Solar (ON)
Sun Volts Unlimited Inc. (ON)
Suntime Enterprises (NS)
Taylor Munro Energy Systems (BC)
Thermo Dynamics Ltd (NS)
Thermomax Industries Ltd (BC)
Wenvor Technologies Inc. (ON)

Heat Storage

Echo Systeme (QC)
Enerconcept Inc. (QC)
Enerfree Enterprises (MB)
Energy Alternatives Ltd. (BC)
Jack's All Trades Inc. (ON)
Matrix Energy Inc. (QC)
Renewable Energy of Plum Hollow (ON)
Rocky Mountain Solar (CO)
Solcan Ltd. (ON)
Sun Ross Solar Systems Ltd. (NS)
Sun Stream Solar (ON)
Sun Volts Unlimited Inc. (ON)
Taylor Munro Energy Systems (BC)
The Electric Sun & SH Engineering (AB)
Thermomax Industries Ltd (BC)
Wenvor Technologies Inc. (ON)

Pumps

David Nixon & Sun (ON)
Echo Systeme (QC)
Enerconcept Inc. (QC)
Energy Alternatives Ltd. (BC)
Jack's All Trades Inc. (ON)
Kelln Consulting Ltd. (SK)
Northern Lights Energy Systems (ON)
Rocky Mountain Solar (CO)
Solar Utilites Ltd. (AB)
Solcan Ltd. (ON)
Sun Ross Solar Systems Ltd. (NS)
Sun Stream Solar (ON)
Sun Volts Unlimited Inc. (ON)
Suntime Enterprises (NS)
Taylor Munro Energy Systems (BC)
The Electric Sun & SH Engineering (AB)
Thermo Dynamics Ltd (NS)
Wenvor Technologies Inc. (ON)

Storage Tanks

Echo Systeme (QC)
Enerconcept Inc. (QC)
Energy Alternatives Ltd. (BC)
Jack's All Trades Inc. (ON)
Northern Lights Energy Systems (ON)
Renewable Energy of Plum Hollow (ON)
Rocky Mountain Solar (CO)
Solcan Ltd. (ON)
Sun Ross Solar Systems Ltd. (NS)
Sun Stream Solar (ON)
Sun Volts Unlimited Inc. (ON)
Suntime Enterprises (NS)
Taylor Munro Energy Systems (BC)
The Electric Sun & SH Engineering (AB)
Thermomax Industries Ltd (BC)
Wenvor Technologies Inc. (ON)

Swimming Pool Heating

Sovran Energy Inc. (BC)

Valves

Echo Systeme (QC)
Enerconcept Inc. (QC)
Jack's All Trades Inc. (ON)
Solcan Ltd. (ON)
Sun Ross Solar Systems Ltd. (NS)
Taylor Munro Energy Systems (BC)
Wenvor Technologies Inc. (ON)

Bio-Mass Applications

Biothermica International Inc. (QC)

Biofuel petrol/diesel

IOGEN Corporation (ON)
Resource Efficient Agricultural Production (QC)

District Heat by Biogas

KMW Systems Inc. (ON)
Trigen Energy Canada Inc. (ON)

Energy from Waste Products

Biothermica International Inc. (QC)
Kemestrie inc. et CQVB (QC)
KMW Systems Inc. (ON)
Trigen Energy Canada Inc. (ON)

Ethanol Production

Kemestrie inc. et CQVB (QC)
Resource Efficient Agricultural Production (QC)

Gasification

Biothermica International Inc. (QC)
Kemestrie inc. et CQVB (QC)

Landfill Gas Utilization

Biothermica International Inc. (QC)

Wood Waste Utilization

Biothermica International Inc. (QC)
Kemestrie inc. et CQVB (QC)
KMW Systems Inc. (ON)
Resource Efficient Agricultural Production (QC)

Bio-Mass Products

Anaerobic Digesters

Allen Kani Associates (ON)

Boilers

Allen Kani Associates (ON)
KMW Systems Inc. (ON)
North Central Heating (BC)
Trigen Energy Canada Inc. (ON)

Combusters

KMW Systems Inc. (ON)

Cooking Stoves & Biomass Fired

Allen Kani Associates (ON)
Suntime Enterprises (NS)

Engines – Gas

Trigen Energy Canada Inc. (ON)

Feeding Equipment

KMW Systems Inc. (ON)

Gas Compressors

Buchanan & Hall Ltd. (ON)

Generators

Jack's All Trades Inc. (ON)

Heating

Echo Systeme (QC)
Suntime Enterprises (NS)

Pellets

Renewable Energy of Plum Hollow (ON)
Resource Efficient Agricultural Production (QC)
Suntime Enterprises (NS)

Storage Containers – Solid Biomass

Resource Efficient Agricultural Production (QC)

Turbines Gas

Buchanan & Hall Ltd. (ON)
North Central Heating (BC)
Trigen Energy Canada Inc. (ON)

Turbines Steam

Trigen Energy Canada Inc. (ON)

Wood Stoves

Allen Kani Associates (ON)
Compact Appliances Limited (NB)
Jack's All Trades Inc. (ON)
North Central Heating (BC)
Renewable Energy of Plum Hollow (ON)
Sunpower Co. (ON)

Earth Energy Applications

Cooling

Allen Kani Associates (ON)
Boonstra Union Energy (ON)
Buchanan & Hall Ltd. (ON)
Canada Heat Pumps (ON)
DNS Geothermal (MB)
Gravenhurst Plumbing & Electrical Company Ltd. (ON)
Jack's All Trades Inc. (ON)
Lockhart Industries (BC)
North Central Heating (BC)
Oshawa Refrigeration Service Ltd. (ON)
Paradise Plumbing & Heating (MB)
Reed & Schaab Plumbing & Heating (ON)
Sunnyside Furnaces Ltd. (ON)
Sustainable Home Solutions (QC)
Trigen Energy Canada Inc. (ON)

Heating

Allen Kani Associates (ON)
Boonstra Union Energy (ON)
Buchanan & Hall Ltd. (ON)
Canada Heat Pumps (ON)
DNS Geothermal (MB)
Gravenhurst Plumbing & Electrical Company Ltd. (ON)
Jack's All Trades Inc. (ON)
Lockhart Industries (BC)
North Central Heating (BC)
Oshawa Refrigeration Service Ltd. (ON)
Paradise Plumbing & Heating (MB)
Reed & Schaab Plumbing & Heating (ON)
Sunnyside Furnaces Ltd. (ON)
Sustainable Home Solutions (QC)

Thermal Storage

Trigen Energy Canada Inc. (ON)

Power Generation

Acres International (ON)
ICP Global Technologies (QC)
Jack's All Trades Inc. (ON)
R. Nair & Associates (QC)
Solar Plus (BC)
Solar Power Systems (ON)
Statpower Technologies Corporation (BC)
Westrock Battery Ltd. (Solar Division) (MB)
Trigen Energy Canada Inc. (ON)

Earth Energy Products

Heat Exchanger & Heat Pump – Forced Air Closed Loop

Boonstra Union Energy (ON)
Buchanan & Hall Ltd. (ON)
Canada Heat Pumps (ON)
Dectron Internationale (QC)
DNS Geothermal (MB)
Gravenhurst Plumbing & Electrical Company Ltd. (ON)
Jack's All Trades Inc. (ON)
Lockhart Industries (BC)
Longhill Energy Products (Eastern) Ltd (ON)
North Central Heating (BC)
Oshawa Refrigeration Service Ltd. (ON)
Paradise Plumbing & Heating (MB)
Polar Refrigeration Sales (BC)
Reed & Schaab Plumbing & Heating (ON)
Shute Plumbing & Heating Ltd. (ON)
Sunnyside Furnaces Ltd. (ON)

Heat Exchanger & Heat Pump – Forced Air Open Loop

Boonstra Union Energy (ON)
Buchanan & Hall Ltd. (ON)
Canada Heat Pumps (ON)
Gravenhurst Plumbing & Electrical Company Ltd. (ON)
Jack's All Trades Inc. (ON)
Lockhart Industries (BC)
Longhill Energy Products (Eastern) Ltd (ON)
North Central Heating (BC)
Oshawa Refrigeration Service Ltd. (ON)
Paradise Plumbing & Heating (MB)
Reed & Schaab Plumbing & Heating (ON)
Shute Plumbing & Heating Ltd. (ON)
Sunnyside Furnaces Ltd. (ON)
Trigen Energy Canada Inc. (ON)

Heat Exchanger & Heat Pump - Water to Water Closed Loop

Boonstra Union Energy (ON)
Buchanan & Hall Ltd. (ON)
Canada Heat Pumps (ON)
DNS Geothermal (MB)
Gravenhurst Plumbing & Electrical Company Ltd. (ON)
Jack's All Trades Inc. (ON)
Lockhart Industries (BC)
Longhill Energy Products (Eastern) Ltd (ON)
North Central Heating (BC)
Oshawa Refrigeration Service Ltd. (ON)
Paradise Plumbing & Heating (MB)
Polar Refrigeration Sales (BC)
Shute Plumbing & Heating Ltd. (ON)
Sunnyside Furnaces Ltd. (ON)
Trigen Energy Canada Inc. (ON)

Heat Exchanger & Heat Pump – Water to Water Open Loop

Boonstra Union Energy (ON)
Buchanan & Hall Ltd. (ON)
Canada Heat Pumps (ON)
Gravenhurst Plumbing & Electrical Company Ltd. (ON)
Jack's All Trades Inc. (ON)
Lockhart Industries (BC)
Longhill Energy Products (Eastern) Ltd (ON)
North Central Heating (BC)
Oshawa Refrigeration Service Ltd. (ON)
Paradise Plumbing & Heating (MB)
Shute Plumbing & Heating Ltd. (ON)
Sunnyside Furnaces Ltd. (ON)

Thermostats

Boonstra Union Energy (ON)
Buchanan & Hall Ltd. (ON)
Jack's All Trades Inc. (ON)
Longhill Energy Products (Eastern) Ltd (ON)
North Central Heating (BC)
Oshawa Refrigeration Service Ltd. (ON)
Reed & Schaab Plumbing & Heating (ON)
Sunnyside Furnaces Ltd. (ON)

General Services

Architects

Autonomous Sustainable Housing Inc. (AB)
Echo Systeme (QC)
Northeast Sustainable Energy Association (MA)
Richard Kadulski Architect (BC)
Solar Utilites Ltd. (AB)
Stephen F. Pope Architect (ON)

Associations

Canadian Solar Industries Association (ON)
Earth Energy Society of Canada (ON)
Northeast Sustainable Energy Association (MA)
Solar Utilites Ltd. (AB)

Certification

Bodycote - ORTECH Inc. (ON)
Howell-Mayhew Engineering (AB)
Solar Plus (BC)
Solar Utilites Ltd. (AB)
Westrock Battery Ltd. (Solar Division) (MB)

Consultancy

Acres International (ON)
Adkinson & Associates (ON)
Allen Kani Associates (ON)
Autonomous Sustainable Housing Inc. (AB)
Biothermica International Inc. (QC)
Bodycote - ORTECH Inc. (ON)
CIMAT Power Systems Inc. (QC)
Conserval Engineering (ON)
Cousins Consulting Services (AB)
Energy Alternatives Ltd. (BC)
Enermodal Engineering Ltd. (ON)
Enersite International Inc. (ON)
Howell-Mayhew Engineering (AB)
Kelln Consulting Ltd. (SK)
Lapp-Hancock Associates Limited (ON)
March Solar (ON)
Midnight Sun Energy Ltd. (NT)
NewSun Technologies, Ltd. (ON)
R. Nair & Associates (QC)
Resource Efficient Agricultural Production (QC)
Richard Kadulski Architect (BC)
Rocky Mountain Solar (CO)
Rozon Batteries Inc. (QC)
Solar Freedom International (SK)
Solar Plus (BC)
Solar Systems (ON)
Solar Utilites Ltd. (AB)
Solener Inc. (QC)
Soltek Solar Energy Ltd (BC)
Sovran Energy Inc. (BC)
Stephen F. Pope Architect (ON)
Sustainable Home Solutions (QC)
Thermomax Industries Ltd (BC)
TN Conseil (QC)
Wenvor Technologies Inc. (ON)
Westrock Battery Ltd. (Solar Division) (MB)
Zephyr North (ON)
Zerowatt Energy (BC)

Contract R & D

Acres International (ON)
Adkinson & Associates (ON)
Allen Kani Associates (ON)
Autonomous Sustainable Housing Inc. (AB)
Bodycote - ORTECH Inc. (ON)
CANMET Energy Diversification Research Laboratory (QC)
CIMAT Power Systems Inc. (QC)
Conserval Engineering (ON)
Echo Systeme (QC)
Enermodal Engineering Ltd. (ON)
GPCo Inc. (QC)
Howell-Mayhew Engineering (AB)
Kemestrie inc. et CQVB (QC)
Lapp-Hancock Associates Limited (ON)
NewSun Technologies, Ltd. (ON)
Northern Communication & Navigation Systems Ltd. (NT)
Phantom Electronics Inc (ON)
R. Nair & Associates (QC)
Resource Efficient Agricultural Production (QC)
Rozon Batteries Inc. (QC)
Solar Utilites Ltd. (AB)
Soltek Solar Energy Ltd (BC)
Sovran Energy Inc. (BC)
TN Conseil (QC)
Westrock Battery Ltd. (Solar Division) (MB)
Zephyr North (ON)

Data Analysis

Acres International (ON)
Bodycote - ORTECH Inc. (ON)
Cousins Consulting Services (AB)
DNS Geothermal (MB)
Enermodal Engineering Ltd. (ON)
GPCo Inc. (QC)
Howell-Mayhew Engineering (AB)
Lapp-Hancock Associates Limited (ON)
Resource Efficient Agricultural Production (QC)
Solar Plus (BC)
Solar Utilites Ltd. (AB)
Soltek Solar Energy Ltd (BC)
Sovran Energy Inc. (BC)
Sustainable Home Solutions (QC)
TN Conseil (QC)
Wenvor Technologies Inc. (ON)
Zephyr North (ON)

Datalogging

Acres International (ON)
Adkinson & Associates (ON)
Bodycote - ORTECH Inc. (ON)
CIMAT Power Systems Inc. (QC)
Echo Systeme (QC)
Enermodal Engineering Ltd. (ON)
Geneq Inc. (QC)
GPCo Inc. (QC)
Howell-Mayhew Engineering (AB)
Solar Utilites Ltd. (AB)
Soltek Solar Energy Ltd (BC)
Wenvor Technologies Inc. (ON)
West Kootenay Power Ltd. (BC)
Zephyr North (ON)

Design

Acres International (ON)
Adkinson & Associates (ON)
Allen Kani Associates (ON)
Autonomous Sustainable Housing Inc. (AB)
Biothermica International Inc. (QC)
CIMAT Power Systems Inc. (QC)
David Nixon & Sun (ON)
DNS Geothermal (MB)
Echo Systeme (QC)
Enerdata Systems Inc. (BC)
Enerfree Enterprises (MB)
Energy Alternatives Ltd. (BC)
Enermodal Engineering Ltd. (ON)
Enviro Store Ltd The (AB)
Frank's Alternate Energy (ON)
Jack's All Trades Inc. (ON)
Kemestrie inc. et CQVB (QC)
Lenbrook Energy Systems (ON)
March Solar (ON)
Matrix Energy Inc. (QC)
Midnight Sun Energy Ltd. (NT)
Northeast Sustainable Energy Association (MA)
Northern Communication & Navigation Systems Ltd. (NT)
Northern Lights Energy Systems (ON)
Phantom Electronics Inc (ON)
Richard Kadulski Architect (BC)
Rocky Mountain Solar (CO)
Solar Freedom International (SK)
Solar Plus (BC)
Solar Power Systems (ON)
Solar Systems (ON)
Solar Utilites Ltd. (AB)
Solener Inc. (QC)
Soltek Solar Energy Ltd (BC)
Sovran Energy Inc. (BC)
Statpower Technologies Corporation (BC)
Stephen F. Pope Architect (ON)
Sun Stream Solar (ON)
Sunpower Co. (ON)
Sustainable Home Solutions (QC)
The Electric Sun & SH Engineering (AB)
Thermomax Industries Ltd (BC)
TN Conseil (QC)
Trigen Energy Canada Inc. (ON)
Wenvor Technologies Inc. (ON)
Westrock Battery Ltd. (Solar Division) (MB)
Zephyr North (ON)
Zerowatt Energy (BC)

Designers

Acres International (ON)
Autonomous Sustainable Housing Inc. (AB)
Energy Alternatives Ltd. (BC)
Kelln Consulting Ltd. (SK)
Phantom Electronics Inc (ON)
Prometheus Energy (ON)
Richard Kadulski Architect (BC)
Solar Systems (ON)
Solar Utilites Ltd. (AB)
Solener Inc. (QC)
Sovran Energy Inc. (BC)
Wenvor Technologies Inc. (ON)
Westrock Battery Ltd. (Solar Division) (MB)

Distributor

Argus Telecom International Inc. (QC)
David Nixon & Sun (ON)
Echo Systeme (QC)
Energy Alternatives Ltd. (BC)
Enviro Store Ltd The (AB)
Geneq Inc. (QC)
Lenbrook Energy Systems (ON)
March Solar (ON)
Matrix Energy Inc. (QC)
Midnight Sun Energy Ltd. (NT)
Prometheus Energy (ON)
Rozon Batteries Inc. (QC)
Solar Freedom International (SK)
Solar Plus (BC)
Solar Utilites Ltd. (AB)
Soltek Solar Energy Ltd (BC)
Sovran Energy Inc. (BC)
Sun Ross Solar Systems Ltd. (NS)
Suncurrent Industries (AB)
Sunpower Co. (ON)
Thermomax Industries Ltd (BC)
Trigen Energy Canada Inc. (ON)
Wenvor Technologies Inc. (ON)
Westrock Battery Ltd. (Solar Division) (MB)
Zerowatt Energy (BC)

Documentation

Acres International (ON)
Howell-Mayhew Engineering (AB)
Rozon Batteries Inc. (QC)
Solar Plus (BC)
Solar Utilites Ltd. (AB)
Statpower Technologies Corporation (BC)
The Electric Sun & SH Engineering (AB)
Westrock Battery Ltd. (Solar Division) (MB)

Environmental Impact Assessments

Acres International (ON)
Enviro Store Ltd The (AB)
Lapp-Hancock Associates Limited (ON)
Resource Efficient Agricultural Production (QC)
Solar Utilites Ltd. (AB)
Solener Inc. (QC)
TN Conseil (QC)
Zephyr North (ON)

Feasibility Studies

Acres International (ON)
Adkinson & Associates (ON)
Allen Kani Associates (ON)
Autonomous Sustainable Housing Inc. (AB)
Biothermica International Inc. (QC)
Cousins Consulting Services (AB)
DNS Geothermal (MB)
Echo Systeme (QC)
Enerdata Systems Inc. (BC)
Energy Alternatives Ltd. (BC)
Enermodal Engineering Ltd. (ON)
GPCo Inc. (QC)
Kemestrie inc. et CQVB (QC)
Lapp-Hancock Associates Limited (ON)
Matrix Energy Inc. (QC)
NewSun Technologies, Ltd. (ON)
Resource Efficient Agricultural Production (QC)
Solar Plus (BC)
Solar Utilites Ltd. (AB)
Solener Inc. (QC)
Soltek Solar Energy Ltd (BC)
Sovran Energy Inc. (BC)
Stephen F. Pope Architect (ON)
Sustainable Home Solutions (QC)
The Electric Sun & SH Engineering (AB)
TN Conseil (QC)
Trigen Energy Canada Inc. (ON)
Wenvor Technologies Inc. (ON)
West Kootenay Power Ltd. (BC)
Westrock Battery Ltd. (Solar Division) (MB)
Zephyr North (ON)

Financial Advice

Acres International (ON)
DNS Geothermal (MB)
Enermodal Engineering Ltd. (ON)
GPCo Inc. (QC)
Resource Efficient Agricultural Production (QC)
Rocky Mountain Solar (CO)
Solar Plus (BC)
Soltek Solar Energy Ltd (BC)
Sovran Energy Inc. (BC)
Sustainable Home Solutions (QC)
The Electric Sun & SH Engineering (AB)
TN Conseil (QC)
West Kootenay Power Ltd. (BC)
Westrock Battery Ltd. (Solar Division) (MB)

Information

Acres International (ON)
Canadian Solar Industries Association (ON)
Citizens for Renewable Energy (ON)
Conserval Engineering (ON)
DNS Geothermal (MB)
Earth Energy Society of Canada (ON)
Echo Systeme (QC)
Enerfree Enterprises (MB)
Enermodal Engineering Ltd. (ON)
Enviro Store Ltd The (AB)
Frank's Alternate Energy (ON)
Howell-Mayhew Engineering (AB)
Jack's All Trades Inc. (ON)
Lenbrook Energy Systems (ON)
March Solar (ON)
Matrix Energy Inc. (QC)

Midnight Sun Energy Ltd. (NT)
Northern Communication & Navigation Systems Ltd. (NT)
Rozon Batteries Inc. (QC)
Shute Plumbing & Heating Ltd. (ON)
Solar Plus (BC)
Solar Power Systems (ON)
Solar Utilites Ltd. (AB)
Soltek Solar Energy Ltd (BC)
Sovran Energy Inc. (BC)
Statpower Technologies Corporation (BC)
Sun Ross Solar Systems Ltd. (NS)
Sunpower Co. (ON)
Sustainable Home Solutions (QC)
The Electric Sun & SH Engineering (AB)
Trigen Energy Canada Inc. (ON)
Wenvor Technologies Inc. (ON)
West Kootenay Power Ltd. (BC)
Westrock Battery Ltd. (Solar Division) (MB)

Installation

Argus Telecom International Inc. (QC)
Boonstra Union Energy (ON)
Conserval Engineering (ON)
David Nixon & Sun (ON)
DNS Geothermal (MB)
Echo Systeme (QC)
Enerfree Enterprises (MB)
Energy Alternatives Ltd. (BC)
Enviro Store Ltd The (AB)
Frank's Alternate Energy (ON)
Jack's All Trades Inc. (ON)
March Solar (ON)
Matrix Energy Inc. (QC)
Midnight Sun Energy Ltd. (NT)
North Central Heating (BC)
Northern Communication & Navigation Systems Ltd. (NT)
Phantom Electronics Inc (ON)
Prometheus Energy (ON)
Shute Plumbing & Heating Ltd. (ON)
Solar Plus (BC)
Solar Power Systems (ON)
Solar Utilites Ltd. (AB)
Soltek Solar Energy Ltd (BC)
Sovran Energy Inc. (BC)
Sun Ross Solar Systems Ltd. (NS)
Sun Stream Solar (ON)
Sunpower Co. (ON)
Suntime Enterprises (NS)
The Electric Sun & SH Engineering (AB)
Wenvor Technologies Inc. (ON)
Westrock Battery Ltd. (Solar Division) (MB)
Zerowatt Energy (BC)

Maintenance

Acres International (ON)
Argus Telecom International Inc. (QC)
Boonstra Union Energy (ON)
David Nixon & Sun (ON)
DNS Geothermal (MB)
Echo Systeme (QC)
Frank's Alternate Energy (ON)
Jack's All Trades Inc. (ON)
March Solar (ON)
Midnight Sun Energy Ltd. (NT)
North Central Heating (BC)
Northern Lights Energy Systems (ON)
Prometheus Energy (ON)
Shute Plumbing & Heating Ltd. (ON)
Solar Plus (BC)
Solar Power Systems (ON)
Solar Systems (ON)
Soltek Solar Energy Ltd (BC)
Sovran Energy Inc. (BC)
Sun Ross Solar Systems Ltd. (NS)
Sunpower Co. (ON)
Suntime Enterprises (NS)
The Electric Sun & SH Engineering (AB)
Trigen Energy Canada Inc. (ON)
Wenvor Technologies Inc. (ON)
Westrock Battery Ltd. (Solar Division) (MB)

Monitoring

Acres International (ON)
Adkinson & Associates (ON)
Argus Telecom International Inc. (QC)
Bodycote - ORTECH Inc. (ON)
CIMAT Power Systems Inc. (QC)
Echo Systeme (QC)
Enermodal Engineering Ltd. (ON)
Frank's Alternate Energy (ON)
Geneq Inc. (QC)
Howell-Mayhew Engineering (AB)
Jack's All Trades Inc. (ON)
March Solar (ON)
Northern Lights Energy Systems (ON)
Phantom Electronics Inc (ON)
Solar Plus (BC)
Solar Systems (ON)
Solar Utilites Ltd. (AB)
Soltek Solar Energy Ltd (BC)
Sovran Energy Inc. (BC)
Sun Ross Solar Systems Ltd. (NS)
The Electric Sun & SH Engineering (AB)
TN Conseil (QC)
Trigen Energy Canada Inc. (ON)
Wenvor Technologies Inc. (ON)
West Kootenay Power Ltd. (BC)
Westrock Battery Ltd. (Solar Division) (MB)
Zephyr North (ON)

Planning Permission

Acres International (ON)
GPCo Inc. (QC)
Kemestrie inc. et CQVB (QC)
Solar Plus (BC)
Stephen F. Pope Architect (ON)
The Electric Sun & SH Engineering (AB)
Westrock Battery Ltd. (Solar Division) (MB)

Product Sourcing

Acres International (ON)
Adkinson & Associates (ON)
Argus Telecom International Inc. (QC)
Energy Alternatives Ltd. (BC)
Enviro Store Ltd The (AB)
Frank's Alternate Energy (ON)
Howell-Mayhew Engineering (AB)
Jack's All Trades Inc. (ON)
March Solar (ON)
Matrix Energy Inc. (QC)
Prometheus Energy (ON)
Renewable Energy of Plum Hollow (ON)
Solar Plus (BC)
Solar Power Systems (ON)
Solar Utilites Ltd. (AB)
Soltek Solar Energy Ltd (BC)
Sovran Energy Inc. (BC)
Sustainable Home Solutions (QC)
The Electric Sun & SH Engineering (AB)
Wenvor Technologies Inc. (ON)
Westrock Battery Ltd. (Solar Division) (MB)

Project Management

Acres International (ON)
Argus Telecom International Inc. (QC)
Autonomous Sustainable Housing Inc. (AB)
Biothermica International Inc. (QC)
CIMAT Power Systems Inc. (QC)
DNS Geothermal (MB)
Echo Systeme (QC)
Enerdata Systems Inc. (BC)
Energy Alternatives Ltd. (BC)
Enviro Store Ltd The (AB)
GPCo Inc. (QC)
Howell-Mayhew Engineering (AB)
Jack's All Trades Inc. (ON)
Kemestrie inc. et CQVB (QC)
Lapp-Hancock Associates Limited (ON)
March Solar (ON)
Matrix Energy Inc. (QC)
NewSun Technologies, Ltd. (ON)
Prometheus Energy (ON)
Renewable Energy of Plum Hollow (ON)
Resource Efficient Agricultural Production (QC)
Rozon Batteries Inc. (QC)
Solar Plus (BC)
Solar Utilites Ltd. (AB)
Soltek Solar Energy Ltd (BC)
Sovran Energy Inc. (BC)
Statpower Technologies Corporation (BC)
Suncurrent Industries (AB)
Suntime Enterprises (NS)
Sustainable Home Solutions (QC)
The Electric Sun & SH Engineering (AB)
TN Conseil (QC)
Wenvor Technologies Inc. (ON)
Westrock Battery Ltd. (Solar Division) (MB)

Repair

Acres International (ON)
Argus Telecom International Inc. (QC)
David Nixon & Sun (ON)
DNS Geothermal (MB)
Echo Systeme (QC)
Enviro Store Ltd The (AB)
Frank's Alternate Energy (ON)
Jack's All Trades Inc. (ON)
Northern Lights Energy Systems (ON)
Shute Plumbing & Heating Ltd. (ON)
Solar Plus (BC)
Solar Power Systems (ON)
Solar Systems (ON)
Soltek Solar Energy Ltd (BC)
Statpower Technologies Corporation (BC)
Sun Ross Solar Systems Ltd. (NS)
Sunpower Co. (ON)
Suntime Enterprises (NS)
Wenvor Technologies Inc. (ON)
Westrock Battery Ltd. (Solar Division) (MB)

Site Evaluation

Acres International (ON)
Argus Telecom International Inc. (QC)
Autonomous Sustainable Housing Inc. (AB)
Biothermica International Inc. (QC)
CIMAT Power Systems Inc. (QC)
DNS Geothermal (MB)
Echo Systeme (QC)
Energy Alternatives Ltd. (BC)
Enermodal Engineering Ltd. (ON)
Enviro Store Ltd The (AB)
Frank's Alternate Energy (ON)
GPCo Inc. (QC)
Howell-Mayhew Engineering (AB)
March Solar (ON)
Midnight Sun Energy Ltd. (NT)
Northern Lights Energy Systems (ON)
Prometheus Energy (ON)
Richard Kadulski Architect (BC)
Solar Plus (BC)
Solar Power Systems (ON)
Solar Systems (ON)
Solar Utilites Ltd. (AB)
Solener Inc. (QC)
Soltek Solar Energy Ltd (BC)
Stephen F. Pope Architect (ON)
Sun Ross Solar Systems Ltd. (NS)
Sunpower Co. (ON)
Suntime Enterprises (NS)
Sustainable Home Solutions (QC)
Wenvor Technologies Inc. (ON)
Westrock Battery Ltd. (Solar Division) (MB)
Zephyr North (ON)
Zerowatt Energy (BC)

Software

Acres International (ON)
Argus Telecom International Inc. (QC)
Echo Systeme (QC)
Enermodal Engineering Ltd. (ON)
GPCo Inc. (QC)
Howell-Mayhew Engineering (AB)
Numerical Logics Inc (ON)
Solar Plus (BC)
Solar Utilites Ltd. (AB)
Sustainable Home Solutions (QC)
Wenvor Technologies Inc. (ON)
Zephyr North (ON)

System Design

Acres International (ON)
Echo Systeme (QC)
Solar Plus (BC)
Sustainable Home Solutions (QC)
Wenvor Technologies Inc. (ON)

Testing

Acres International (ON)
Argus Telecom International Inc. (QC)
Bodycote - ORTECH Inc. (ON)
CANMET Energy Diversification Research Laboratory (QC)
Echo Systeme (QC)
Enermodal Engineering Ltd. (ON)
Frank's Alternate Energy (ON)
Howell-Mayhew Engineering (AB)
Solar Plus (BC)
Solar Utilites Ltd. (AB)
Sovran Energy Inc. (BC)
The Electric Sun & SH Engineering (AB)
Wenvor Technologies Inc. (ON)
Westrock Battery Ltd. (Solar Division) (MB)

Training

Acres International (ON)
Argus Telecom International Inc. (QC)
Autonomous Sustainable Housing Inc. (AB)
Citizens for Renewable Energy (ON)
David Nixon & Sun (ON)
Echo Systeme (QC)
Energy Alternatives Ltd. (BC)
Enermodal Engineering Ltd. (ON)
Enersite International Inc. (ON)
Lenbrook Energy Systems (ON)
NewSun Technologies, Ltd. (ON)
Rozon Batteries Inc. (QC)
Solar Plus (BC)
Solar Utilites Ltd. (AB)
Solener Inc. (QC)
Sovran Energy Inc. (BC)
Sunpower Co. (ON)
Thermomax Industries Ltd (BC)
TN Conseil (QC)
Wenvor Technologies Inc. (ON)
Westrock Battery Ltd. (Solar Division) (MB)

Upgrading

Acres International (ON)
Argus Telecom International Inc. (QC)
March Solar (ON)
Solar Plus (BC)
Solar Utilites Ltd. (AB)
Solener Inc. (QC)
Sunpower Co. (ON)
Trigen Energy Canada Inc. (ON)
Wenvor Technologies Inc. (ON)
Westrock Battery Ltd. (Solar Division) (MB)

Wind Farm Development

Allen Kani Associates (ON)
Argus Telecom International Inc. (QC)
Echo Systeme (QC)
GPCo Inc. (QC)
Solar Utilites Ltd. (AB)
Wenvor Technologies Inc. (ON)

Hybrid Systems

Hybrid Systems - Solar/Biomass

Allen Kani Associates (ON)
Argus Telecom International Inc. (QC)
Jack's All Trades Inc. (ON)
Solar Power Systems (ON)
Solar Utilites Ltd. (AB)
Sunpower Co. (ON)
Suntime Enterprises (NS)
Sustainable Home Solutions (QC)

Hybrid Systems - Solar/Diesel

Argus Telecom International Inc. (QC)
CIMAT Power Systems Inc. (QC)
Enerfree Enterprises (MB)
Energy Alternatives Ltd. (BC)
Frank's Alternate Energy (ON)
Kelln Consulting Ltd. (SK)
Lenbrook Energy Systems (ON)
March Solar (ON)
Matrix Energy Inc. (QC)
Midnight Sun Energy Ltd. (NT)
Nith Valley Solar (ON)
Northern Communication & Navigation Systems Ltd. (NT)
Phantom Electronics Inc (ON)
Rocky Mountain Solar (CO)
Solar Plus (BC)
Solar Power Systems (ON)
Solar Utilites Ltd. (AB)
Solener Inc. (QC)
Sovran Energy Inc. (BC)
Sun Stream Solar (ON)
Sun Volts Unlimited Inc. (ON)
Sunpower Co. (ON)
The Electric Sun & SH Engineering (AB)
Westrock Battery Ltd. (Solar Division) (MB)

Hybrid Systems - Solar/Other

CIMAT Power Systems Inc. (QC)
Conserval Engineering (ON)
Echo Systeme (QC)
Jack's All Trades Inc. (ON)
Lenbrook Energy Systems (ON)
Light Solar Wind Mfg. (ON)
March Solar (ON)
Matrix Energy Inc. (QC)
Nith Valley Solar (ON)
Northern Lights Energy Systems (ON)
Powersource Energy Systems Inc. (BC)
Rocky Mountain Solar (CO)
Solar Plus (BC)
Solar Power Systems (ON)
Solar Utilites Ltd. (AB)
Sun Ross Solar Systems Ltd. (NS)
Sun Stream Solar (ON)
Sun Volts Unlimited Inc. (ON)
Sunpower Co. (ON)
Wenvor Technologies Inc. (ON)
Westrock Battery Ltd. (Solar Division) (MB)

Hybrid Systems - Solar/Wind

Allen Kani Associates (ON)
Argus Telecom International Inc. (QC)
CIMAT Power Systems Inc. (QC)
David Nixon & Sun (ON)
Echo Systeme (QC)
Enerfree Enterprises (MB)
Energy Alternatives Ltd. (BC)
Enviro Store Ltd The (AB)
Frank's Alternate Energy (ON)
Golden Genesis Co. Ltd. (AB)
Jack's All Trades Inc. (ON)
Kelln Consulting Ltd. (SK)
Lenbrook Energy Systems (ON)
Light Solar Wind Mfg. (ON)
March Solar (ON)
Matrix Energy Inc. (QC)
Midnight Sun Energy Ltd. (NT)
Nith Valley Solar (ON)
Nor'wester Energy Systems (AB)
Northern Communication & Navigation Systems Ltd. (NT)
Northern Lights Energy Systems (ON)
Powersource Energy Systems Inc. (BC)
Prometheus Energy (ON)
Renewable Energy of Plum Hollow (ON)
Rocky Mountain Solar (CO)
Rozon Batteries Inc. (QC)
Sambrabec Inc. (QC)
Solar Plus (BC)
Solar Power Systems (ON)
Solar Utilites Ltd. (AB)
Solener Inc. (QC)
Soltek Solar Energy Ltd (BC)
Sovran Energy Inc. (BC)
Sun Stream Solar (ON)
Sun Volts Unlimited Inc. (ON)
Sunpower Co. (ON)
Sustainable Home Solutions (QC)
The Electric Sun & SH Engineering (AB)
Trillium Windmills Inc. (ON)
Wenvor Technologies Inc. (ON)
Westrock Battery Ltd. (Solar Division) (MB)

Hybrid Systems - Wind/Biomass

Enviro Store Ltd The (AB)
Jack's All Trades Inc. (ON)
March Solar (ON)
Rocky Mountain Solar (CO)
Solar Utilites Ltd. (AB)
Sustainable Home Solutions (QC)

Hybrid Systems - Wind/Diesel

CIMAT Power Systems Inc. (QC)
Enviro Store Ltd The (AB)
GPCo Inc. (QC)
Lenbrook Energy Systems (ON)
Matrix Energy Inc. (QC)
Midnight Sun Energy Ltd. (NT)
Nith Valley Solar (ON)
Nor'wester Energy Systems (AB)
Prometheus Energy (ON)
Rocky Mountain Solar (CO)
Sambrabec Inc. (QC)
Solar Plus (BC)
Solar Utilites Ltd. (AB)
Sun Stream Solar (ON)
Suncurrent Industries (AB)
Wenvor Technologies Inc. (ON)

Hybrid Systems - Wind/Other

Allen Kani Associates (ON)
Argus Telecom International Inc. (QC)
Energy Alternatives Ltd. (BC)
Enviro Store Ltd The (AB)
GPCo Inc. (QC)
Lenbrook Energy Systems (ON)
Light Solar Wind Mfg. (ON)
March Solar (ON)
Nith Valley Solar (ON)
Rocky Mountain Solar (CO)
Solar Plus (BC)
Solar Utilites Ltd. (AB)
Sunpower Co. (ON)

Micro Hydro Applications

Battery Based Systems

Argus Telecom International Inc. (QC)
Echo Systeme (QC)
Energy Alternatives Ltd. (BC)
Frank's Alternate Energy (ON)
Light Solar Wind Mfg. (ON)
March Solar (ON)
Matrix Energy Inc. (QC)
Nor'wester Energy Systems (AB)
Prometheus Energy (ON)
R. Nair & Associates (QC)
Rocky Mountain Solar (CO)
Solar Power Systems (ON)
Sun Stream Solar (ON)
Sun Volts Unlimited Inc. (ON)
Sunpower Co. (ON)
Westrock Battery Ltd. (Solar Division) (MB)
Zerowatt Energy (BC)

DC Only Systems

Echo Systeme (QC)
Energy Alternatives Ltd. (BC)
Frank's Alternate Energy (ON)
Light Solar Wind Mfg. (ON)
March Solar (ON)
Matrix Energy Inc. (QC)
Nor'wester Energy Systems (AB)
Prometheus Energy (ON)
Rocky Mountain Solar (CO)
Solar Power Systems (ON)
Sun Stream Solar (ON)
Sun Volts Unlimited Inc. (ON)
Sunpower Co. (ON)
Westrock Battery Ltd. (Solar Division) (MB)
Zerowatt Energy (BC)

Stand Alone Systems

R. Nair & Associates (QC)

Micro Hydro Products

Northern Lights Energy Systems (ON)
Solar Plus (BC)
Wenvor Technologies Inc. (ON)

Diversions/Intakes

Acres International (ON)
Energy Alternatives Ltd. (BC)
Frank's Alternate Energy (ON)
Rocky Mountain Solar (CO)
Ward & Associates (BC)

Generators

Acres International (ON)
Energy Alternatives Ltd. (BC)
Frank's Alternate Energy (ON)
Matrix Energy Inc. (QC)
Northern Lights Energy Systems (ON)
Rocky Mountain Solar (CO)
Solar Plus (BC)
Sun Volts Unlimited Inc. (ON)
Sunpower Co. (ON)

Load Control Governing

Acres International (ON)
Echo Systeme (QC)
Energy Alternatives Ltd. (BC)
Frank's Alternate Energy (ON)
Northern Lights Energy Systems (ON)

High-Head Turbines

Acres International (ON)
Echo Systeme (QC)
Energy Alternatives Ltd. (BC)
Frank's Alternate Energy (ON)
Light Solar Wind Mfg. (ON)
Matrix Energy Inc. (QC)
Solar Plus (BC)
Solar Power Systems (ON)
Ward & Associates (BC)

Medium-Head Turbines

Acres International (ON)
Echo Systeme (QC)
Energy Alternatives Ltd. (BC)
Frank's Alternate Energy (ON)
Matrix Energy Inc. (QC)
Nor'wester Energy Systems (AB)
Solar Plus (BC)
Solar Power Systems (ON)
Sunpower Co. (ON)
Ward & Associates (BC)

Low-Head Turbines

Acres International (ON)
Energy Alternatives Ltd. (BC)
Enviro Store Ltd The (AB)
Frank's Alternate Energy (ON)
Matrix Energy Inc. (QC)
Nor'wester Energy Systems (AB)
Northern Lights Energy Systems (ON)
Solar Plus (BC)
Solar Power Systems (ON)
Sunpower Co. (ON)

Pipelines

Acres International (ON)
Echo Systeme (QC)
Energy Alternatives Ltd. (BC)
Frank's Alternate Energy (ON)
Nor'wester Energy Systems (AB)
Solar Plus (BC)

Power Houses

Acres International (ON)
Energy Alternatives Ltd. (BC)
Frank's Alternate Energy (ON)

Water Pumping

Ward & Associates (BC)

Passive Solar Applications

Commercial Buildings

Allen Kani Associates (ON)
Autonomous Sustainable Housing Inc. (AB)
Cousins Consulting Services (AB)
Enermodal Engineering Ltd. (ON)
Jack's All Trades Inc. (ON)
Stephen F. Pope Architect (ON)
The Centre for Ecological Building Design (ON)
West Kootenay Power Ltd. (BC)

Cooling

Allen Kani Associates (ON)
Autonomous Sustainable Housing Inc. (AB)
Enermodal Engineering Ltd. (ON)
Jack's All Trades Inc. (ON)
Solar Utilites Ltd. (AB)
Stephen F. Pope Architect (ON)
Sustainable Home Solutions (QC)
The Centre for Ecological Building Design (ON)

Daylighting

Advanced Glazings Ltd. (NS)
Allen Kani Associates (ON)
Autonomous Sustainable Housing Inc. (AB)
Echo Systeme (QC)
Enermodal Engineering Ltd. (ON)
Rocky Mountain Solar (CO)
Solar Utilites Ltd. (AB)
Stephen F. Pope Architect (ON)
Sustainable Home Solutions (QC)
The Centre for Ecological Building Design (ON)
Zerowatt Energy (BC)

Heating

Advanced Glazings Ltd. (NS)
Allen Kani Associates (ON)
Autonomous Sustainable Housing Inc. (AB)
Echo Systeme (QC)
Enermodal Engineering Ltd. (ON)
Jack's All Trades Inc. (ON)
Rocky Mountain Solar (CO)
Solar Utilites Ltd. (AB)
Stephen F. Pope Architect (ON)
Sun Ross Solar Systems Ltd. (NS)
Sustainable Home Solutions (QC)
The Centre for Ecological Building Design (ON)
Zerowatt Energy (BC)

Home & Cottage

Advanced Glazings Ltd. (NS)
Allen Kani Associates (ON)
Autonomous Sustainable Housing Inc. (AB)
Echo Systeme (QC)
Enermodal Engineering Ltd. (ON)
Jack's All Trades Inc. (ON)
Rocky Mountain Solar (CO)
Solar Utilites Ltd. (AB)
Stephen F. Pope Architect (ON)
Sun Ross Solar Systems Ltd. (NS)
Sustainable Home Solutions (QC)
The Centre for Ecological Building Design (ON)
Zerowatt Energy (BC)

Industrial Applications

Advanced Glazings Ltd. (NS)
Cousins Consulting Services (AB)
Echo Systeme (QC)
Enermodal Engineering Ltd. (ON)
Solar Utilites Ltd. (AB)
Sustainable Home Solutions (QC)

Information

The Centre for Ecological Building Design (ON)

Multi-Unit Residential

Allen Kani Associates (ON)
Autonomous Sustainable Housing Inc. (AB)
Enermodal Engineering Ltd. (ON)
Jack's All Trades Inc. (ON)
Solar Utilites Ltd. (AB)
Stephen F. Pope Architect (ON)
Sustainable Home Solutions (QC)
Zerowatt Energy (BC)

Passive Solar Products

Domestic Hot Water

Bodycote - ORTECH Inc. (ON)

Envelope Systems & Components

Advanced Glazings Ltd. (NS)
Bodycote - ORTECH Inc. (ON)
Echo Systeme (QC)
Enermodal Engineering Ltd. (ON)
Jack's All Trades Inc. (ON)
Richard Kadulski Architect (BC)

Heating

Richard Kadulski Architect (BC)

Insulations & Airsealing products

Advanced Glazings Ltd. (NS)
Enermodal Engineering Ltd. (ON)
Richard Kadulski Architect (BC)

Mechanical Systems & Components

Echo Systeme (QC)
Enermodal Engineering Ltd. (ON)
Jack's All Trades Inc. (ON)
Richard Kadulski Architect (BC)
Solar Utilites Ltd. (AB)

Solar Control, Daylighting & Shading Devices

Advanced Glazings Ltd. (NS)
Bodycote - ORTECH Inc. (ON)
Enermodal Engineering Ltd. (ON)
Richard Kadulski Architect (BC)
Solar Utilites Ltd. (AB)

Thermal Storage

Echo Systeme (QC)
Enermodal Engineering Ltd. (ON)
Richard Kadulski Architect (BC)

Windows, High Performance Glazing & Components

Advanced Glazings Ltd. (NS)
Enermodal Engineering Ltd. (ON)
Solar Utilites Ltd. (AB)

Photovoltaic Applications

Battery Charging

Analytic Systems (BC)
Analytic Systems (BC)
Argus Telecom International Inc. (QC)
Canrom Photovoltaics Inc. (ON)
CIMAT Power Systems Inc. (QC)
David Nixon & Sun (ON)
Echo Systeme (QC)
Eco Energy (ON)
Enerfree Enterprises (MB)
Energy Alternatives Ltd. (BC)
Frank's Alternate Energy (ON)
Geneq Inc. (QC)
Golden Genesis Co. Ltd. (AB)
ICP Global Technologies (QC)
Jack's All Trades Inc. (ON)
Kelln Consulting Ltd. (SK)
Lenbrook Energy Systems (ON)
Light Solar Wind Mfg. (ON)
March Solar (ON)
Matrix Energy Inc. (QC)
Midnight Sun Energy Ltd. (NT)
Nith Valley Solar (ON)
Nor'wester Energy Systems (AB)
Northern Communication & Navigation Systems Ltd. (NT)
Northern Lights Energy Systems (ON)
Phantom Electronics Inc (ON)
Powersource Energy Systems Inc. (BC)
Prometheus Energy (ON)
Renewable Energy of Plum Hollow (ON)
Rocky Mountain Solar (CO)
Rozon Batteries Inc. (QC)
Solar Plus (BC)
Solar Power Systems (ON)
Solar Systems (ON)
Solar Utilites Ltd. (AB)
Soltek Solar Energy Ltd (BC)
Sovran Energy Inc. (BC)
Statpower Technologies Corporation (BC)
Statpower Technologies Corporation (BC)
Sun Ross Solar Systems Ltd. (NS)
Sun Stream Solar (ON)
Sun Volts Unlimited Inc. (ON)
Sunpower Co. (ON)
Suntime Enterprises (NS)
Sustainable Home Solutions (QC)
The Electric Sun & SH Engineering (AB)
Tideland Signal Canada Ltd. (BC)
Total North Communications Ltd. (YT)
Wenvor Technologies Inc. (ON)
Westrock Battery Ltd. (Solar Division) (MB)
Zerowatt Energy (BC)

Battery Charging – RV & Marine

Argus Telecom International Inc. (QC)
CIMAT Power Systems Inc. (QC)
David Nixon & Sun (ON)
Echo Systeme (QC)
Eco Energy (ON)
Enerfree Enterprises (MB)
Energy Alternatives Ltd. (BC)
Frank's Alternate Energy (ON)
Geneq Inc. (QC)
Golden Genesis Co. Ltd. (AB)
ICP Global Technologies (QC)
Jack's All Trades Inc. (ON)
Kelln Consulting Ltd. (SK)
Lenbrook Energy Systems (ON)
Light Solar Wind Mfg. (ON)
March Solar (ON)
Matrix Energy Inc. (QC)
Midnight Sun Energy Ltd. (NT)
Nith Valley Solar (ON)
Northern Lights Energy Systems (ON)
Prometheus Energy (ON)
Rocky Mountain Solar (CO)
Rozon Batteries Inc. (QC)
Solar Plus (BC)
Solar Power Systems (ON)
Solar Systems (ON)
Solar Utilites Ltd. (AB)
Soltek Solar Energy Ltd (BC)
Sovran Energy Inc. (BC)
Statpower Technologies Corporation (BC)
Sun Ross Solar Systems Ltd. (NS)
Sun Stream Solar (ON)
Sun Volts Unlimited Inc. (ON)
Sunpower Co. (ON)
The Electric Sun & SH Engineering (AB)
Tideland Signal Canada Ltd. (BC)
Total North Communications Ltd. (YT)
Wenvor Technologies Inc. (ON)
Westrock Battery Ltd. (Solar Division) (MB)

Cathodic Protection

CIMAT Power Systems Inc. (QC)
David Nixon & Sun (ON)
Frank's Alternate Energy (ON)
Golden Genesis Co. Ltd. (AB)
Kelln Consulting Ltd. (SK)
Lenbrook Energy Systems (ON)
Light Solar Wind Mfg. (ON)
March Solar (ON)
Matrix Energy Inc. (QC)
Nor'wester Energy Systems (AB)
Powersource Energy Systems Inc. (BC)
Rozon Batteries Inc. (QC)
Solar Plus (BC)
Solar Utilites Ltd. (AB)
Soltek Solar Energy Ltd (BC)
Sovran Energy Inc. (BC)
The Electric Sun & SH Engineering (AB)
Wenvor Technologies Inc. (ON)

Consumer Products

Amsdell Inc. (ON)
David Nixon & Sun (ON)
Echo Systeme (QC)
Energy Alternatives Ltd. (BC)
Frank's Alternate Energy (ON)
Golden Genesis Co. Ltd. (AB)
ICP Global Technologies (QC)
Jack's All Trades Inc. (ON)
Lenbrook Energy Systems (ON)
March Solar (ON)
Matrix Energy Inc. (QC)
Midnight Sun Energy Ltd. (NT)
Northern Lights Energy Systems (ON)
Powersource Energy Systems Inc. (BC)
Prometheus Energy (ON)
Renewable Energy of Plum Hollow (ON)
Rozon Batteries Inc. (QC)
Solar Freedom International (SK)
Solar Plus (BC)
Solar Utilites Ltd. (AB)
Soltek Solar Energy Ltd (BC)
Sovran Energy Inc. (BC)
Statpower Technologies Corporation (BC)
Sun Stream Solar (ON)
Sun Volts Unlimited Inc. (ON)
Sunpower Co. (ON)
The Electric Sun & SH Engineering (AB)
Wenvor Technologies Inc. (ON)
Westrock Battery Ltd. (Solar Division) (MB)

Cooling

Conserval Engineering (ON)
Golden Genesis Co. Ltd. (AB)
Jack's All Trades Inc. (ON)
Kelln Consulting Ltd. (SK)

Matrix Energy Inc. (QC)
Midnight Sun Energy Ltd. (NT)
Northern Communication & Navigation Systems Ltd. (NT)
Powersource Energy Systems Inc. (BC)
Renewable Energy of Plum Hollow (ON)
Rozon Batteries Inc. (QC)
Solar Plus (BC)
Solar Utilites Ltd. (AB)
Soltek Solar Energy Ltd (BC)
Sun Volts Unlimited Inc. (ON)
Wenvor Technologies Inc. (ON)
Westrock Battery Ltd. (Solar Division) (MB)

Desalination

Golden Genesis Co. Ltd. (AB)
Matrix Energy Inc. (QC)
Midnight Sun Energy Ltd. (NT)
NewSun Technologies, Ltd. (ON)
Powersource Energy Systems Inc. (BC)
Solar Plus (BC)
Solar Utilites Ltd. (AB)
Westrock Battery Ltd. (Solar Division) (MB)

Domestic Hot Water

Midnight Sun Energy Ltd. (NT)

Grid-Connected Systems

Allen Kani Associates (ON)
Argus Telecom International Inc. (QC)
Canrom Photovoltaics Inc. (ON)
Energy Alternatives Ltd. (BC)
Enersite International Inc. (ON)
Golden Genesis Co. Ltd. (AB)
Howell-Mayhew Engineering (AB)
Kelln Consulting Ltd. (SK)
Lenbrook Energy Systems (ON)
March Solar (ON)
Matrix Energy Inc. (QC)
Midnight Sun Energy Ltd. (NT)
Nith Valley Solar (ON)
Nor'wester Energy Systems (AB)
Phantom Electronics Inc (ON)
Powersource Energy Systems Inc. (BC)
Prometheus Energy (ON)
Renewable Energy of Plum Hollow (ON)
Rocky Mountain Solar (CO)
Rozon Batteries Inc. (QC)
Solar Plus (BC)
Solar Power Systems (ON)
Solar Utilites Ltd. (AB)
Soltek Solar Energy Ltd (BC)
Sovran Energy Inc. (BC)
Sun Stream Solar (ON)
Sun Volts Unlimited Inc. (ON)
Sustainable Home Solutions (QC)
Wenvor Technologies Inc. (ON)
Westrock Battery Ltd. (Solar Division) (MB)
Zerowatt Energy (BC)

Home & Cottage

Solar Freedom International (SK)
Sun Volts Unlimited Inc. (ON)
The Electric Sun & SH Engineering (AB)

Industrial Applications

Conserval Engineering (ON)

Integration into Buildings

Allen Kani Associates (ON)
Canrom Photovoltaics Inc. (ON)
CIMAT Power Systems Inc. (QC)
Conserval Engineering (ON)
Echo Systeme (QC)
Energy Alternatives Ltd. (BC)
Golden Genesis Co. Ltd. (AB)
Howell-Mayhew Engineering (AB)
Jack's All Trades Inc. (ON)
Lenbrook Energy Systems (ON)
March Solar (ON)
Matrix Energy Inc. (QC)
Midnight Sun Energy Ltd. (NT)
Northern Communication & Navigation Systems Ltd. (NT)
Phantom Electronics Inc (ON)
Powersource Energy Systems Inc. (BC)
Renewable Energy of Plum Hollow (ON)
Rocky Mountain Solar (CO)
Solar Plus (BC)
Solar Utilites Ltd. (AB)
Sovran Energy Inc. (BC)
Sun Volts Unlimited Inc. (ON)
The Electric Sun & SH Engineering (AB)
Wenvor Technologies Inc. (ON)
Westrock Battery Ltd. (Solar Division) (MB)
Zerowatt Energy (BC)

Irrigation

CIMAT Power Systems Inc. (QC)
David Nixon & Sun (ON)
Eco Energy (ON)
Enerfree Enterprises (MB)
Energy Alternatives Ltd. (BC)
Frank's Alternate Energy (ON)
Lenbrook Energy Systems (ON)
Light Solar Wind Mfg. (ON)
March Solar (ON)
Matrix Energy Inc. (QC)
Midnight Sun Energy Ltd. (NT)
NewSun Technologies, Ltd. (ON)
Nith Valley Solar (ON)
Nor'wester Energy Systems (AB)
Northern Lights Energy Systems (ON)
Powersource Energy Systems Inc. (BC)
Renewable Energy of Plum Hollow (ON)
Rocky Mountain Solar (CO)
Rozon Batteries Inc. (QC)
Solar Plus (BC)
Solar Power Systems (ON)
Soltek Solar Energy Ltd (BC)
Sovran Energy Inc. (BC)
Sun Stream Solar (ON)
Sun Volts Unlimited Inc. (ON)
Sunpower Co. (ON)
Wenvor Technologies Inc. (ON)
Westrock Battery Ltd. (Solar Division) (MB)

Lighting

Allen Kani Associates (ON)
Argus Telecom International Inc. (QC)
CIMAT Power Systems Inc. (QC)
David Nixon & Sun (ON)
Echo Systeme (QC)
Enerfree Enterprises (MB)
Energy Alternatives Ltd. (BC)
Frank's Alternate Energy (ON)
Golden Genesis Co. Ltd. (AB)
Kelln Consulting Ltd. (SK)
Lenbrook Energy Systems (ON)
Light Solar Wind Mfg. (ON)
March Solar (ON)
Matrix Energy Inc. (QC)
Midnight Sun Energy Ltd. (NT)
NewSun Technologies, Ltd. (ON)
Nith Valley Solar (ON)
Northern Lights Energy Systems (ON)
Phantom Electronics Inc (ON)
Powersource Energy Systems Inc. (BC)
Renewable Energy of Plum Hollow (ON)
Rocky Mountain Solar (CO)
Rozon Batteries Inc. (QC)
Solar Freedom International (SK)
Solar Plus (BC)
Solar Power Systems (ON)
Solar Utilites Ltd. (AB)
Solener Inc. (QC)
Soltek Solar Energy Ltd (BC)
Sovran Energy Inc. (BC)
Sun Ross Solar Systems Ltd. (NS)
Sun Stream Solar (ON)
Sun Volts Unlimited Inc. (ON)
Sunpower Co. (ON)
Suntime Enterprises (NS)
The Electric Sun & SH Engineering (AB)
Wenvor Technologies Inc. (ON)
Westrock Battery Ltd. (Solar Division) (MB)

Monitoring Systems

Argus Telecom International Inc. (QC)
CIMAT Power Systems Inc. (QC)
David Nixon & Sun (ON)
Echo Systeme (QC)
Enerfree Enterprises (MB)
Frank's Alternate Energy (ON)
Geneq Inc. (QC)
Golden Genesis Co. Ltd. (AB)
Lenbrook Energy Systems (ON)
March Solar (ON)
Matrix Energy Inc. (QC)

Midnight Sun Energy Ltd. (NT)
Phantom Electronics Inc (ON)
Powersource Energy Systems Inc. (BC)
Rocky Mountain Solar (CO)
Rozon Batteries Inc. (QC)
Solar Plus (BC)
Solar Utilites Ltd. (AB)
Soltek Solar Energy Ltd (BC)
Sovran Energy Inc. (BC)
Sun Volts Unlimited Inc. (ON)
Sunpower Co. (ON)
Wenvor Technologies Inc. (ON)
Westrock Battery Ltd. (Solar Division) (MB)

Navigational Aids

Argus Telecom International Inc. (QC)
CIMAT Power Systems Inc. (QC)
Echo Systeme (QC)
Enerfree Enterprises (MB)
Golden Genesis Co. Ltd. (AB)
Lenbrook Energy Systems (ON)
Light Solar Wind Mfg. (ON)
March Solar (ON)
Matrix Energy Inc. (QC)
Midnight Sun Energy Ltd. (NT)
NewSun Technologies, Ltd. (ON)
Nith Valley Solar (ON)
Powersource Energy Systems Inc. (BC)
Rozon Batteries Inc. (QC)
Solar Plus (BC)
Solar Utilites Ltd. (AB)
Soltek Solar Energy Ltd (BC)
Sovran Energy Inc. (BC)
Sun Ross Solar Systems Ltd. (NS)
Sun Stream Solar (ON)
Sun Volts Unlimited Inc. (ON)
Sunpower Co. (ON)
Tideland Signal Canada Ltd. (BC)
Wenvor Technologies Inc. (ON)
Westrock Battery Ltd. (Solar Division) (MB)

Refrigeration

Argus Telecom International Inc. (QC)
CIMAT Power Systems Inc. (QC)
Echo Systeme (QC)
Energy Alternatives Ltd. (BC)
Frank's Alternate Energy (ON)
Golden Genesis Co. Ltd. (AB)
Jack's All Trades Inc. (ON)
Kelln Consulting Ltd. (SK)
Lenbrook Energy Systems (ON)
Light Solar Wind Mfg. (ON)
March Solar (ON)
Matrix Energy Inc. (QC)
Midnight Sun Energy Ltd. (NT)
Powersource Energy Systems Inc. (BC)
Rocky Mountain Solar (CO)
Rozon Batteries Inc. (QC)
Solar Plus (BC)
Solar Utilites Ltd. (AB)
Sovran Energy Inc. (BC)
Sun Volts Unlimited Inc. (ON)
Suntime Enterprises (NS)
Wenvor Technologies Inc. (ON)
Westrock Battery Ltd. (Solar Division) (MB)

Remote Instrumentation

Geneq Inc. (QC)

Roof Shingles

Allen Kani Associates (ON)
Kelln Consulting Ltd. (SK)
Lenbrook Energy Systems (ON)
March Solar (ON)
Matrix Energy Inc. (QC)
Midnight Sun Energy Ltd. (NT)
Phantom Electronics Inc (ON)
Powersource Energy Systems Inc. (BC)
Prometheus Energy (ON)
Rocky Mountain Solar (CO)
Solar Plus (BC)
Solar Utilites Ltd. (AB)
Soltek Solar Energy Ltd (BC)
Sovran Energy Inc. (BC)
Sun Stream Solar (ON)
Sun Volts Unlimited Inc. (ON)
Sustainable Home Solutions (QC)
Westrock Battery Ltd. (Solar Division) (MB)

Rural Electrification

Amsdell Inc. (ON)
Canrom Photovoltaics Inc. (ON)
CIMAT Power Systems Inc. (QC)
David Nixon & Sun (ON)
Echo Systeme (QC)
Enerfree Enterprises (MB)
Energy Alternatives Ltd. (BC)
Enersite International Inc. (ON)
Golden Genesis Co. Ltd. (AB)
Jack's All Trades Inc. (ON)
Kelln Consulting Ltd. (SK)
March Solar (ON)
Matrix Energy Inc. (QC)
Midnight Sun Energy Ltd. (NT)
NewSun Technologies, Ltd. (ON)
Nith Valley Solar (ON)
Nor'wester Energy Systems (AB)
Northern Lights Energy Systems (ON)
Powersource Energy Systems Inc. (BC)
Renewable Energy of Plum Hollow (ON)
Rocky Mountain Solar (CO)
Rozon Batteries Inc. (QC)
Solar Plus (BC)
Solar Power Systems (ON)
Solar Utilites Ltd. (AB)
Solener Inc. (QC)
Soltek Solar Energy Ltd (BC)
Sovran Energy Inc. (BC)
Sun Ross Solar Systems Ltd. (NS)
Sun Stream Solar (ON)
Sun Volts Unlimited Inc. (ON)
The Electric Sun & SH Engineering (AB)
Wenvor Technologies Inc. (ON)
Westrock Battery Ltd. (Solar Division) (MB)

Solar Control, Daylighting & Shading

Rocky Mountain Solar (CO)

Stand Alone Systems

Amsdell Inc. (ON)
Argus Telecom International Inc. (QC)
Canrom Photovoltaics Inc. (ON)
CIMAT Power Systems Inc. (QC)
David Nixon & Sun (ON)
Echo Systeme (QC)
Enerfree Enterprises (MB)
Energy Alternatives Ltd. (BC)
Enersite International Inc. (ON)
Frank's Alternate Energy (ON)
Golden Genesis Co. Ltd. (AB)
Jack's All Trades Inc. (ON)
Lenbrook Energy Systems (ON)
Light Solar Wind Mfg. (ON)
March Solar (ON)
Matrix Energy Inc. (QC)
Midnight Sun Energy Ltd. (NT)
Nith Valley Solar (ON)
Nor'wester Energy Systems (AB)
Northern Lights Energy Systems (ON)
Phantom Electronics Inc (ON)
Powersource Energy Systems Inc. (BC)
Prometheus Energy (ON)
Renewable Energy of Plum Hollow (ON)
Rocky Mountain Solar (CO)
Solar Plus (BC)
Solar Power Systems (ON)
Solar Utilites Ltd. (AB)
Soltek Solar Energy Ltd (BC)
Sovran Energy Inc. (BC)
Sun Ross Solar Systems Ltd. (NS)
Sun Stream Solar (ON)
Sun Volts Unlimited Inc. (ON)
Sunpower Co. (ON)
The Electric Sun & SH Engineering (AB)
Wenvor Technologies Inc. (ON)
Westrock Battery Ltd. (Solar Division) (MB)
Zerowatt Energy (BC)

Telecommunications

Argus Telecom International Inc. (QC)
Canrom Photovoltaics Inc. (ON)
CIMAT Power Systems Inc. (QC)
Echo Systeme (QC)
Energy Alternatives Ltd. (BC)
Geneq Inc. (QC)
Golden Genesis Co. Ltd. (AB)
Kelln Consulting Ltd. (SK)
Lenbrook Energy Systems (ON)
Light Solar Wind Mfg. (ON)
March Solar (ON)

Matrix Energy Inc. (QC)
Midnight Sun Energy Ltd. (NT)
Nor'wester Energy Systems (AB)
Northern Communication & Navigation Systems Ltd. (NT)
Phantom Electronics Inc (ON)
Powersource Energy Systems Inc. (BC)
Renewable Energy of Plum Hollow (ON)
Rocky Mountain Solar (CO)
Rozon Batteries Inc. (QC)
Solar Plus (BC)
Solar Utilites Ltd. (AB)
Solener Inc. (QC)
Soltek Solar Energy Ltd (BC)
Sovran Energy Inc. (BC)
Sun Stream Solar (ON)
Sun Volts Unlimited Inc. (ON)
Sunpower Co. (ON)
Total North Communications Ltd. (YT)
Wenvor Technologies Inc. (ON)
Westrock Battery Ltd. (Solar Division) (MB)

Vaccine/Medical Refrigeration

Lenbrook Energy Systems (ON)
March Solar (ON)
Matrix Energy Inc. (QC)
Midnight Sun Energy Ltd. (NT)
Powersource Energy Systems Inc. (BC)
Rocky Mountain Solar (CO)
Solar Plus (BC)
Solar Utilites Ltd. (AB)
Sovran Energy Inc. (BC)
Wenvor Technologies Inc. (ON)
Westrock Battery Ltd. (Solar Division) (MB)

Ventilation

Argus Telecom International Inc. (QC)
CIMAT Power Systems Inc. (QC)
Echo Systeme (QC)
Enerfree Enterprises (MB)
Energy Alternatives Ltd. (BC)
Golden Genesis Co. Ltd. (AB)
Jack's All Trades Inc. (ON)
Kelln Consulting Ltd. (SK)
March Solar (ON)
Matrix Energy Inc. (QC)
Midnight Sun Energy Ltd. (NT)
Powersource Energy Systems Inc. (BC)
Rocky Mountain Solar (CO)
Solar Plus (BC)
Soltek Solar Energy Ltd (BC)
Sovran Energy Inc. (BC)
Sun Stream Solar (ON)
Sun Volts Unlimited Inc. (ON)
The Electric Sun & SH Engineering (AB)
Wenvor Technologies Inc. (ON)
Westrock Battery Ltd. (Solar Division) (MB)

Water Pumping

Argus Telecom International Inc. (QC)
CIMAT Power Systems Inc. (QC)
David Nixon & Sun (ON)
Echo Systeme (QC)
Eco Energy (ON)
Enerfree Enterprises (MB)
Energy Alternatives Ltd. (BC)
Golden Genesis Co. Ltd. (AB)
Jack's All Trades Inc. (ON)
Kelln Consulting Ltd. (SK)
Lenbrook Energy Systems (ON)
March Solar (ON)
Matrix Energy Inc. (QC)
NewSun Technologies, Ltd. (ON)
Nith Valley Solar (ON)
Nor'wester Energy Systems (AB)
Northern Lights Energy Systems (ON)
Powersource Energy Systems Inc. (BC)
Renewable Energy of Plum Hollow (ON)
Rozon Batteries Inc. (QC)
Solar Plus (BC)
Solar Power Systems (ON)
Solar Utilites Ltd. (AB)
Solener Inc. (QC)
Soltek Solar Energy Ltd (BC)
Sovran Energy Inc. (BC)
Sun Ross Solar Systems Ltd. (NS)
Sun Stream Solar (ON)
Sunpower Co. (ON)
Wenvor Technologies Inc. (ON)
Westrock Battery Ltd. (Solar Division) (MB)

Photovoltaic Products

Staticon Ltd. (ON)

Arrays

Argus Telecom International Inc. (QC)
Bodycote - ORTECH Inc. (ON)
Canrom Photovoltaics Inc. (ON)
CIMAT Power Systems Inc. (QC)
David Nixon & Sun (ON)
Echo Systeme (QC)
Enerfree Enterprises (MB)
Energy Alternatives Ltd. (BC)
Enersite International Inc. (ON)
Enviro Store Ltd The (AB)
Frank's Alternate Energy (ON)
Geneq Inc. (QC)
Golden Genesis Co. Ltd. (AB)
Kelln Consulting Ltd. (SK)
Lenbrook Energy Systems (ON)
Light Solar Wind Mfg. (ON)
March Solar (ON)
Matrix Energy Inc. (QC)
Midnight Sun Energy Ltd. (NT)
NewSun Technologies, Ltd. (ON)
Nith Valley Solar (ON)
Nor'wester Energy Systems (AB)
Northern Communication & Navigation Systems Ltd. (NT)
Northern Lights Energy Systems (ON)
Phantom Electronics Inc (ON)
Powersource Energy Systems Inc. (BC)
Prometheus Energy (ON)
Renewable Energy of Plum Hollow (ON)
Rocky Mountain Solar (CO)
Rozon Batteries Inc. (QC)
Solar Freedom International (SK)
Solar Plus (BC)
Solar Power Systems (ON)
Solar Systems (ON)
Solar Utilites Ltd. (AB)
Soltek Solar Energy Ltd (BC)
Sovran Energy Inc. (BC)
Sun Ross Solar Systems Ltd. (NS)
Sun Stream Solar (ON)
Sun Volts Unlimited Inc. (ON)
Sunpower Co. (ON)
The Electric Sun & SH Engineering (AB)
Thermo Dynamics Ltd (NS)
Tideland Signal Canada Ltd. (BC)
Total North Communications Ltd. (YT)
Wenvor Technologies Inc. (ON)
Westrock Battery Ltd. (Solar Division) (MB)

Batteries

GNB Technologies (QC)
Surrette Battery Company Limited (NS)

Batteries – Lead Acid

Argus Telecom International Inc. (QC)
Canrom Photovoltaics Inc. (ON)
CIMAT Power Systems Inc. (QC)
David Nixon & Sun (ON)
Echo Systeme (QC)
Enerfree Enterprises (MB)
Energy Alternatives Ltd. (BC)
Enersite International Inc. (ON)
Enviro Store Ltd The (AB)
Frank's Alternate Energy (ON)
Geneq Inc. (QC)
Jack's All Trades Inc. (ON)
Kelln Consulting Ltd. (SK)
Lenbrook Energy Systems (ON)
Light Solar Wind Mfg. (ON)
March Solar (ON)
Matrix Energy Inc. (QC)
Midnight Sun Energy Ltd. (NT)
NewSun Technologies, Ltd. (ON)
Nith Valley Solar (ON)
Nor'wester Energy Systems (AB)
Northern Communication & Navigation Systems Ltd. (NT)
Northern Lights Energy Systems (ON)
Phantom Electronics Inc (ON)
Powersource Energy Systems Inc. (BC)
Prometheus Energy (ON)
Renewable Energy of Plum Hollow (ON)
Rocky Mountain Solar (CO)
Rozon Batteries Inc. (QC)

Solar Plus (BC)
Solar Power Systems (ON)
Solar Utilites Ltd. (AB)
Soltek Solar Energy Ltd (BC)
Sovran Energy Inc. (BC)
Sun Ross Solar Systems Ltd. (NS)
Sun Stream Solar (ON)
Sun Volts Unlimited Inc. (ON)
Sunpower Co. (ON)
Suntime Enterprises (NS)
The Electric Sun & SH Engineering (AB)
Tideland Signal Canada Ltd. (BC)
Total North Communications Ltd. (YT)
Wenvor Technologies Inc. (ON)
Westrock Battery Ltd. (Solar Division) (MB)

Batteries – NiCad

Argus Telecom International Inc. (QC)
CIMAT Power Systems Inc. (QC)
Echo Systeme (QC)
Enerfree Enterprises (MB)
Enviro Store Ltd The (AB)
Frank's Alternate Energy (ON)
Golden Genesis Co. Ltd. (AB)
Grassroots Environmental Products (ON)
Jack's All Trades Inc. (ON)
Kelln Consulting Ltd. (SK)
Lenbrook Energy Systems (ON)
Matrix Energy Inc. (QC)
Midnight Sun Energy Ltd. (NT)
Northern Communication & Navigation Systems Ltd. (NT)
Powersource Energy Systems Inc. (BC)
Rocky Mountain Solar (CO)
Rozon Batteries Inc. (QC)
Solar Plus (BC)
Solar Power Systems (ON)
Solar Utilites Ltd. (AB)
Sovran Energy Inc. (BC)
Sun Volts Unlimited Inc. (ON)
Sunpower Co. (ON)
Tideland Signal Canada Ltd. (BC)
Total North Communications Ltd. (YT)
Wenvor Technologies Inc. (ON)
Westrock Battery Ltd. (Solar Division) (MB)

Cabling

Argus Telecom International Inc. (QC)
CIMAT Power Systems Inc. (QC)
Energy Alternatives Ltd. (BC)
Frank's Alternate Energy (ON)
Geneq Inc. (QC)
Jack's All Trades Inc. (ON)
Lenbrook Energy Systems (ON)
Light Solar Wind Mfg. (ON)
March Solar (ON)
Matrix Energy Inc. (QC)
Midnight Sun Energy Ltd. (NT)
Nor'wester Energy Systems (AB)
Northern Lights Energy Systems (ON)
Phantom Electronics Inc (ON)
Powersource Energy Systems Inc. (BC)
Prometheus Energy (ON)
Renewable Energy of Plum Hollow (ON)
Rozon Batteries Inc. (QC)
Solar Plus (BC)
Solar Power Systems (ON)
Solar Utilites Ltd. (AB)
Soltek Solar Energy Ltd (BC)
Sovran Energy Inc. (BC)
Sun Stream Solar (ON)
Sun Volts Unlimited Inc. (ON)
Sunpower Co. (ON)
The Electric Sun & SH Engineering (AB)
Total North Communications Ltd. (YT)
Wenvor Technologies Inc. (ON)
Westrock Battery Ltd. (Solar Division) (MB)

Cathodic Protection

TN Conseil (QC)

Cells - Monocrystalline

Argus Telecom International Inc. (QC)
Canrom Photovoltaics Inc. (ON)
CIMAT Power Systems Inc. (QC)
Energy Alternatives Ltd. (BC)
Enviro Store Ltd The (AB)
Golden Genesis Co. Ltd. (AB)
Jack's All Trades Inc. (ON)
Kelln Consulting Ltd. (SK)
Lenbrook Energy Systems (ON)
Light Solar Wind Mfg. (ON)
March Solar (ON)
Matrix Energy Inc. (QC)
Midnight Sun Energy Ltd. (NT)
NewSun Technologies, Ltd. (ON)
Nor'wester Energy Systems (AB)
Northern Lights Energy Systems (ON)
Powersource Energy Systems Inc. (BC)
Renewable Energy of Plum Hollow (ON)
Rocky Mountain Solar (CO)
Rozon Batteries Inc. (QC)
Solar Freedom International (SK)
Solar Plus (BC)
Solar Systems (ON)
Solar Utilites Ltd. (AB)
Soltek Solar Energy Ltd (BC)
Sovran Energy Inc. (BC)
Sun Stream Solar (ON)
Sun Volts Unlimited Inc. (ON)
Sunpower Co. (ON)
Wenvor Technologies Inc. (ON)

Cells – Polycrystalline

Argus Telecom International Inc. (QC)
Canrom Photovoltaics Inc. (ON)
Energy Alternatives Ltd. (BC)
Enviro Store Ltd The (AB)
Golden Genesis Co. Ltd. (AB)
Kelln Consulting Ltd. (SK)
Lenbrook Energy Systems (ON)
March Solar (ON)
Matrix Energy Inc. (QC)
Midnight Sun Energy Ltd. (NT)
NewSun Technologies, Ltd. (ON)
Nor'wester Energy Systems (AB)
Northern Lights Energy Systems (ON)
Powersource Energy Systems Inc. (BC)
Renewable Energy of Plum Hollow (ON)
Rocky Mountain Solar (CO)
Rozon Batteries Inc. (QC)
Solar Freedom International (SK)
Solar Plus (BC)
Solar Systems (ON)
Solar Utilites Ltd. (AB)
Soltek Solar Energy Ltd (BC)
Sovran Energy Inc. (BC)
Sun Stream Solar (ON)
Sun Volts Unlimited Inc. (ON)
Sunpower Co. (ON)
Wenvor Technologies Inc. (ON)

Cells - Thin Film

Argus Telecom International Inc. (QC)
Energy Alternatives Ltd. (BC)
Enviro Store Ltd The (AB)
Golden Genesis Co. Ltd. (AB)
ICP Global Technologies (QC)
Kelln Consulting Ltd. (SK)
Lenbrook Energy Systems (ON)
Light Solar Wind Mfg. (ON)
March Solar (ON)
Midnight Sun Energy Ltd. (NT)
Northern Lights Energy Systems (ON)
Powersource Energy Systems Inc. (BC)
Rocky Mountain Solar (CO)
Solar Plus (BC)
Solar Utilites Ltd. (AB)
Sovran Energy Inc. (BC)
Sun Stream Solar (ON)
Sun Volts Unlimited Inc. (ON)
Sunpower Co. (ON)
Wenvor Technologies Inc. (ON)

Charge Controllers

Argus Telecom International Inc. (QC)
CIMAT Power Systems Inc. (QC)
David Nixon & Sun (ON)
Diversitel Communications (ON)
Echo Systeme (QC)
Eco Energy (ON)
Enerfree Enterprises (MB)
Energy Alternatives Ltd. (BC)
Enersite International Inc. (ON)
Enviro Store Ltd The (AB)
Frank's Alternate Energy (ON)
Geneq Inc. (QC)
Golden Genesis Co. Ltd. (AB)
ICP Global Technologies (QC)
Jack's All Trades Inc. (ON)
Lenbrook Energy Systems (ON)
Light Solar Wind Mfg. (ON)
March Solar (ON)

Matrix Energy Inc. (QC)
Midnight Sun Energy Ltd. (NT)
Nith Valley Solar (ON)
Nor'wester Energy Systems (AB)
Northern Communication & Navigation Systems Ltd. (NT)
Northern Lights Energy Systems (ON)
Phantom Electronics Inc (ON)
Powersource Energy Systems Inc. (BC)
Prometheus Energy (ON)
Renewable Energy of Plum Hollow (ON)
Rocky Mountain Solar (CO)
Rozon Batteries Inc. (QC)
Solar Freedom International (SK)
Solar Plus (BC)
Solar Power Systems (ON)
Solar Systems (ON)
Solar Utilites Ltd. (AB)
Soltek Solar Energy Ltd (BC)
Sovran Energy Inc. (BC)
Sun Ross Solar Systems Ltd. (NS)
Sun Stream Solar (ON)
Sun Volts Unlimited Inc. (ON)
Sunpower Co. (ON)
The Electric Sun & SH Engineering (AB)
Thermo Dynamics Ltd (NS)
Total North Communications Ltd. (YT)
Wenvor Technologies Inc. (ON)
Westrock Battery Ltd. (Solar Division) (MB)

Charge Kits (RV & Marine)

Argus Telecom International Inc. (QC)
David Nixon & Sun (ON)
Echo Systeme (QC)
Enerfree Enterprises (MB)
Energy Alternatives Ltd. (BC)
Enviro Store Ltd The (AB)
Frank's Alternate Energy (ON)
Geneq Inc. (QC)
Golden Genesis Co. Ltd. (AB)
ICP Global Technologies (QC)
Jack's All Trades Inc. (ON)
Kelln Consulting Ltd. (SK)
Lenbrook Energy Systems (ON)
Light Solar Wind Mfg. (ON)
March Solar (ON)
Matrix Energy Inc. (QC)
Midnight Sun Energy Ltd. (NT)
Nith Valley Solar (ON)
Northern Lights Energy Systems (ON)
Powersource Energy Systems Inc. (BC)
Prometheus Energy (ON)
Renewable Energy of Plum Hollow (ON)
Rocky Mountain Solar (CO)
Rozon Batteries Inc. (QC)
Solar Plus (BC)
Solar Power Systems (ON)
Solar Systems (ON)
Solar Utilites Ltd. (AB)
Soltek Solar Energy Ltd (BC)
Sovran Energy Inc. (BC)
Sun Ross Solar Systems Ltd. (NS)
Sun Stream Solar (ON)
Sun Volts Unlimited Inc. (ON)
Sunpower Co. (ON)
The Electric Sun & SH Engineering (AB)
Total North Communications Ltd. (YT)
Wenvor Technologies Inc. (ON)
Westrock Battery Ltd. (Solar Division) (MB)

Electrolysers

Echo Systeme (QC)
Kelln Consulting Ltd. (SK)
Midnight Sun Energy Ltd. (NT)
Powersource Energy Systems Inc. (BC)
Solar Utilites Ltd. (AB)
Sun Volts Unlimited Inc. (ON)
Westrock Battery Ltd. (Solar Division) (MB)

Generator Controllers

Argus Telecom International Inc. (QC)
CIMAT Power Systems Inc. (QC)
Enerfree Enterprises (MB)
Energy Alternatives Ltd. (BC)
Enviro Store Ltd The (AB)
Jack's All Trades Inc. (ON)
Kelln Consulting Ltd. (SK)
March Solar (ON)
Matrix Energy Inc. (QC)
Midnight Sun Energy Ltd. (NT)
Nith Valley Solar (ON)
Powersource Energy Systems Inc. (BC)
Prometheus Energy (ON)
Renewable Energy of Plum Hollow (ON)
Rocky Mountain Solar (CO)
Rozon Batteries Inc. (QC)
Sambrabec Inc. (QC)
Solar Plus (BC)
Solar Utilites Ltd. (AB)
Soltek Solar Energy Ltd (BC)
Sun Stream Solar (ON)
Sun Volts Unlimited Inc. (ON)
Wenvor Technologies Inc. (ON)
Westrock Battery Ltd. (Solar Division) (MB)

Generators

Argus Telecom International Inc. (QC)
Enerfree Enterprises (MB)
Energy Alternatives Ltd. (BC)
Enviro Store Ltd The (AB)
Golden Genesis Co. Ltd. (AB)
Jack's All Trades Inc. (ON)
Kelln Consulting Ltd. (SK)
Lenbrook Energy Systems (ON)
Light Solar Wind Mfg. (ON)
March Solar (ON)
Matrix Energy Inc. (QC)
Midnight Sun Energy Ltd. (NT)
Nith Valley Solar (ON)
Powersource Energy Systems Inc. (BC)
Prometheus Energy (ON)
Renewable Energy of Plum Hollow (ON)
Rocky Mountain Solar (CO)
Solar Plus (BC)
Solar Power Systems (ON)
Solar Utilites Ltd. (AB)
Soltek Solar Energy Ltd (BC)
Sovran Energy Inc. (BC)
Sun Stream Solar (ON)
Sun Volts Unlimited Inc. (ON)
The Electric Sun & SH Engineering (AB)
Wenvor Technologies Inc. (ON)
Westrock Battery Ltd. (Solar Division) (MB)

Home & Cottage

Enviro Store Ltd The (AB)

Inverters

Solar Converters Inc (ON)

Inverters — Modified Sine Wave

Amsdell Inc. (ON)
Analytic Systems (BC)
Argus Telecom International Inc. (QC)
Canrom Photovoltaics Inc. (ON)
CIMAT Power Systems Inc. (QC)
David Nixon & Sun (ON)
Echo Systeme (QC)
Enerfree Enterprises (MB)
Energy Alternatives Ltd. (BC)
Enersite International Inc. (ON)
Enviro Store Ltd The (AB)
Frank's Alternate Energy (ON)
Golden Genesis Co. Ltd. (AB)
Kelln Consulting Ltd. (SK)
Lenbrook Energy Systems (ON)
Light Solar Wind Mfg. (ON)
March Solar (ON)
Matrix Energy Inc. (QC)
Midnight Sun Energy Ltd. (NT)
NewSun Technologies, Ltd. (ON)
Nith Valley Solar (ON)
Nor'wester Energy Systems (AB)
Northern Communication & Navigation Systems Ltd. (NT)
Northern Lights Energy Systems (ON)
Phantom Electronics Inc (ON)
Powersource Energy Systems Inc. (BC)
Prometheus Energy (ON)
Renewable Energy of Plum Hollow (ON)
Rocky Mountain Solar (CO)
Rozon Batteries Inc. (QC)
Sambrabec Inc. (QC)
Solar Freedom International (SK)
Solar Plus (BC)
Solar Power Systems (ON)
Solar Systems (ON)
Solar Utilites Ltd. (AB)
Soltek Solar Energy Ltd (BC)
Sovran Energy Inc. (BC)
Sun Ross Solar Systems Ltd. (NS)
Sun Stream Solar (ON)
Sun Volts Unlimited Inc. (ON)

Sunpower Co. (ON)
The Electric Sun & SH Engineering (AB)
Total North Communications Ltd. (YT)
Wenvor Technologies Inc. (ON)
Westrock Battery Ltd. (Solar Division) (MB)

Inverters – Sine Wave

Amsdell Inc. (ON)
Analytic Systems (BC)
Argus Telecom International Inc. (QC)
Canrom Photovoltaics Inc. (ON)
CIMAT Power Systems Inc. (QC)
David Nixon & Sun (ON)
Echo Systeme (QC)
Enerfree Enterprises (MB)
Energy Alternatives Ltd. (BC)
Enersite International Inc. (ON)
Enviro Store Ltd The (AB)
Frank's Alternate Energy (ON)
Geneq Inc. (QC)
Golden Genesis Co. Ltd. (AB)
Kelln Consulting Ltd. (SK)
Lenbrook Energy Systems (ON)
Light Solar Wind Mfg. (ON)
March Solar (ON)
Matrix Energy Inc. (QC)
Midnight Sun Energy Ltd. (NT)
NewSun Technologies, Ltd. (ON)
Nith Valley Solar (ON)
Nor'wester Energy Systems (AB)
Northern Communication & Navigation Systems Ltd. (NT)
Northern Lights Energy Systems (ON)
Phantom Electronics Inc (ON)
Powersource Energy Systems Inc. (BC)
Prometheus Energy (ON)
Renewable Energy of Plum Hollow (ON)
Rocky Mountain Solar (CO)
Rozon Batteries Inc. (QC)
Sambrabec Inc. (QC)
Solar Freedom International (SK)
Solar Plus (BC)
Solar Power Systems (ON)
Solar Systems (ON)
Solar Utilites Ltd. (AB)
Soltek Solar Energy Ltd (BC)
Sovran Energy Inc. (BC)
Staticon Ltd. (ON)
Sun Ross Solar Systems Ltd. (NS)
Sun Stream Solar (ON)
Sun Volts Unlimited Inc. (ON)
Sunpower Co. (ON)
Suntime Enterprises (NS)
The Electric Sun & SH Engineering (AB)
Total North Communications Ltd. (YT)
Wenvor Technologies Inc. (ON)
Westrock Battery Ltd. (Solar Division) (MB)

Inverters - Square Wave

Amsdell Inc. (ON)
Argus Telecom International Inc. (QC)
Echo Systeme (QC)
Enerfree Enterprises (MB)
Enersite International Inc. (ON)
Enviro Store Ltd The (AB)
Lenbrook Energy Systems (ON)
March Solar (ON)
Matrix Energy Inc. (QC)
Nith Valley Solar (ON)
Rocky Mountain Solar (CO)
Solar Freedom International (SK)
Solar Plus (BC)
Solar Power Systems (ON)
Solar Systems (ON)
Solar Utilites Ltd. (AB)
Sovran Energy Inc. (BC)
Sun Volts Unlimited Inc. (ON)
Sunpower Co. (ON)
Wenvor Technologies Inc. (ON)
Westrock Battery Ltd. (Solar Division) (MB)

Low Voltage DC Equipment

Analytic Systems (BC)
Argus Telecom International Inc. (QC)
CIMAT Power Systems Inc. (QC)
Diversitel Communications (ON)
Echo Systeme (QC)
Enerfree Enterprises (MB)
Energy Alternatives Ltd. (BC)
Enviro Store Ltd The (AB)
Frank's Alternate Energy (ON)
Geneq Inc. (QC)
Jack's All Trades Inc. (ON)
Kelln Consulting Ltd. (SK)
Lenbrook Energy Systems (ON)
Light Solar Wind Mfg. (ON)
March Solar (ON)
Matrix Energy Inc. (QC)
Midnight Sun Energy Ltd. (NT)
Nith Valley Solar (ON)
Northern Lights Energy Systems (ON)
Phantom Electronics Inc (ON)
Powersource Energy Systems Inc. (BC)
Prometheus Energy (ON)
Rocky Mountain Solar (CO)
Rozon Batteries Inc. (QC)
Solar Freedom International (SK)
Solar Plus (BC)
Solar Power Systems (ON)
Solar Utilites Ltd. (AB)
Soltek Solar Energy Ltd (BC)
Sovran Energy Inc. (BC)
Staticon Ltd. (ON)
Sun Ross Solar Systems Ltd. (NS)
Sun Stream Solar (ON)
Sun Volts Unlimited Inc. (ON)
Sunpower Co. (ON)
Tideland Signal Canada Ltd. (BC)
Total North Communications Ltd. (YT)
Wenvor Technologies Inc. (ON)
Westrock Battery Ltd. (Solar Division) (MB)

Modules

Argus Telecom International Inc. (QC)
Bodycote - ORTECH Inc. (ON)
Canrom Photovoltaics Inc. (ON)
CIMAT Power Systems Inc. (QC)
Enerfree Enterprises (MB)
Energy Alternatives Ltd. (BC)
Enersite International Inc. (ON)
Enviro Store Ltd The (AB)
Frank's Alternate Energy (ON)
Geneq Inc. (QC)
Golden Genesis Co. Ltd. (AB)
Kelln Consulting Ltd. (SK)
Lenbrook Energy Systems (ON)
Light Solar Wind Mfg. (ON)
March Solar (ON)
Matrix Energy Inc. (QC)
Midnight Sun Energy Ltd. (NT)
NewSun Technologies, Ltd. (ON)
Nith Valley Solar (ON)
Nor'wester Energy Systems (AB)
Northern Lights Energy Systems (ON)
Phantom Electronics Inc (ON)
Powersource Energy Systems Inc. (BC)
Prometheus Energy (ON)
Renewable Energy of Plum Hollow (ON)
Rocky Mountain Solar (CO)
Rozon Batteries Inc. (QC)
Solar Freedom International (SK)
Solar Plus (BC)
Solar Power Systems (ON)
Solar Systems (ON)
Solar Utilites Ltd. (AB)
Soltek Solar Energy Ltd (BC)
Sovran Energy Inc. (BC)
Sun Ross Solar Systems Ltd. (NS)
Sun Stream Solar (ON)
Sun Volts Unlimited Inc. (ON)
Sunpower Co. (ON)
The Electric Sun & SH Engineering (AB)
Thermo Dynamics Ltd (NS)
Wenvor Technologies Inc. (ON)
Westrock Battery Ltd. (Solar Division) (MB)

Monitoring Equipment

Argus Telecom International Inc. (QC)
Bodycote - ORTECH Inc. (ON)
CIMAT Power Systems Inc. (QC)
Echo Systeme (QC)
Eco Energy (ON)
Enerfree Enterprises (MB)
Energy Alternatives Ltd. (BC)
Enersite International Inc. (ON)
Enviro Store Ltd The (AB)
Frank's Alternate Energy (ON)
Golden Genesis Co. Ltd. (AB)
Howell-Mayhew Engineering (AB)
Light Solar Wind Mfg. (ON)
March Solar (ON)
Matrix Energy Inc. (QC)
Midnight Sun Energy Ltd. (NT)
NewSun Technologies, Ltd. (ON)

Nith Valley Solar (ON)
Northern Lights Energy Systems (ON)
Phantom Electronics Inc (ON)
Powersource Energy Systems Inc. (BC)
Prometheus Energy (ON)
Rocky Mountain Solar (CO)
Rozon Batteries Inc. (QC)
Solar Plus (BC)
Solar Power Systems (ON)
Solar Systems (ON)
Solar Utilites Ltd. (AB)
Soltek Solar Energy Ltd (BC)
Sovran Energy Inc. (BC)
Sun Ross Solar Systems Ltd. (NS)
Sun Stream Solar (ON)
Sun Volts Unlimited Inc. (ON)
Sunpower Co. (ON)
The Electric Sun & SH Engineering (AB)
Tideland Signal Canada Ltd. (BC)
Total North Communications Ltd. (YT)
Wenvor Technologies Inc. (ON)
Westrock Battery Ltd. (Solar Division) (MB)

Mounts

Argus Telecom International Inc. (QC)
Canrom Photovoltaics Inc. (ON)
CIMAT Power Systems Inc. (QC)
Echo Systeme (QC)
Energy Alternatives Ltd. (BC)
Enersite International Inc. (ON)
Enviro Store Ltd The (AB)
Frank's Alternate Energy (ON)
Geneq Inc. (QC)
Golden Genesis Co. Ltd. (AB)
Lenbrook Energy Systems (ON)
Light Solar Wind Mfg. (ON)
March Solar (ON)
Matrix Energy Inc. (QC)
Midnight Sun Energy Ltd. (NT)
Nith Valley Solar (ON)
Nor'wester Energy Systems (AB)
Northern Lights Energy Systems (ON)
Powersource Energy Systems Inc. (BC)
Prometheus Energy (ON)
Renewable Energy of Plum Hollow (ON)
Rocky Mountain Solar (CO)
Rozon Batteries Inc. (QC)
Solar Plus (BC)
Solar Power Systems (ON)
Solar Systems (ON)
Solar Utilites Ltd. (AB)
Soltek Solar Energy Ltd (BC)
Sovran Energy Inc. (BC)
Sun Ross Solar Systems Ltd. (NS)
Sun Stream Solar (ON)
Sun Volts Unlimited Inc. (ON)
Sunpower Co. (ON)
The Electric Sun & SH Engineering (AB)
Wenvor Technologies Inc. (ON)
Westrock Battery Ltd. (Solar Division) (MB)

Pumps

Argus Telecom International Inc. (QC)
CIMAT Power Systems Inc. (QC)
Echo Systeme (QC)
Enerfree Enterprises (MB)
Energy Alternatives Ltd. (BC)
Enviro Store Ltd The (AB)
Frank's Alternate Energy (ON)
Geneq Inc. (QC)
Golden Genesis Co. Ltd. (AB)
Jack's All Trades Inc. (ON)
Kelln Consulting Ltd. (SK)
Lenbrook Energy Systems (ON)
Light Solar Wind Mfg. (ON)
March Solar (ON)
Matrix Energy Inc. (QC)
Midnight Sun Energy Ltd. (NT)
Nith Valley Solar (ON)
Nor'wester Energy Systems (AB)
Northern Lights Energy Systems (ON)
Powersource Energy Systems Inc. (BC)
Prometheus Energy (ON)
Renewable Energy of Plum Hollow (ON)
Rocky Mountain Solar (CO)
Rozon Batteries Inc. (QC)
Solar Freedom International (SK)
Solar Plus (BC)
Solar Power Systems (ON)
Solar Systems (ON)
Solar Utilites Ltd. (AB)
Soltek Solar Energy Ltd (BC)
Sovran Energy Inc. (BC)
Sun Ross Solar Systems Ltd. (NS)
Sun Stream Solar (ON)
Sun Volts Unlimited Inc. (ON)
Sunpower Co. (ON)
The Electric Sun & SH Engineering (AB)
Wenvor Technologies Inc. (ON)
Westrock Battery Ltd. (Solar Division) (MB)

Telecommunications

Diversitel Communications (ON)

Trackers

Argus Telecom International Inc. (QC)
Canrom Photovoltaics Inc. (ON)
Echo Systeme (QC)
Energy Alternatives Ltd. (BC)
Golden Genesis Co. Ltd. (AB)
Kelln Consulting Ltd. (SK)
Lenbrook Energy Systems (ON)
March Solar (ON)
Matrix Energy Inc. (QC)
Midnight Sun Energy Ltd. (NT)
Nor'wester Energy Systems (AB)
Northern Lights Energy Systems (ON)
Powersource Energy Systems Inc. (BC)
Renewable Energy of Plum Hollow (ON)
Rozon Batteries Inc. (QC)
Solar Plus (BC)
Solar Power Systems (ON)
Soltek Solar Energy Ltd (BC)
Sovran Energy Inc. (BC)
Sun Stream Solar (ON)
Sun Volts Unlimited Inc. (ON)
Sunpower Co. (ON)
The Electric Sun & SH Engineering (AB)
Wenvor Technologies Inc. (ON)
Westrock Battery Ltd. (Solar Division) (MB)

Water Pumping

Thermo Dynamics Ltd (NS)

Wind Energy Applications

Howell-Mayhew Engineering (AB)

Battery Charging

Argus Telecom International Inc. (QC)
CIMAT Power Systems Inc. (QC)
David Nixon & Sun (ON)
Echo Systeme (QC)
Enerfree Enterprises (MB)
Energy Alternatives Ltd. (BC)
Enviro Store Ltd The (AB)
Frank's Alternate Energy (ON)
Jack's All Trades Inc. (ON)
Kelln Consulting Ltd. (SK)
Lenbrook Energy Systems (ON)
Light Solar Wind Mfg. (ON)
March Solar (ON)
Matrix Energy Inc. (QC)
Midnight Sun Energy Ltd. (NT)
Nith Valley Solar (ON)
Nor'wester Energy Systems (AB)
Northern Lights Energy Systems (ON)
Prometheus Energy (ON)
Renewable Energy of Plum Hollow (ON)
Rocky Mountain Solar (CO)
Rozon Batteries Inc. (QC)
Sambrabec Inc. (QC)
Solar Plus (BC)
Solar Power Systems (ON)
Solar Systems (ON)
Soltek Solar Energy Ltd (BC)
Sun Stream Solar (ON)
Sun Volts Unlimited Inc. (ON)
Sunpower Co. (ON)
Sustainable Home Solutions (QC)
The Electric Sun & SH Engineering (AB)
Trillium Windmills Inc. (ON)
Wenvor Technologies Inc. (ON)
Westrock Battery Ltd. (Solar Division) (MB)

Grid-Connected Systems

Allen Kani Associates (ON)
Argus Telecom International Inc. (QC)
CIMAT Power Systems Inc. (QC)

Echo Systeme (QC)
Energy Alternatives Ltd. (BC)
Enviro Store Ltd The (AB)
GPCo Inc. (QC)
Jack's All Trades Inc. (ON)
Kelln Consulting Ltd. (SK)
Lenbrook Energy Systems (ON)
Light Solar Wind Mfg. (ON)
Matrix Energy Inc. (QC)
NEG Micon Canada Inc. (QC)
Nith Valley Solar (ON)
Nor'wester Energy Systems (AB)
Prometheus Energy (ON)
Rocky Mountain Solar (CO)
Rozon Batteries Inc. (QC)
Solar Plus (BC)
Solar Power Systems (ON)
Soltek Solar Energy Ltd (BC)
Sun Stream Solar (ON)
Sunpower Co. (ON)
Wenvor Technologies Inc. (ON)
Westrock Battery Ltd. (Solar Division) (MB)

Lighting

Argus Telecom International Inc. (QC)
CIMAT Power Systems Inc. (QC)
Echo Systeme (QC)
Enviro Store Ltd The (AB)
Frank's Alternate Energy (ON)
Jack's All Trades Inc. (ON)
Kelln Consulting Ltd. (SK)
Lenbrook Energy Systems (ON)
Light Solar Wind Mfg. (ON)
March Solar (ON)
Matrix Energy Inc. (QC)
Midnight Sun Energy Ltd. (NT)
Nith Valley Solar (ON)
Northern Lights Energy Systems (ON)
Prometheus Energy (ON)
Rocky Mountain Solar (CO)
Rozon Batteries Inc. (QC)
Sambrabec Inc. (QC)
Solar Freedom International (SK)
Solar Plus (BC)
Solar Power Systems (ON)
Solar Systems (ON)
Solar Utilites Ltd. (AB)
Soltek Solar Energy Ltd (BC)
Sun Stream Solar (ON)
Sunpower Co. (ON)
The Electric Sun & SH Engineering (AB)
Trillium Windmills Inc. (ON)
Wenvor Technologies Inc. (ON)
Westrock Battery Ltd. (Solar Division) (MB)

Marine

Argus Telecom International Inc. (QC)
CIMAT Power Systems Inc. (QC)
Echo Systeme (QC)
Energy Alternatives Ltd. (BC)
Enviro Store Ltd The (AB)
Frank's Alternate Energy (ON)
Jack's All Trades Inc. (ON)
Kelln Consulting Ltd. (SK)
Lenbrook Energy Systems (ON)
March Solar (ON)
Matrix Energy Inc. (QC)
Midnight Sun Energy Ltd. (NT)
Nith Valley Solar (ON)
Prometheus Energy (ON)
Rocky Mountain Solar (CO)
Rozon Batteries Inc. (QC)
Sambrabec Inc. (QC)
Solar Plus (BC)
Solar Systems (ON)
Sun Stream Solar (ON)
Sunpower Co. (ON)
Trillium Windmills Inc. (ON)
Wenvor Technologies Inc. (ON)
Westrock Battery Ltd. (Solar Division) (MB)

Pumping

Argus Telecom International Inc. (QC)
CIMAT Power Systems Inc. (QC)
David Nixon & Sun (ON)
Echo Systeme (QC)
Enviro Store Ltd The (AB)
Frank's Alternate Energy (ON)
Jack's All Trades Inc. (ON)
Kelln Consulting Ltd. (SK)
Lenbrook Energy Systems (ON)
Light Solar Wind Mfg. (ON)
March Solar (ON)
Matrix Energy Inc. (QC)
Midnight Sun Energy Ltd. (NT)
Nith Valley Solar (ON)
Nor'wester Energy Systems (AB)
Northern Lights Energy Systems (ON)
Prometheus Energy (ON)
Renewable Energy of Plum Hollow (ON)
Rocky Mountain Solar (CO)
Rozon Batteries Inc. (QC)
Sambrabec Inc. (QC)
Solar Freedom International (SK)
Solar Plus (BC)
Solar Power Systems (ON)
Solar Systems (ON)
Solar Utilites Ltd. (AB)
Soltek Solar Energy Ltd (BC)
Sun Stream Solar (ON)
Suncurrent Industries (AB)
Sunpower Co. (ON)
The Electric Sun & SH Engineering (AB)
Trillium Windmills Inc. (ON)
Wenvor Technologies Inc. (ON)
Westrock Battery Ltd. (Solar Division) (MB)

Remote Instrumentation

Argus Telecom International Inc. (QC)
CIMAT Power Systems Inc. (QC)
Echo Systeme (QC)
Enviro Store Ltd The (AB)
GPCo Inc. (QC)
Lenbrook Energy Systems (ON)
March Solar (ON)
Midnight Sun Energy Ltd. (NT)
Northern Communication & Navigation Systems Ltd. (NT)
Northern Lights Energy Systems (ON)
Rocky Mountain Solar (CO)
Rozon Batteries Inc. (QC)
Solar Plus (BC)
Solar Power Systems (ON)
Soltek Solar Energy Ltd (BC)
Trillium Windmills Inc. (ON)
Wenvor Technologies Inc. (ON)
Zephyr North (ON)

Security Systems

CIMAT Power Systems Inc. (QC)
Echo Systeme (QC)
Enviro Store Ltd The (AB)
Lenbrook Energy Systems (ON)
Light Solar Wind Mfg. (ON)
March Solar (ON)
Prometheus Energy (ON)
Rozon Batteries Inc. (QC)
Solar Plus (BC)
Trillium Windmills Inc. (ON)
Wenvor Technologies Inc. (ON)
Westrock Battery Ltd. (Solar Division) (MB)

Stand Alone Systems

Argus Telecom International Inc. (QC)
CIMAT Power Systems Inc. (QC)
David Nixon & Sun (ON)
Echo Systeme (QC)
Energy Alternatives Ltd. (BC)
Enviro Store Ltd The (AB)
Frank's Alternate Energy (ON)
Jack's All Trades Inc. (ON)
Kelln Consulting Ltd. (SK)
Lenbrook Energy Systems (ON)
March Solar (ON)
Matrix Energy Inc. (QC)
Midnight Sun Energy Ltd. (NT)
NEG Micon Canada Inc. (QC)
Nith Valley Solar (ON)
Nor'wester Energy Systems (AB)
Northern Lights Energy Systems (ON)
Prometheus Energy (ON)
Renewable Energy of Plum Hollow (ON)
Rozon Batteries Inc. (QC)
Sambrabec Inc. (QC)
Solar Plus (BC)
Solar Power Systems (ON)
Solar Systems (ON)
Solar Utilites Ltd. (AB)
Soltek Solar Energy Ltd (BC)
Sun Stream Solar (ON)
Sunpower Co. (ON)
The Electric Sun & SH Engineering (AB)
Trillium Windmills Inc. (ON)
Wenvor Technologies Inc. (ON)
Westrock Battery Ltd. (Solar Division) (MB)

Telecommunications

Argus Telecom International Inc. (QC)
CIMAT Power Systems Inc. (QC)
Echo Systeme (QC)
Enviro Store Ltd The (AB)
Lenbrook Energy Systems (ON)
Light Solar Wind Mfg. (ON)
March Solar (ON)
Matrix Energy Inc. (QC)
Midnight Sun Energy Ltd. (NT)
Nor'wester Energy Systems (AB)
Northern Communication & Navigation Systems Ltd. (NT)
Prometheus Energy (ON)
Rocky Mountain Solar (CO)
Rozon Batteries Inc. (QC)
Sambrabec Inc. (QC)
Solar Plus (BC)
Soltek Solar Energy Ltd (BC)
Sunpower Co. (ON)
Trillium Windmills Inc. (ON)
Wenvor Technologies Inc. (ON)
Westrock Battery Ltd. (Solar Division) (MB)
Zephyr North (ON)

Wellhead Automation

Enviro Store Ltd The (AB)
Lenbrook Energy Systems (ON)
March Solar (ON)
Solar Plus (BC)
Trillium Windmills Inc. (ON)

Wind Farm Systems

Argus Telecom International Inc. (QC)
Echo Systeme (QC)
Enviro Store Ltd The (AB)
GPCo Inc. (QC)
NEG Micon Canada Inc. (QC)
Nor'wester Energy Systems (AB)
Rocky Mountain Solar (CO)
Rozon Batteries Inc. (QC)
Solar Plus (BC)
Sun Stream Solar (ON)
Wenvor Technologies Inc. (ON)

Wind Monitoring

Enviro Store Ltd The (AB)
Solar Plus (BC)
Soltek Solar Energy Ltd (BC)

Wind Energy Products

Batteries

Argus Telecom International Inc. (QC)
CIMAT Power Systems Inc. (QC)
David Nixon & Sun (ON)
Echo Systeme (QC)
Energy Alternatives Ltd. (BC)
Enviro Store Ltd The (AB)
Frank's Alternate Energy (ON)
Jack's All Trades Inc. (ON)
Kelln Consulting Ltd. (SK)
Lenbrook Energy Systems (ON)
Light Solar Wind Mfg. (ON)
March Solar (ON)
Matrix Energy Inc. (QC)
Midnight Sun Energy Ltd. (NT)
Nith Valley Solar (ON)
Nor'wester Energy Systems (AB)
Northern Lights Energy Systems (ON)
Prometheus Energy (ON)
Renewable Energy of Plum Hollow (ON)
Rocky Mountain Solar (CO)
Rozon Batteries Inc. (QC)
Sambrabec Inc. (QC)
Solar Plus (BC)
Solar Power Systems (ON)
Solar Systems (ON)
Solar Utilites Ltd. (AB)
Soltek Solar Energy Ltd (BC)
Sun Stream Solar (ON)
Sun Volts Unlimited Inc. (ON)
Sunpower Co. (ON)
The Electric Sun & SH Engineering (AB)
Trillium Windmills Inc. (ON)
Wenvor Technologies Inc. (ON)
Westrock Battery Ltd. (Solar Division) (MB)

Blade Protection

Enviro Store Ltd The (AB)
NEG Micon Canada Inc. (QC)
Nor'wester Energy Systems (AB)
Prometheus Energy (ON)
Solar Plus (BC)
Solar Systems (ON)
Soltek Solar Energy Ltd (BC)
Sun Stream Solar (ON)
Sun Volts Unlimited Inc. (ON)
Wenvor Technologies Inc. (ON)

Blades

Echo Systeme (QC)
Energy Alternatives Ltd. (BC)
Enviro Store Ltd The (AB)
Jack's All Trades Inc. (ON)
Light Solar Wind Mfg. (ON)
Matrix Energy Inc. (QC)
NEG Micon Canada Inc. (QC)
Nor'wester Energy Systems (AB)
Northern Lights Energy Systems (ON)
Prometheus Energy (ON)
Renewable Energy of Plum Hollow (ON)
Rozon Batteries Inc. (QC)
Solar Plus (BC)
Solar Systems (ON)
Soltek Solar Energy Ltd (BC)
Sun Stream Solar (ON)
Sun Volts Unlimited Inc. (ON)
Sunpower Co. (ON)
Trillium Windmills Inc. (ON)
Wenvor Technologies Inc. (ON)

Brakes

Enviro Store Ltd The (AB)
Frank's Alternate Energy (ON)
Light Solar Wind Mfg. (ON)
Nor'wester Energy Systems (AB)
Prometheus Energy (ON)
Solar Plus (BC)
Solar Power Systems (ON)
Soltek Solar Energy Ltd (BC)
Sun Stream Solar (ON)
Sun Volts Unlimited Inc. (ON)
Wenvor Technologies Inc. (ON)

Cabling

Argus Telecom International Inc. (QC)
Echo Systeme (QC)
Energy Alternatives Ltd. (BC)
Frank's Alternate Energy (ON)
Lenbrook Energy Systems (ON)
Light Solar Wind Mfg. (ON)
Nor'wester Energy Systems (AB)
Northern Lights Energy Systems (ON)
Prometheus Energy (ON)
Renewable Energy of Plum Hollow (ON)
Rozon Batteries Inc. (QC)
Sambrabec Inc. (QC)
Solar Plus (BC)
Solar Power Systems (ON)
Sun Volts Unlimited Inc. (ON)
Sunpower Co. (ON)
The Electric Sun & SH Engineering (AB)
Wenvor Technologies Inc. (ON)
Westrock Battery Ltd. (Solar Division) (MB)

Controllers

Argus Telecom International Inc. (QC)
CIMAT Power Systems Inc. (QC)
Echo Systeme (QC)
Energy Alternatives Ltd. (BC)
Enviro Store Ltd The (AB)
Jack's All Trades Inc. (ON)
Lenbrook Energy Systems (ON)
Light Solar Wind Mfg. (ON)
March Solar (ON)
Matrix Energy Inc. (QC)
Midnight Sun Energy Ltd. (NT)
NEG Micon Canada Inc. (QC)
Nith Valley Solar (ON)
Nor'wester Energy Systems (AB)
Northern Lights Energy Systems (ON)
Prometheus Energy (ON)
Rocky Mountain Solar (CO)
Rozon Batteries Inc. (QC)
Sambrabec Inc. (QC)
Solar Plus (BC)
Solar Power Systems (ON)

Soltek Solar Energy Ltd (BC)
Sun Stream Solar (ON)
Sun Volts Unlimited Inc. (ON)
Sunpower Co. (ON)
The Electric Sun & SH Engineering (AB)
Trillium Windmills Inc. (ON)
Wenvor Technologies Inc. (ON)
Westrock Battery Ltd. (Solar Division) (MB)

Coupling

Light Solar Wind Mfg. (ON)
Nor'wester Energy Systems (AB)
Prometheus Energy (ON)
Solar Plus (BC)
Soltek Solar Energy Ltd (BC)
Sun Volts Unlimited Inc. (ON)
Wenvor Technologies Inc. (ON)

Direction Vanes

Jack's All Trades Inc. (ON)
Nor'wester Energy Systems (AB)
Prometheus Energy (ON)
Renewable Energy of Plum Hollow (ON)
Solar Plus (BC)
Sun Volts Unlimited Inc. (ON)
Wenvor Technologies Inc. (ON)
Zephyr North (ON)

Foundations

Echo Systeme (QC)
NEG Micon Canada Inc. (QC)
Nor'wester Energy Systems (AB)
Prometheus Energy (ON)
Renewable Energy of Plum Hollow (ON)
Solar Plus (BC)
Soltek Solar Energy Ltd (BC)
Sun Volts Unlimited Inc. (ON)
The Electric Sun & SH Engineering (AB)
Wenvor Technologies Inc. (ON)

Gear Units

Jack's All Trades Inc. (ON)
Nor'wester Energy Systems (AB)
Prometheus Energy (ON)
Renewable Energy of Plum Hollow (ON)
Solar Plus (BC)
Soltek Solar Energy Ltd (BC)
Sun Volts Unlimited Inc. (ON)
Wenvor Technologies Inc. (ON)

Generators

Argus Telecom International Inc. (QC)
CIMAT Power Systems Inc. (QC)
Echo Systeme (QC)
Energy Alternatives Ltd. (BC)
Jack's All Trades Inc. (ON)
Lenbrook Energy Systems (ON)
Light Solar Wind Mfg. (ON)
Matrix Energy Inc. (QC)
Nith Valley Solar (ON)
Nor'wester Energy Systems (AB)
Northern Lights Energy Systems (ON)
Prometheus Energy (ON)
Renewable Energy of Plum Hollow (ON)
Sambrabec Inc. (QC)
Solar Plus (BC)
Sun Volts Unlimited Inc. (ON)
Sunpower Co. (ON)
The Electric Sun & SH Engineering (AB)
Trillium Windmills Inc. (ON)
Wenvor Technologies Inc. (ON)
Westrock Battery Ltd. (Solar Division) (MB)

Horizontal Axis Turbines – Large – < 1MW

Light Solar Wind Mfg. (ON)
NEG Micon Canada Inc. (QC)
Nor'wester Energy Systems (AB)
Rocky Mountain Solar (CO)

Horizontal Axis Turbines – Medium 100 kW - 1 MW

Enviro Store Ltd The (AB)
Lenbrook Energy Systems (ON)
Light Solar Wind Mfg. (ON)
NEG Micon Canada Inc. (QC)
Nor'wester Energy Systems (AB)
Northern Communication & Navigation Systems Ltd. (NT)
Rocky Mountain Solar (CO)
Solar Utilites Ltd. (AB)
Sunpower Co. (ON)

Horizontal Axis Turbines – Small 5 - 95 kW

Argus Telecom International Inc. (QC)
CIMAT Power Systems Inc. (QC)
David Nixon & Sun (ON)
Energy Alternatives Ltd. (BC)
Enviro Store Ltd The (AB)
Jack's All Trades Inc. (ON)
Kelln Consulting Ltd. (SK)
Lenbrook Energy Systems (ON)
Light Solar Wind Mfg. (ON)
Matrix Energy Inc. (QC)
Nor'wester Energy Systems (AB)
Northern Communication & Navigation Systems Ltd. (NT)
Northern Lights Energy Systems (ON)
Prometheus Energy (ON)
Renewable Energy of Plum Hollow (ON)
Rocky Mountain Solar (CO)
Rozon Batteries Inc. (QC)
Solar Plus (BC)
Solar Utilites Ltd. (AB)
Soltek Solar Energy Ltd (BC)
Sun Stream Solar (ON)
Sun Volts Unlimited Inc. (ON)
Sunpower Co. (ON)
Wenvor Technologies Inc. (ON)
Westrock Battery Ltd. (Solar Division) (MB)

Horizontal Axis Turbines – Very Small – < 5kW

Nor'wester Energy Systems (AB)
Renewable Energy of Plum Hollow (ON)
Soltek Solar Energy Ltd (BC)
Sun Volts Unlimited Inc. (ON)

Hubs

Light Solar Wind Mfg. (ON)
NEG Micon Canada Inc. (QC)
Nor'wester Energy Systems (AB)
Northern Lights Energy Systems (ON)
Renewable Energy of Plum Hollow (ON)
Solar Plus (BC)
Soltek Solar Energy Ltd (BC)
Wenvor Technologies Inc. (ON)

Hydraulic Systems

Echo Systeme (QC)
Jack's All Trades Inc. (ON)
Solar Plus (BC)

Inverters

Argus Telecom International Inc. (QC)
CIMAT Power Systems Inc. (QC)
Echo Systeme (QC)
Energy Alternatives Ltd. (BC)
Enviro Store Ltd The (AB)
Frank's Alternate Energy (ON)
Jack's All Trades Inc. (ON)
Kelln Consulting Ltd. (SK)
Lenbrook Energy Systems (ON)
Light Solar Wind Mfg. (ON)
March Solar (ON)
Matrix Energy Inc. (QC)
Midnight Sun Energy Ltd. (NT)
Nith Valley Solar (ON)
Nor'wester Energy Systems (AB)
Northern Lights Energy Systems (ON)
Prometheus Energy (ON)
Renewable Energy of Plum Hollow (ON)
Rocky Mountain Solar (CO)
Rozon Batteries Inc. (QC)
Sambrabec Inc. (QC)
Solar Plus (BC)
Soltek Solar Energy Ltd (BC)
Sun Stream Solar (ON)
Sun Volts Unlimited Inc. (ON)
Sunpower Co. (ON)
The Electric Sun & SH Engineering (AB)
Trillium Windmills Inc. (ON)
Westrock Battery Ltd. (Solar Division) (MB)

Lubricants

Nor'wester Energy Systems (AB)
Solar Plus (BC)
Wenvor Technologies Inc. (ON)

Masts

Echo Systeme (QC)
Enviro Store Ltd The (AB)
NEG Micon Canada Inc. (QC)
Nor'wester Energy Systems (AB)
Renewable Energy of Plum Hollow (ON)
Solar Plus (BC)
Soltek Solar Energy Ltd (BC)
Sun Stream Solar (ON)
The Electric Sun & SH Engineering (AB)
Wenvor Technologies Inc. (ON)
Zephyr North (ON)

Monitoring Equipment

Argus Telecom International Inc. (QC)
Echo Systeme (QC)
Enviro Store Ltd The (AB)
Frank's Alternate Energy (ON)
Jack's All Trades Inc. (ON)
Light Solar Wind Mfg. (ON)
Nith Valley Solar (ON)
Nor'wester Energy Systems (AB)
Northern Lights Energy Systems (ON)
Renewable Energy of Plum Hollow (ON)
Rocky Mountain Solar (CO)
Rozon Batteries Inc. (QC)
Solar Plus (BC)
Sun Stream Solar (ON)
Sun Volts Unlimited Inc. (ON)
Trillium Windmills Inc. (ON)
Wenvor Technologies Inc. (ON)
Westrock Battery Ltd. (Solar Division) (MB)
Zephyr North (ON)

Towers

Echo Systeme (QC)
Energy Alternatives Ltd. (BC)
Enviro Store Ltd The (AB)
Frank's Alternate Energy (ON)
Jack's All Trades Inc. (ON)
Lenbrook Energy Systems (ON)
Light Solar Wind Mfg. (ON)
Midnight Sun Energy Ltd. (NT)
NEG Micon Canada Inc. (QC)
Nor'wester Energy Systems (AB)
Prometheus Energy (ON)
Rocky Mountain Solar (CO)
Solar Plus (BC)
Soltek Solar Energy Ltd (BC)
Sun Stream Solar (ON)
Sun Volts Unlimited Inc. (ON)
Sunpower Co. (ON)
Wenvor Technologies Inc. (ON)
Westrock Battery Ltd. (Solar Division) (MB)

Vertical Axis Turbines – Large > 1 MW

Rocky Mountain Solar (CO)

Vertical Axis Turbines – Medium 100 kW - 1 MW

Rocky Mountain Solar (CO)
Sambrabec Inc. (QC)

Vertical Axis Turbines – Small 5 -95 kW

Echo Systeme (QC)
Midnight Sun Energy Ltd. (NT)
Rocky Mountain Solar (CO)
Sambrabec Inc. (QC)
Solar Plus (BC)
Sun Volts Unlimited Inc. (ON)

Vertical Axis Turbines – Very Small < 5kW

David Nixon & Sun (ON)
Echo Systeme (QC)
Frank's Alternate Energy (ON)
Kelln Consulting Ltd. (SK)
March Solar (ON)
Midnight Sun Energy Ltd. (NT)
Prometheus Energy (ON)
Rocky Mountain Solar (CO)
Sambrabec Inc. (QC)
Solar Plus (BC)
Sun Volts Unlimited Inc. (ON)
Westrock Battery Ltd. (Solar Division) (MB)

CHAPTER 11

A-Z Directory

A

ABB - Asea Brown Boveri Inc.
1410 Blair Place Suite 600
Gloucester, ON, K1J 9B9
Tel: (613) 747-5816
Fax: (613) 747-5851
Contact: Stefan Krajewski

Acres International
4342 Queen St
Box 1001
Niagara Falls, ON, L2E 6W1
Tel: (905) 374-5200
Fax: (905) 374-1157
Contact: Hans de Meel
E-Mail: postmaster@acres.com
Website: www.acres.com
No of Employees: 600; Years in Operation: 75
Acres International has earned a world-wide reputation for providing innovative and cost effective engineering services to the power industry (hydroelectric, thermal and transmission) as well as to the transportation, mining and industrial sectors

Adkinson & Associates
4044 Aberdeen Road
Beamsville, ON, L0R 1B6
Tel: (905) 563-0232
Fax: (905) 563-0455
Contact: Don Adkinson
E-Mail: dadkinso@npiec.on.ca
No of Employees: 2; Years in Operation: 20
Assisting PV users, manufacturers, developers and funding agencies in their use of PV systems and products. Answers provided to your questions quickly, accurately and reliably.

Advanced Glazings Ltd.
65E Memorial Drive
PO Box 277
North Sydney, NS, B2A 3M3
Tel: (902) 794-2899
Fax: (902) 794-1869
Contact: Doug Milburn
E-Mail: info@advancedglazings.com
No of Employees: 10; Years in Operation: 3
Advanced Glazings Ltd. is a manufacturer of cost-effective high performance transparent insulation for passive solar hot water, passive solar heating, trombe walls and energy efficient green houses.

Alberta Renewable Energy Test Site
3000 College Drive South
Lethbridge, AB, T1K 1L6
Tel: (403) 329-1212
Fax: (403) 328-5562
Contact: Rick Atkins
E-Mail: rick.atkins@agric.gov.ab.ca

Allen Kani Associates
292 Merton St. Suite 2
Toronto, ON, M4S 1A9
Tel: (416) 488-4425
Fax: (416) 488-7608
Contact: Greg Allen
E-Mail: alive@inforamp.net
No of Employees: 4; Years in Operation: 10
Allen Kani Associates specializes in energy efficient and environmentally appropriate design and engineering of residential, institutional & commercial buildings. Our designs also stress ecological building practices and a healthy indoor environment.

Amazing Solar Energy
56 The Esplanade, #405
Toronto, ON, M5E 1A7
Tel: (416) 867-8969
Fax: (416) 941-9038
Contact: Nick de Pencier

Amsdell Inc.
45 Mural St., Unit 5
Richmond Hill, ON, L4B 1J4
Tel: (905) 881-3020 Ext: 326
Fax: (905) 881-3023
Contact: Benedict Chan
E-Mail: bchan@amsdell.com
No of Employees: 5; Years in Operation: 6
Amsdell Inc. of Richmond Hill is a technology research and development company specialising in power inverters. Established in 1993, Amsdell has developed numerous solutions to lead and to compliment the ceaseless changes in power electronics and information technology development.

Analytic Systems
#207, 12448-82nd Avenue
Surrey, BC, V3W 3E9
Tel: (604) 543-7378
Fax: (604) 543-7354
Contact: Jim Hargrove
E-Mail: jim@analyticsystems.com
No of Employees: 8; Years in Operation: 22
Analytic Systems designs and manufactures inverters (pure sine & quasi sine), battery chargers (from DC or AC), DC to DC voltage converters and power supplies from 60 to 1000 watts.

Argus Telecom International Inc.
809 Deslauriers
St-Laurent, QC, H4N 1X3
Tel: (514) 331-0840
Fax: (514) 331-0843
Contact: John McCabe
No of Employees: 86; Years in Operation: 14

Arise Technologies Corporation
156 Columbia Street West Suite 4
Waterloo, ON, N2L 3L3
Tel: (519) 725-2244
Fax: (519) 725-8907
Contact: Ian MacLellan
E-Mail: ian.maclellan@arisetech.com

Atlantic Orient Canada Inc.
780 Windmill Road Suite 302
Dartmouth, NS, B3B 1T7
Tel: (902) 468 1621
Fax: (902) 468-2424
Contact: David Lombardi

Atlantic Wind Test Site
R.R. 4, Tignish, PEI, C0B 2B0
Tel: (902) 882-2746
Fax: (902) 882-3823
Contact: Carl Brothers
No of Employees: 5; Years in Operation: 10+
Canada's wind energy testing & development facility.

Aurora Research Institute
Box 1450
Inuvik, NT, Y0E 0T0
Tel: (403) 979-3298
Fax: (403) 979-4264
Contact: David Malcolm

Aurora Technologies
PO Box 275
Norman Wells, NT, X0E 0V0
Tel: (867) 5873039
Fax: (867) 5873038
Contact: Todd McCauley

Autonomous & Sustainable Housing Inc.
9211 Scurfield Dr. NW
Calgary, AB, T3L 1V9
Tel: (403) 2391882
Fax: (403) 5472671
Contact: Jorg Ostrowski
E-Mail: jdo@ucalgary.ca
Website: www.ucalgary.ca/jdo/ecotecture.htm

ASH Inc. is a Canadian company whose partners have been in professional practice since 1976 in: Architectural & Environmental Design; Ecological Planning, Community Design & Master Planning; Educational Seminars, Workshops & Lectures; Consulting & Plan Reviews; Research & Development

B

Ballard Power Systems Ltd.
4000 Glenlyon Parkway
Burnaby, BC, V5J 5J9
Tel: (604) 454-0900
Fax: (604) 412-4702
Contact: Toby Little
E-Mail: toblit@ballard.com
Website: www.ballard.com
No of Employees: 500; Years in Operation: 15
Ballard is the world leader in Proton exchange membrane fuel cells. These cells have been incorporated into light and heavy duty engines and also for stationary power generation, portable and marine power applications.

BC IT Technology Centre
3700 Willingdon Avenue
Burnaby, BC, V5G 3H2
Tel: (604) 451-6934
Fax: (604) 436-0286
Contact: Ljubisav Stamenic
E-Mail: lstameni@BCIT.BC.ca

Biothermica International Inc.
3333 Blvd. Cavendish Suite 440
Montréal , QC, H4B 2M5
Tel: (514) 488-3881
Fax: (514) 488-3125
Contact: Marie Leroux
E-Mail: lerouxmj@total.net
Website: www.biothermica.com
No of Employees: 20; Years in Operation: 10
Biothermica specialises in sanitary landfill gas management and utilisation and the building of small power plants fueled by biomass waste.

Bodycote - ORTECH Inc.
National Solar Test Facility-Canada
2395 Speakman Drive
Mississauga, ON, L5K 1B3
Tel: (905) 822-4111 Ext: 544
Fax: (905) 823-1446
Contact: Alfred Brunger
E-Mail: abrunger@ortech.on.ca
No of Employees: 90; Years in Operation: 20
Bodycote Ortech, one of 31 facilities worldwide that are part of Bodycote Materials Testing Group, the world's largest independent group of testing laboratories, operates the NSTF in co-operation with Natural Resources Canada.

Boonstra Union Energy
21 Highway #5 West
Flamborough, ON, L9H 7L5
Tel: (905) 689-7951
Fax: (905) 689-3030
Contact: Kevin Mitchell
E-Mail: boonstra.on@sympatico.ca
No of Employees: 60; Years in Operation: 37
In business since 1961, Boonstra's licensed experts take care of all of your indoor comfort needs including, heating/cooling equipment, ventilation, planned service & preventative maintenance.

Boreal Alternate Energy
32 Sunset Drive North
Whitehorse, YT, Y1A 4M8
Tel: (403) 668-5744
Contact: Doug Craig

Brace Research Institute
Macdonald Campus Of McGill University
P O Box 900, 3 Stewart Park
Ste Anne De Bellevue, QC, H9X 3V9
Tel: (514) 398-7833
Fax: (514) 398-7767
Contact: Tom A. Lawand
E-Mail: ae12000@musica.mcgill.ca

Buchanan & Hall Ltd.
RR#5
Stratford, ON, N5A 6S6
Tel: (519) 271-4793
Fax: (519) 271-8022
Contact: Bob Buchanan
No of Employees: 7; Years in Operation: 48
Commercial refrigeration, residential air, ground source heat pumps - installation & service.

C

Canada Heat Pumps
5488 Main St.
Box 219
Osgoode, ON, K0A 2W0
Tel: (613) 826-0762
Fax: (613) 826-0808
Contact: William Hamilton
E-Mail: geo2000@magma.ca
No of Employees: ; Years in Operation: 16
Manufacturer of ground source heat pumps. Custom design and construction.

Canada Mortgage and Housing Corporation
700 Montreal Road
Ottawa, ON, K1A 0P7
Tel: (613) 7482802
Fax: (613) 7484072
Contact: Paul Sauve

Canadian Gas Association
Assistant Manager, Environmental Affairs
243 Consumers Road, Suite 1200
North York, ON, M2J 5E3
Tel: (416) 498-1994 Ext: 332
Fax: (416) 498-7465
Contact: Jennifer Keyes
E-Mail: jkeyes@cga.ca

Canadian Geo-Solar Ltd.
Box 249 Suite 640 Gartshore Road
Fergus, ON, N1M 2W8
Tel: (519) 843-3393
Fax: (519) 843-6944
Contact: Jeffrey Markle

Canadian Hydro Developers Inc.
200, 622 - 5th Ave. SW
Calgary, AB, T2P 0M6
Tel: (403) 269-9379
Fax: (403) 262-8786
Contact: John Keating
E-Mail: canhydro@canhydro.com
No of Employees: 12; Years in Operation: 10
Canadian Hydro (TSE "KHD") owns and operates 8 hydro electric plants in B.C., Alberta and Ontario providing over 30 megawatts of power. Additional facilities are planned for these provinces in the near future.

Canadian Solar Industries Association
2415 Holly Lane Suite 250
Ottawa, ON, K1V 7P2
Tel: (613) 7369077
Fax: (613) 7368938
Contact: Alexa Pritchard
E-Mail: cansia@magmacom.com

Canadian Sunflower Enterprises Inc.
General Delivery
Rosseau, ON, P0C 1J0
Tel: (705) 732-4363
Fax: (705) 732-4363
Contact: Stephen Gammage

Canbar Inc.
Box 280, One Canbar Street
Waterloo, ON, N2J 4A7
Tel: (519) 886-2880
Fax: (519) 886-5546
Contact: Jim Smith

CANMET Energy Diversification Research Laboratory
1615 Lionel-Boulet
Box 4800
Varennes, QC, J3X 1S6
Tel: (450) 652-4821
Fax: (450) 652-5177
Contact: Alexandre Monarque
E-Mail: amonarqu@nrcan.gc.ca

Website: cedrl.mets.nrcan.gc.ca
No of Employees: 75; Years in Operation: 7
The Renewable and Hybrid Systems Group at the CANMET Energy Diversification Research Laboratory (CEDRL) works in close partnership with industry, consultants, academia and end-users to help develop, commercialize or acquire advanced PV technologies.

Canrom Photovoltaics Inc.
108 Aikman Avenue
Hamilton, ON, L8M 1P9
Tel: (905) 526-7634
Fax: (905) 526-9341
Contact: Nick Dalacu
E-Mail: ndalacu@fox.mstn.ca

Carrie Associates Inc.
609 Indian Road
Toronto, ON, M6P 2C4
Tel: (416) 534-9945
Fax: (416) 534-1420
Contact: Peter Carrie
E-Mail: carpe@interlog.com
No of Employees: 7; Years in Operation: 2
Provide a range of consulting and facilitation services for renewable energy projects, from assessment of opportunity to "turnkey" delivery of completed projects.

Cen Can Marketing
Unit H, 1420 Clarence Ave.
Winnipeg, MB, R3T 1T6
Tel: (204) 452-3800
Fax: (204) 452-1080
Contact: Richard McConachy

CIMAT Power Systems Inc.
PO Box 422
Pointe-Claire, QC, H9R 4P3
Tel: (514) 695-1546
Fax: (514) 630-8938
Contact: Raymond Lalonde
E-Mail: ray@cimat.com
Website: www.cimat.com
No of Employees: 5; Years in Operation: 5
Cimat Power Systems specialises in designing and supplying remote site power solutions to individuals as well as industry.

Citizens for Renewable Energy
462 East Road, RR#4
Lion's Head, ON, N0H 1W0
Tel: (519) 795-7725
Fax: (519) 795-7725
Contact: Ziggy Kleinau
E-Mail: cfre@web.net
Website: www.web.net/~cfre
No of Employees: ; Years in Operation: 3
CFRE is an incorporated non-profit group organisation dedicated to promoting clean and safe renewable energy in small scale decentralised installations. We lobby and inform governments and the public about new developments.

Compact Appliances Limited
142 Crescent Street
Box 6222
Sackville, NB, E4L 1G6
Tel: (506) 364-8002
Fax: (506) 536-3578
Contact: Ned Fisher
No of Employees: 10; Years in Operation: 15
Compact Appliances is a wholesale distributor serving retailers and contractors with supplementary equipment using wood, woodpellets, oil, propane & natural gas and solar thermal products throughout the Atlantic Provinces.

Conserval Engineering
200 Wildcat Road
North York, ON, M3J 2N5
Tel: (416) 661-7057
Fax: (416) 661-7146
Contact: John Hollick
E-Mail: conserval@globalserve.net
No of Employees: 10; Years in Operation: 21
Conserval manufactures Solarwall air heaters for industrial, commercial and crop drying applications nationally and internationally. Solarwall is the world's leading solar air heating system. When combined with PV panels, Solarwall removes heat, thereby improving PV cell efficiency.

Cousins Consulting Services
12524-42nd Avenue
Edmonton, AB, T6J 0X3
Tel: (403) 430-0669
Fax: (403) 430-9894
Contact: David Cousins
No of Employees: 1; Years in Operation: 9
We offer energy auditing services and design of building automation systems for commercial/ institutional sector, including office buildings, schools, hospitals and government buildings.

Cowley Ridge Wind Services Inc.
Box 1408
Pincher Creek, AB, T0K 1W0
Tel: (403) 627-5811
Fax: (403) 627-5814
Contact: Gavin Lowe
E-Mail: cwind@telusplanet.net.

D

David Nixon & Sun
70 Dufferin St.
Guelph, ON, N1H 4A1
Tel: (519) 822-2037
Fax: (519) 822-2037
Contact: David Nixon
E-Mail: nixon@freespace.net
No of Employees: 1; Years in Operation: 3
Design, build, install, maintain and repair small stand-alone, single source/ hybrid/ portable renewable energy systems. Develop and conduct renewable energy hands-on workshops.

DBA Solar Electric
RR #3
Hunt River, PE, C0A 1N0
Tel: (902) 9643006
Contact: Aubrey McLeod

Dectron Internationale
4300 Poirier Blvd.
Montreal, QC,
Tel: (514) 334-9609
Fax: (514) 334-9184
Contact: Roxanne Landawale
E-Mail: mail@dectron.com
No of Employees: ; Years in Operation: 23
Manufacturers of dehumidification equipment & ventilation, air conditioners & refrigeration products. Leaders in indoor air quality (IAQ) & pool dehumification. Pioneers in the HVAC market with in house engineering & service technicians.

Denoco Energy Systems Ltd.
RR#1
Smiths Falls, ON, K7A 5B8
Tel: (613) 283-0574
Fax: (613) 283-3593
Contact: Dennis O'Connor

Diversitel Communications
8-21 Concourse Gate
Nepean, ON, K2E 7S4
Tel: (613) 727-0083
Fax: (613) 727-5433
Contact: John Strickland
E-Mail: information@diversitel.ca
No of Employees: 2; Years in Operation: 16
Diversitel has developed and installed several radio repeater power supply systems in the High Arctic which integrate solar power and primary batteries. They have also conceived, designed, built and installed an experimental solar power system at CFB Alert providing electrical power year round.

DNS Geothermal
Box 314
Swan River, MB, R0L 1Z0
Tel: (204) 734-2567
Fax: (204) 734-3911
Contact: James Brown
No of Employees: 3; Years in Operation: 6
DNS Geothermal promotes energy conservation through installation of geothermal heating, cooling and hot water systems. We also custom install vertical and horizontal ground loops as well as lake loops.

Doc's Maintenance Ltd.
Box 121 Hebron
Yarmouth, NS, B0W 1X0
Tel: (902) 7428604
Fax: (902) 7429403
Contact: Chuck Outhouse
E-Mail: charles@klis.com

Dutch Industries Ltd.
PO Box 568
Pilot Butte, SK, S0G 3Z0
Tel: (306) 781-4818
Fax: (306) 781-4877
Contact: Izaak Cruson

E

Earth Energy Society of Canada
130 Slater Street Suite 1050
Ottawa, ON, K1P 6E2
Tel: (613) 2302332
Fax: (613) 2371480
Contact: Bill Eggertson
E-Mail: ceea@earthenergy.org
Website: www.earthenergy.org

Echo Systeme
3400, 5e Avenue
Shawinigan-Sud, QC, G9P 1S1
Tel: (819) 537-0560
Fax: (819) 537-0560
Contact: Jean-Pierre Belisle
E-Mail: jean-pierre_belisle@marche.com
Website: www.echosysteme.qc.ca
No of Employees: 5; Years in Operation: 14
Network of local dealers & installers, group buying, training and workshop for professionals and consumers. Research facilities & demonstration centre (PV, Wind, Hydro & the Autonomous House)

Eco Energy
57 Alexander Street
Belleville, ON, K8N 2H4
Tel: (613) 962-9889
Fax: (613) 962-9889
Contact: Jan King
E-Mail: eco.energy@sympatico.ca
Website: www3.sympatico.ca/eco.energy
No of Employees: 2; Years in Operation: 7
Canada's #1 manufacturer of solar electric controls. Innovative and efficient electronics for photovoltaic systems that are easy to install, rugged and reliable. All our products have a 5 year warrenty. See our excellent products on the web at www3.sympatico.ca/eco.energy

EcoPathways Consulting Inc.
57 Allayden Drive
Whitby, ON, L1P 1L5
Tel: (905) 430-2172
Fax: (905) 430-4574
Contact: Brian Kelly
E-Mail: ecopathways@sympatico.ca

Enerconcept Inc.
10 Park St.
Lennoxville, QC, J1M 1H1
Tel: (819) 829-1690
Fax: (819) 829-9146
Contact: Christian Vachon
E-Mail: cvachon@ican.net
No of Employees: 2; Years in Operation: 1
Consultancy and product distribution of solar air heaters and solar water heating systems, photovoltaics on demand.

Enerdata Systems Inc.
PO Box 15353
Vancouver, BC, V6B 5B1
Tel: (604) 649-2736
Fax: (604) 535-4154
Contact: Tom Knox
E-Mail: enerdata@yahoo.com
No of Employees: 4; Years in Operation: 12
More than 12 years experience in resource conservation, auditing and recommendations for institutional, industrial and commercial facilities. Over 800 on-site audits completed totalling more than 30 million square feet.

Enerfree Enterprises
Box 49, GRP 336, RR#3
Winnipeg, MB, R3C 2E7
Tel: (204) 667-8080
Contact: Neil Gilley
No of Employees: 1; Years in Operation: 7
Custom design of independent power systems for cottages and RVs.

Energy Alternatives Ltd.
825 Laval Crescent
Kamloops, BC, V2C 5P2
Tel: (800) 265-8898
Contact: Robert Mathews
E-Mail: energyalternatives@bc.sympatico.ca
No of Employees: 2; Years in Operation: 13
Energy Alternatives supplies all aspects of renewable energy for remote and grid connected applications, including site assessment/feasibility studies and reports. Our 2-day micro-hydro course is proven popular over 6 years.

Energy Council of Canada
30 Colonnade Road, Suite 400
Ottawa, ON, K2E 7J6
Tel: (613) 952-6469
Fax: (613) 952-6470
Contact: E.P. Cockshutt

Enerkem Technologies Inc.
a subsidiary of Kemestrie inc.
4245, rue Garlock
Sherbrooke, QC, J1L 2C8
Tel: (819) 569-4888
Fax: (819) 569-8411
Contact: Nicholas Abatzoglou
E-Mail: kem@interlinx.qc.ca
No of Employees: 16; Years in Operation: 5
Enerkem commercalises biomass and waste-to-energy technologies. Gasification/reforming and Steam Aqueous Fractionation/Hydrolysis to sugars/ethanol as gas as liquid conditioning modules are all covered by patents.

Enermodal Engineering Ltd.
650 Riverbend Drive
Kitchener, ON, N2K 3S2
Tel: (519) 743-8777
Fax: (519) 743-8778
Contact: Stephen C. Carpenter
E-Mail: office@enermodal.com
Website: www.enermodal.com
No of Employees: 14; Years in Operation: 18
Enermodal provides consulting and building design services in the energy conservation and renewable energy fields. Enermodal has designed some of the best examples of "green" buildings including the Waterloo Green Home. Other services include building energy simulation, detailed monitoring and IAQ testing.

Enersite International Inc.
74 Auriga Drive - Unit 1
Nepean, ON, K2E 7X7
Tel: (613) 723-5714
Fax: (613) 723-5980
Contact: Raye Thomas
E-Mail: rthomas@newsun.ca
No of Employees: 4; Years in Operation: 3
Enersite provides consulting services and its patented pre-payment Enerkey technology for rural electrification based on PV systems and local grid systems.

ENMAX - Calgary's Electric System
PO Box 2100 Stn M
2808 Spiller Road SE
Calgary, AB, T2P 2M5
Tel: (403) 268-1227
Contact: Leon Burn

Enviro Store Ltd The
9845-99th Ave
Towne Centre Mall
Grande Prairie, AB, T8V 4B2
Tel: (877) 955-0552
Contact: Leon Pendleton
No of Employees: 4; Years in Operation: 5
We believe in the health and enviromental integity of the planet. We offer products that lessen our impact and raise our consciousness in health, nutrition and the environment. "Healthful living - naturally!"

Envirometrex Corporation
131 Bloor Street West Suite 903
Toronto, ON, M5S 1S3
Tel: (416) 928-0917
Fax: (416) 928-0714
Contact: Richard Kolomeychuk
E-Mail: b20037@accesspt.north.net

Environergie Québec S.E.N.C.
561 rue Pacifique
Lac-Saint-Charles, QC, G0A 2H0
Tel: (418) 849-5221
Fax: (418) 849-5221
Contact: Clement Bergeron
E-Mail: envirorj@qbc.clic.net

Environment Canada
4905 Dufferin Street
Downsview, ON, M3H 5T4
Tel: (416) 739-4361
Fax: (416) 739-4446
Contact: Bob Morris

Environmental Building Design Network
Box 20005 Suite 431 Boler Road
London, ON, N6K 2K8
Tel: (519) 471-2225
Fax: (519) 471-9542
Contact: Roger Good

EPCOR Group of Companies
Manager, Sustainable Development
16th Floor, 10065 Jasper Avenue
Edmonton, AB, T5J 3B1
Tel: (403) 412-3487
Fax: (403) 412-3435
Contact: Joel Nodelman
E-Mail: jnodelma@epcor-group.com
jnodelma@edpower.com

F

Force 10 Consulting
R.R. 2
Cochrane, AB, T0L 0W0
Tel: (403) 932-6299
Fax: (403) 932-6299
Contact: Mike Bourns

Forest Engineering Institute of Canada - Eastern Division
580 boul, Saint-Jean
Pointe Claire, QC, H3R 3J9
Tel: (514) 6941140
Fax: (514) 6944351
Contact: Jean-Francois Gingras

Frank's Alternate Energy
RR 3
Thunder Bay, ON, P7C 4V2
Tel: (807) 964-2050
Fax: (807) 964-2050
Contact: Frank Ilczyszyn
E-Mail: sunwind@norlink.net
Website: www.sunwindwater.com
No of Employees: 3; Years in Operation: 5
Frank's Alternate Energy specializes in the custom design and installation of off-grid solar and wind power systems. Frank's also provides wind-powered pond aerators.

G

Geneq Inc.
8047 Jarry Street East
Montreal, QC, H1J 1H6
Tel: (514) 354-2511
Fax: (514) 354-6948
Contact: Marcel Doucet
E-Mail: mdoucet@geneq.com
No of Employees: 18; Years in Operation: 26
Systems integrator and distributor of remote data acquisition systems and telemetry systems.

GNB Technologies
10,500 Cote-de-liesse Suite 100
Lachine, QC, H8T 1A4
Tel: (514) 636-9081
Fax: (514) 633-1131
Contact: Louis Forcucci
E-Mail: lforcucci@gnb.com
No of Employees: 22; Years in Operation: 50
GNB Technologies is reknowned worldwide as a quality manufacturer of batteries for many applications. GNB's "Absolyte" technology was developed in 1982 as the first deep cycle battery for alternate energy applications.

Golden Genesis Co. Ltd.
122 Queen Street
Trenton, ON
Tel: (888) 765-2797
Fax: (613) 392-2263
Contact: Mark Paddison
E-Mail: paddison@goldengenesis.com

GPCo Inc.
2516, rue Saint-Jacques
Montreal, QC, H3J 2T4
Tel: (514) 932-1511
Fax: (514) 879-3415
Contact: Guy Painchaud
E-Mail: guyp@cam.org
GPCo provides professional services to the international wind power industry. Its areas of expertise covers all aspects of wind project development. Project management for Le Nordaire project and lead consultant for the development of the RETScreen software are among its latest mandates.

Grassroots Environmental Products
372 Danforth Avenue
Toronto, ON, M4K 1N8
Tel: (416) 466-2841
Contact: Rob Grand
No of Employees: 10; Years in Operation: 6
Grassroots offers environmentally responsible alternatives for a low impact lifestyle. Products include biodgredable household cleaners, recycled office supplies, hemp clothing, non-toxic personal care products, environmental books & magazines and solar powered equipment.

Gravenhurst Plumbing & Electrical Company Ltd.
210 Brock Street
Gravenhurst, ON, P1P 1H4
Tel: (705) 687-3402
Fax: (705) 687-7945
Contact: Chip Lougheed
E-Mail: c_lougheed@hotmail.com
No of Employees: 11; Years in Operation: 54
Since 1945, excellent, reliable, mechanical electrical contractor serving Muskoka. Regional Water Furnace authorised dealer.

H

Haliburton North Conststruction
P.O.Box 215
Dorset, ON, P0A 1E0
Tel: (705) 766-2881
Fax: (705) 766-2881
Contact: Sandy MacKay

Helicol Canada
2577, DeMiniac
Ville St-Laurent, QC, H4S 1E5
Tel: (514) 335-3595
Fax: (514) 335-0187
Contact: Ken Tait
E-Mail: sales@sunsolar.com

Heliotech Inc
480 St, Hubert
Granby, QC, J2G 5P4
Tel: (514) 379-7505
Fax: (514) 375-5417
Contact: Andrew Van Doorn

Howell-Mayhew Engineering
15006 - 103rd Ave.
Edmonton, AB, T5P 0N8
Tel: (780) 484-0476
Fax: (780) 484-3956
Contact: Gordon Howell
E-Mail: ghowell@compusmart.ab.ca
No of Employees: 2; Years in Operation: 13
We specialise in solar electric power systems, R-2000 energy efficient housing & indoor air quality. We provide complete perfomance monitoring, evalulation, commissioning & inspection services.

I

ICP Global Technologies
9511 Cote-de-Liesse
Montreal, QC, H9P 2N9
Tel: (514) 420-0700
Fax: (514) 420-0701
Contact: Dan Kruger
E-Mail: customers@icpglobal.com
No of Employees: 20; Years in Operation: 10
We specialixe in consumer related products using solar energy. We have offices worldwide and offer solutions to all your 12 volt charging needs.

Indal Technologies
3570 Hawkestone Rd
Mississauga, ON, L5C 2V8
Tel: (905) 275-5300
Fax: (905) 275-2418
Contact: Vincent Lacey

IOGEN Corporation
400 Hunt Club Road
Ottawa, ON, K1V 1C1
Tel: (613) 733-9830 Ext: 407
Fax: (613) 733-5127
Contact: Jeff Passmore
E-Mail: jefp@iogen.ca
No of Employees: 60; Years in Operation: 20
Iogen is a leading developer of technology to make a cellulose based clean fuel. Iogen develops, manufactures and markets industrial enzymes to the pulp and paper, textiles and animal feed industries.

IPPSO
163-C Eastbourne Ave.
Toronto, ON, M5P 2G5
Tel: (416) 322-6549
Fax: (416) 481-5785
Contact: Jake Brooks

Island Technologies
Box 832
Charlottetown, PEI, C1A 7L9
Tel: (902) 368-7171
Fax: (902) 368-7139
Contact: Malcolm Lodge

J

Jack's All Trades Inc.
51 Lakeside Lane
Pine Lake
Ardoch, ON, K0H 1C0
Tel: (613) 479-2400
Fax: (613) 479-2330
Contact: Jack Rennie
No of Employees: 2; Years in Operation: 16
Jack's All Trades Inc. is a licensed professional company with over 40 years experience in heating, air conditioning, ventilation & refrigeration. Active in HRAI & CEEA with full membership in both associations.

Jensen Engineering/CAP
5037-50th Street
Olds, AB, T4H 1R8
Tel: (403) 556-8755
Fax: (403) 556-7799
Contact: Eric Jensen

K

Kelln Consulting Ltd.
50 Jane Street
Box 94
Lumsden, SK, S0G 3C0
Tel: (306) 731-2224
Fax: (306) 731-2774
Contact: Ken Kelln
E-Mail: kelln.solar@sympatico.ca
No of Employees: 4; Years in Operation: 15

Ken Church Engineering
240-700 Industrial Ave.
Ottawa, ON, K1G 7V1
Tel: (613) 7393374
Fax: (613) 7389425
Contact: Ken Church
E-Mail: tdcl-kce@istas.ca

KMW Systems Inc.
3330 White Oak Road
London, ON, N6E 1L8
Tel: (519) 6861774
Fax: (519) 686-1132
Contact: Eric Bertil Rosen
E-Mail: brosen@kmwgroup.com
No of Employees: 12; Years in Operation: 12
KMW Systems Inc. specialises in supplying KMW Energy Systems to burn biomass fuel. At the heat of the system is KMW's Combustion Chamber, which together with auxiliary equipment make up a complete energy system.

Kortright Centre
5 Shoreham Drive
Downsview, ON, M3N 1S4
Tel: (905) 832-2289
Fax: (905) 832-8238
Contact: Wendy Heron
E-Mail: kcc@interlog.com
No of Employees: 15; Years in Operation: 22
The Kortright Centre is home to Canada's largest public renewable energy demonstration site. Demonstrations include: solar/wind cottage, 10kW Bergy Wind Turbine, grid connected solar shingles and more.

L

Lapp-Hancock Associates Limited
280 Albert Street Suite 900
Ottawa, ON, K1P 5G8
Tel: (613) 2382483
Fax: (613) 2381734
Contact: Ken Hancock
E-Mail: lapp-han@fox.nstn.ca
No of Employees: 15; Years in Operation: 13
Scientific and Technological consulting company, specialising in strategic planning and research. Retained by the Canadian Space Agency to provide technical, scientific and planning support to the Canadian Space Power Initiative (CSPI).

Lenbrook Energy Systems
633 Granite Court
Pickering, ON, L1W 3K1
Tel: (905) 831-6333
Fax: (905) 831-6356
Contact: Lea Harrison
E-Mail: lharrison@lenbrook.com
Website: www.lenbrooksolar.com
No of Employees: 25; Years in Operation: 20
Full service provider of solar power systems for the residential, recreational, industrial and government markets. Lenbrook has sold over 20,000 systems worldwide.

Light Solar Wind Mfg.
1524 Greenbriar Dr.
Oakville, ON, L6M 1Y5
Tel: (905) 847-8190
Fax: (905) 847-7271
Contact: Barry Boyd
E-Mail: bboyd@cgocable.net
No of Employees: 3; Years in Operation:
We provide components and design facilities to install alternative energy systems. Our team will first look at conservation and then help to develop sustainable systems for power needs of upto 25kw. Larger co-generation systems are also available.

Lockhart Industries
Box 784
Duncan, BC, V9L 3Y1
Tel: (250) 748-1731
Fax: (250) 748-4520
Contact: Doug Lockhart
No of Employees: 6; Years in Operation: 18
Refrigeration/ Air conditioning contractor installing geo-exchange heating/cooling systems in all types of buildings.

Longhill Energy Products (Eastern) Ltd
15 Capella Court
Unit 111
Nepean, ON, K2E 7X1
Tel: (613) 226-3856
Fax: (613) 226-2715
Contact: Chris Harrison
E-Mail: longhill@capitalnet.com
No of Employees: 6; Years in Operation: 20
Longhill is a HVAC distributor which has been servicing Eastern Ontario for over twenty years.

M

Marbek Resource Consultants
1355 Bank Street Suite 500
Ottawa, ON, K1H 8K7
Tel: (613) 5230784
Fax: (613) 5230717
Contact: Martin Adelaar
E-Mail: marbek@achilles.net

March Solar
Box 5338, Station F
Ottawa, ON, K2C 3J1
Tel: (613) 5918639
Fax: (613) 5918639
Contact: Alexander March
E-Mail: sunny@solarpowerstore.com
Website: www.solarpowerstore.com
No of Employees: 3; Years in Operation: 10

Matrix Energy Inc.
16807 Hymus Blvd.
Kirkland, QC, H9H 3L4
Tel: (514) 630-5630
Fax: (514) 426-9123
Contact: B. K. Wilkinson
E-Mail: matrixenergy@sympatico.ca
No of Employees: 3; Years in Operation: 14
Matrix Energy Inc. is a full service systems integrator specialising in remote power PV systems & hybrids; a second division specialises in commercial & industrial solar ventilation using SolarwallTM

McKerlie Solar Systems
138 Garden Cr.
Hamilton, ON, L8V 4T4
Tel: (905) 388-3459
Fax: (905) 388-4335
Contact: David McKerlie

Midnight Sun Energy Ltd.
PO Box 1683
5 Ellesmere Drive
Yellowknife, NT, X1A 2P3
Tel: (867) 873-8760
Fax: (867) 873-8768
Contact: Alexandra Hampson
E-Mail: midsun@internorth.com
Website: www.midnightsunenergy.com
No of Employees: 8; Years in Operation: 6
Midnight Sun Energy Ltd. is a northern remote power company specializing in the design, sales and installation of solar, wind and diesel hybrid power systems for cold climate applications and remote regions world-wide.

N

NEG Micon Canada Inc.
1380 Graham Bell
Boucherville, QC, J4B 6H5
Tel: (450) 4492292
Fax: (450) 4490039
Contact: Sylvie Charette
E-Mail: sylviecharrette@compuserve.com

NewSun Technologies, Ltd.
74 Auriga Dr. # 2
Nepean, ON, K2E 7X7
Tel: (613) 723-5750
Fax: (613) 723-5980
Contact: Raye Thomas
E-Mail: rthomas@newsun.ca
No of Employees: 10; Years in Operation: 19
NewSun manufactures solar cell and photovoltaic production and test equipment, develops technology and carries out technology transfer. International in focus, NewSun also provides materials for cell & module manufacturers and specialised systems for navigation, water purification and lighting.

Nith Valley Solar
P. O. Box 261
Paris, ON, N3L 3G2
Tel: (519) 442-3534
Fax: (519) 442-4828
Contact: Mike Morse
E-Mail: morsun@worldchat.com
No of Employees: 1; Years in Operation: 6
Nith Valley Solar specialises in retail sales & installations of renewable energy related products (PV, wind, Balance-of system components).

Nor'wester Energy Systems
3553 - 31st St. N.W.
Calgary, AB, T2L 2K7
Tel: (403) 249-3337
Fax: (403) 282-1238
Contact: Jason Edworthy

North Central Heating
3835 Broadway Ave.
Box 665
Smithers, BC, VOJ 2N0
Tel: (250) 847-3060
Fax: (250) 847-5255
Contact: Brent R. Weme
E-Mail: brentw@mail.netshop.net
No of Employees: 12; Years in Operation: 30

Northern Communication & Navigation Systems Ltd.
PO Box 2317
Yellowknife, NT, X1A 2P7
Tel: (867) 873-3953
Fax: (867) 920-4282
Contact: Jim Pook
E-Mail: ncns@internorth.com
No of Employees: 4; Years in Operation: 24
Solar, Wind & Diesel power systems with planning and installation. Our machine shop services allow us to customize components to suit your needs.

Northern Lights Energy Systems
R.R. 1
Richards Landing, ON, P0R 1J0
Tel: (705) 246-2073
Fax: (705) 246-2073
Contact: Laurence McKay
E-Mail: northern.lights@sympatico.ca
No of Employees: 2; Years in Operation: 11

Northern Lights Solar Systems
Box 1898, 44 Poplar Ave.
Atikokan, ON, P0T 1C0
Tel: (807) 597-4253
Fax: (807) 597-4253
Contact: J. Fred Kimberley
E-Mail: jkimberl@atikokan.Lakeheadu.ca

Nova Electric
1439 Railway Street
Kenora, ON, P9N 3W8
Tel: (807) 468-3531
Fax: (807) 468-6955
Contact: Jeff Vandenbroek

Nova Independant Resources Ltd.
2050 Bennett Road
Kelowna, BC, V1V 2C2
Tel: (250) 7638484
Fax: (250) 7637620
Contact: Harold Lunner

Numerical Logics Inc
119 University Avenue East
Waterloo, ON, N2L 2W1
Tel: (519) 886-7820
Fax: (519) 747-0881
Contact: Didier Thevenard
E-Mail: numlog@sympatico.ca
Website: www3.sympatico.ca/numlog
No of Employees: 1; Years in Operation: 3
Numerical Logics provides consulting services for solar energy software development, renewable energy systems simulation and analysis, and weather processing.

O

Oshawa Refrigeration Service Ltd.
#35 - 585 Wentworth St. E.
Oshawa, ON, L1H 3V8
Tel: (905) 728-5115
Fax: (905) 728-3897
Contact: Pat Rudka
No of Employees: 6; Years in Operation: 60

P

Paradise Plumbing & Heating
Box 96
Dunrea, MB, R0K 0S0
Tel: (204) 776-2178
Fax: (204) 776-2150
Contact: Gerard Paradis
No of Employees: 5; Years in Operation: 100
Plumbing and heating business for 4 generations. Installation of earth energy systems since 1989.

Pembina Institute
251 Laurier Ave. West Suite 500
Ottawa, ON, K1P 5J6
Tel: (613) 235-6288
Fax: (613) 235-2190
Contact: Robert Hornung
E-Mail: rhornung@web.net

Phantom Electronics Inc
111 Industrial Drive
Unit 4
Whitby, ON, L1N 5Z9
Tel: (905) 430-6512
Fax: (905) 666-1188
Contact: Leonard Allen
E-Mail: sales@phantomelectronics.com
Website: www.phantomelectronics.com
No of Employees: 5; Years in Operation: 13
Specialize in PV applications relating to security and communication systems, wireless video transmission and electronic monitoring. Design and manufacture custom power conversion devices.

Photron Canada Inc.
Box 136
Colinton, AB, T0G 0R0
Tel: (403) 675-2586
Fax: (403) 675-2634
Contact: Ron LaPlace
E-Mail: ronl@rrt.com

Polar Refrigeration Sales
910 3rd Avenue
Prince George, BC, V2L 3C9
Tel: (250) 563-6444
Fax: (250) 563-8893
Contact: Vern Badry
No of Employees: 8; Years in Operation: 39

Power Pool Of Alberta
1800 700 4th Ave SW
MacFarlane Tower
Calgary, AB, T2P 3J4
Tel: (403) 233-6496
Fax: (403) 543-0388
Contact: Wayne St. Amour
E-Mail: wayne.st.amour@powerpool.ab.ca

Powersource Energy Systems Inc.
6809 Kirkpatrick Suite 1
Victoria, BC, V8M 1Z8
Tel: (250) 544-2115
Fax: (250) 544-1773
Contact: Mike Cannon
E-Mail: powersource@vvv.com

Probyn & Company Inc.
2nd Flr., 95 King Street East
Toronto, ON, M5C 1G4
Tel: (416) 777-2800
Fax: (416) 777-1190
Contact: Stephen Probyn

Prometheus Energy
33 - 400 Creditstone
Concord, ON, L4K 3Z3
Tel: (905) 660-7868
Fax: (905) 660-8719
Contact: Rob McMonagle
No of Employees: 3; Years in Operation: 20
Renewable Energy equipment supplier since 1979. Complete systems and components to generate your own electricity form sunlight, wind or falling water. Send 7$ for 130 page design manual and product catalogue.

R

R. Nair & Associates
267 Castle Rd.
Beaconsfield, QC, H9W 2W4
Tel: (514) 428-0115
Contact: R. Nair
No of Employees: 5; Years in Operation: 11
Research and Development, Electric generators and motors. Consulting engineers for small power stations. Design audits, breakdown/maintenance/design problems on small to very large electric machinery.

Raydiant Solar Energy
118 Martindale Rd.
St. Catharines, ON, L2S 2X7
Tel: (905) 688-0799
Fax: (905) 984-8331
Contact: Arthur Jemison

Reed & Schaab Plumbing & Heating
Box 109,
RR #2
Walkerton, ON, N0C 2V0
Tel: (519) 8811250
Fax: (519) 8812616
Contact: Bernie Reed
No of Employees: 14; Years in Operation: 8

REIC
15010 Yonge Street
Aurora, ON, L4G 1M6
Tel: (416) 841-5551
Fax: (416) 841-6744
Contact: Brian Marshall

Renewable Energy of Plum Hollow
93 Counter Street
Kingston, ON, K7K 6C7
Tel: (613) 544-5575
Fax: (613) 544-7050
Contact: Ross Rowat

REONAC Energy Systems
1493 rue Begin
St. Laurent, QC, H4R 1V8
Tel: (514) 331-5871
Fax: (514) 335-3193
Contact: Marc Andre

Resource Efficient Agricultural Production
21, 111 Lakeshore Road
Box 125, Glenaladale House
Ste. Anne de Bellevue, QC, H9X 3V9
Tel: (514) 398-7743
Fax: (514) 398-7972
Contact: Patrick Girouard
E-Mail: reap@interlink.net
No of Employees: 5; Years in Operation: 13
REAP Canada is a non-profit research organisation working in the areas of Agriculture, Bioenergy and Pulp & Paper.

Richard Kadulski Architect
204-1037 West Broadway
Vancouver, BC, V6H 1E3
Tel: (604) 689-1841
Fax: (604) 689-1841
Contact: Richard Kadulski
E-Mail: kadulski@direct.ca
No of Employees: 2; Years in Operation: 25
Architectural services focussing on energy conscious, healthy housing design, passive solar, building energy analysis, building science and indoor air quality consulting.

Rozon Batteries Inc.
228, rue Jean-Talon
Saint-Luc, QC, J2W 1J2
Tel: (514) 3482370
Fax: (514) 3484405
Contact: Claude Charbonneau

S

Sambrabec Inc.
5756, Ave. De'Om
Montreal, QC, H3S 2N4
Tel: (514) 738-7241
Fax: (514) 738-7324
Contact: Louis Beaulieu
E-Mail: louisbeaulieu@videotron.ca
Website: www.sambrabec.qc.ca
No of Employees: 8; Years in Operation: 13
Sambrabec is a manufacturer of energy systems, using a revolutionary turbine, the CATAVENT, which works with all fluids.

SAR Engineering Ltd.
8884 - 15th Avenue
Burnaby, BC, V3N 1Y3
Tel: (604) 525-2239
Fax: (604) 525-2146
Contact: Ken Cooper
E-Mail: kacooper@direct.ca

SEA Systems Ltd.
70 Brookfield Road
PO Box 1508, STation C
St. John's, NF, A1C 5N8
Tel: (709) 7472997
Fax: (709) 3688734
Contact: Leo Butler

$SGA Consulting
1355 Bank Street Suite 500
Ottawa, ON, K1H 8K7
Tel: (613) 7307421
Fax: (613) 5230717
Contact: Stephen Graham

Shepherd, Mondrow
5710 Timberlea Blvd, Suite 103
Mississauga, ON, L4W 4W1
Tel: (905) 629-9035
Fax: (905) 629-0264
Contact: Jay Shepherd
Website: www.sgwmiss.com

Shute Plumbing & Heating Ltd.
7 York St. E.
Box 940
Ridgetown, ON, N0P 2C0
Tel: (519) 674-0778
Fax: (519) 674-0779
Contact: Valerie DeBok
No of Employees: 10; Years in Operation: 79
Installing Geothermal units for 17 years from residential to daycare, medical centre and greenhouse. Currently maintaining 2000 units in SW Ontario.

Solar Converters Inc
490 York Road Suite A104
Guelph, ON, N1E 6V1
Tel: (519) 824-5272
Fax: (519) 823-0325
Contact: Daniel Pringle
E-Mail: sci@sentex.net
Website: www.sentex.net/~sci

Solar Electric Engineering Distributors Canada
Box 31024
Whitehorse, YT, Y1A 5P7
Tel: (250) 495-6183
Fax: (250) 495-7183
Contact: Hon. Erik Nielsen
E-Mail: elecerik@cablelan.net
elecerik@oso.auracom.com
No of Employees: 3; Years in Operation: 7

Solar Energy Society of Canada Inc.
116 Lisgar Street Suite 702
Ottawa, ON, K2P 0C2
Tel: (613) 2344151
Fax: (613) 2342988
E-Mail: sesci@cyberus.ca
Website: www.solarenergysociety.ca

Solar Freedom International
P. O. Box 7103
Saskatoon, SK, S7K 4S1
Tel: (306) 652-1442
Fax: (306) 652-1442
Contact: Craig Shearer
E-Mail: cshearer@manynations.com
No of Employees: 1; Years in Operation: 5
Dedicated to the spread of simple and affordable solar cooking technology. Teaches individuals, organisations and businesses throughout the world to build solar ovens.

Solar Plus
PO Box 76
1755 Shawinigan Mill Bay Road
Shawinigan Lake, BC, V0R 2W0
Tel: (250) 743-9796
Fax: (250) 743-9744
Contact: Gary Giles
E-Mail: solar@pine.com
Website: www.vvv.com/~solar
No of Employees: 20; Years in Operation: 12
Solar Plus is a registered electrical contractor specialising in renewable energy offering engineering design and installation of photovoltaic, wind & microhydro systems.

Solar Power Systems
R.R.#4
Tara, ON, N0H 2N0
Tel: (519) 372-6555
Fax: (519) 934-2005
Contact: Judy Kitto
E-Mail: solarlady@bmts.com
No of Employees: 3; Years in Operation: 10
Solar Power Systems has been in business for 10 years and we know that solar power works because we have been running our home and business off-grid since 1986.

Solar Solutions, AECD Inc
29 Pearce Avenue
Winnipeg, MB, R2V 2K3
Tel: (204) 339-1767
Fax: (204) 339-1034
Contact: Tim Yusishen

Solar Systems
RR #4
Hwy 41
Pembroke, ON,
Tel: (613) 735-1659
Contact: Emile Raymond
No of Employees: 2; Years in Operation: 7
Over 7 years in PV & Wind systems. We carry a wide range of power products ranging from school projects to boat, RV, farm, cottage, backup or full house systems.

Solar Utilites Ltd.
328 - 14th St. NW
Calgary, AB, T2L 2L2
Tel: (403) 283-8542
Fax: (403) 270-3065
Contact: Robin Basu
E-Mail: rnbasu@hotmail.com
No of Employees: 8; Years in Operation: 2
Company promotes sustainable development in Canada and abroad. Linked to organisations in the EU, China, Japan, Africa, India and the USA. We can market Canadian products/services abroad.

Solar & Wind Logics Alternative Energy Systems
Box 31012, 211 Main Street
Whitehorse, YT, Y1A 5P7
Tel: (867) 6683220
Fax: (867) 6683555
Contact: Josee Bonhomme

Solargetics Ltd.
5767 Plantagenet
Montreal, QC, H3S 2K4
Tel: (514) 4576968
Fax: (514) 7385498
Contact: Claudette Marsolais

Solcan Ltd.
126 Wychwood Park
London, ON, N6G 1R7
Tel: (519) 473-0501
Fax: (519) 473-0501
Contact: R.K. Swartman
E-Mail: solcan@lon.hookup.net

Solener Inc.
440, rue Lavigueur
Quebec, QC, G1R 1B5
Tel: (418) 640-7444
Fax: (418) 640-7445
Contact: Jimmy Royer
E-Mail: solener@qbc.clic.net
No of Employees: 1; Years in Operation: 5
Bureau d'ingenieurs - conseils en energie renouvelables specialite en energie photovoltaiques, efficacite, energetique.

Solplan Review
Box 86627
Vancouver, BC, V7L 4L2
Tel: (604) 689-1841
Fax: (604) 689-1841
Contact: Richard Kadulski
E-Mail: solplan@direct.ca
No of Employees: 2; Years in Operation: 15
Solplan Review is an independent journal of energy conservation, building science & construction practice. Includes coverage of solar, healthy house, resource efficiency, R2000 & passive solar issues.

Soltek Solar Energy Ltd
2-745 Vanalman Ave
Victoria, BC, V8Z 3B6
Tel: (250) 727-7720
Fax: (250) 727-2135
Contact: David Egles
E-Mail: soltek@soltek.ca

Sovran Energy Inc.
13187 Trewhitt Rd.
Oyama, BC, V4V 2B1
Tel: (250) 548-3642
Fax: (250) 548-3610
Contact: Nigel Roberts
E-Mail: sovran@sovran.ca
Website: www.sovran.ca/~sovran
No of Employees: 4; Years in Operation: 3
Sovran offers a full range of products and services for solar power products. Our services include system design, equipment selection and supply, installation and maintenance.

Staticon Ltd.
5-390 Tapscott Road
Scarborough, ON, M1B 2Y9
Tel: (416) 291-3723
Fax: (416) 291-3871
Contact: Roland Hase
E-Mail: sales@staticon.ca
Website: www.staticon.ca
No of Employees: 50; Years in Operation: 35

Statpower Technologies Corporation
7725 Lougheed Highway
Burnaby, BC, V5A 4V8
Tel: (604) 420-4820
Fax: (604) 420-1591
Contact: Lindsay Hardie
E-Mail: lhardie@statpower.com
No of Employees: 100; Years in Operation: 9
Statpower Technologies is a leader in the development, manufacture and global marketing of electronic power conversion products for the mobile, recreational, industrial, portable electronics and distributed energy markets.

Stephen F. Pope Architect
68 Clegg St.
Ottawa, ON, K1S 0H8
Tel: (613) 231-5681
Fax: (613) 231-5681
Contact: Stephen F. Pope
E-Mail: sfpope@cyberus.ca
spope@NRCan.gc.ca
No of Employees: 1; Years in Operation: 5
Stephen F. Pope Architect is a culturally oriented architectural practice offering the latest information on passive and active solar energy, building energy simulations, and low energy designs for all clients.

Stephen Lamming Associates Ltd.
162 Old Orchard Road
Burlington, ON, L7T 2G1
Tel: (905) 6816215
Fax: (905) 6328330
Contact: Stephen Lamming

Sun Mar Corp.
5035 N. Service Rd. - C9
Burlington, ON, L7L 5V2
Tel: (905) 3321314

Sun Ross Solar Systems Ltd.
Box 502
Port Hawkesbury, NS, B0E 2V0
Tel: (902) 625-1539
Fax: (902) 625-3444
Contact: John Ross
E-Mail: j.a.ross@ns.sympatico.ca
No of Employees: 2; Years in Operation: 17
Installs, services and offers consultation for solar hot water heating, pool systems and photovoltaics.

Sun Stream Solar
2818 Round Lake Road
Round Lake Centre, ON, K0J 2J0
Tel: (613) 7573428
Fax: (613) 7573428
Contact: Doug Wagner
No of Employees: 1; Years in Operation: 7
In business since 1992, we offer sales on a wide variety of renewable energy products, complete installation and service, as well as educational presentatations.

Sun Volts Unlimited Inc.
1260 Hammond St.
North Bay, ON, P1B 2J3
Tel: (800) 558-7939
Contact: Carl Harding
E-Mail: sunvolts@sympatico.ca
No of Employees: 6; Years in Operation: 12
Sales, service and installation for over 10 years throughout Ontario. Showrooms in North Bay & Huntsville area. Sales representatives are always travelling doing free home visits in the Toronto, Peterborough and Jitchener areas. Family business. Free installation!

Suncurrent Industries
640 - 8th Ave. SW
Calgary, AB, T2H 2L3
Tel: (403) 264-2880
Fax: (403) 264-2881
Contact: Paul Liddy
E-Mail: sunjam@cadvision.com
No of Employees: 5; Years in Operation: 9
Suncurrent offers services and products in the renewable energy industry and consulting services in the fields of management, transfer, and marketing of technologies in Canada and the Caribbean.

Sunergy Renewable Energy
PO Box 31015
Guelph, ON, N1H 8K1
Tel: (519) 7672114
Fax: (519) 7670529
Contact: Al Paulissen

Sunfire Energy Corp
Box 848
5569 Taghum Fr Road
Nelson, BC, V1L 6A5
Tel: (250) 352-2001
Fax: (250) 352-3990
Contact: Marc Dupuis

Sunnyside Furnaces Ltd.
352 Brule Rd.
Sault Ste. Marie, ON, P6A 5K8
Tel: (705) 759-9282
Fax: (705) 759-9284
Contact: William R. Fleet
No of Employees: 5; Years in Operation: 43
Sunnyside has been in HVAC business for over 43 years installing geo-thermal heating & cooling since 1985. Have over 100 systems installed & operating in Sault Ste. Marie and surrounding area.

Sunpower Co.
RR #3
Tweed, ON, K0K 3J0
Tel: (613) 4785555
Fax: (613) 478-5555
Contact: Pat Steele
E-Mail: sunpower@blvl.igs.net
Website: www.blvl.igs.net/sunpower/
No of Employees: 2; Years in Operation: 5

Suntap Electric
PO Box 2393
Smithers, BC, V0J 2N0
Tel: (604) 8465436
Fax: (604) 8465436
Contact: David Cody

Suntime Enterprises
Box 158
New Glasgow, NS, B2H 5E2
Tel: (902) 3963387
Fax: (902) 3961487
Contact: Jim Cruikshank
No of Employees: 5; Years in Operation: 21
Suntime is involved in domeatic hot water systems using solar panels, solar boilers and storage tanks to give the customer and sound, reliable system.

SunWind Solar Industries Inc.
Box 714
Ganges Post Office
Salt Spring Island, BC, V8K 2W3
Tel: (250) 537-1044
Fax: (250) 537-2681
Contact: Elizabeth White
E-Mail: sunwind@web.net

Surrette Battery Company Limited
1 Station Road,
PO Box 2020
Springhill, NS, B0M 1X0
Tel: (902) 5973765
Fax: (902) 5978447
Contact: James Surrette
E-Mail: bat.sales@surrette.com
Website: www.surrette.com
No of Employees: 50; Years in Operation: 37
Leading North American manufacturer of deep-cycle, wet-cell, lead-acid batteries. Surrette actively distributes its batteries throughout North, Central and South America & the Middle East.

T

Taylor Munro Energy Systems
4504 45A Street
Delta, BC, V4K 2L8
Tel: (604) 878-1588
Fax: (604) 946-3804
Contact: Joe Thwaites
E-Mail: thwaites@direct.ca

Temiskaming Solar Power
P. O. Box 1264
New Liskeard, ON, P0J 1P0
Tel: (705) 647-8177
Fax: (705) 647-6642
Contact: Jacques Blais

The Centre for Ecological Building Design
Box 20005
431 Boller Road
London, ON, N6K 2L0
Tel: (519) 238-5686
Fax: (519) 238-6626
Contact: Roger Good
No of Employees: 2; Years in Operation: 2
A charitable organisation dedicated to improving public awareness of the human impact on the environment. Focus is residential building methods and designs that are people-friendly and earth-friendly.

The Electric Sun & SH Engineering
Box 6155
Innisfail, AB, T4G 1S8
Tel: (403) 227-2666
Fax: (403) 227-2606
Contact: Stan Haner
E-Mail: sehaner@telusplanet.net
Website: www.telusplanet.net/public/sehaner
No of Employees: 2; Years in Operation: 2
We are engineering consultants as well as retailers of renewable energy products. Our aim is to provide information and services directly to consumers.

Thermo Dynamics Ltd
44 Borden Avenue
Dartmouth, NS, B3B 1C8
Tel: (902) 468-1001
Fax: (902) 468-1002
Contact: Paul Saiko
E-Mail: solarinfo@thermo-dynamics.com
Website: www.thermo-dynamics.com
No of Employees: 7; Years in Operation: 17
Thermo Dynamics Ltd. (TDL) is a manufacturer of solar thermal equipment. TDL offers a range of products from solar fin for absorber plates and solar collectors to complete solar water heating systems. TDL has developed heat exchangers for solar water heaters and potovoltaic pumps for water pumping and solar water heaters.

Thermomax Industries Ltd
2763 Arbutus Road
Victoria, BC, V8N 5X8
Tel: (250) 721-4360
Fax: (250) 721-4329
Contact: Patrick Spearing
E-Mail: patrick@solarThermal.com
No of Employees: 1; Years in Operation: 9
Thermomax Industries Ltd. is a manufacturers representative, agent and importer for Thermomax (UK) products for Canada ans Western USA. They also supply Canadian made stainless steel tanks and heat exchangers.

Thermotech Windows
109-42 Antares Drive
Ottawa, ON, K2E 7Y4
Tel: (613) 2251101
Fax: (613) 2254630
Contact: Stephen Thwaites

Thyssen Canada
2560 Matheson Blvd. East, Suite 425
Mississauga, ON, L4W4Y9
Tel: (905) 602-1300
Fax: (905) 602-7668
Contact: J. Marschall

Tideland Signal Canada Ltd.
105-3650 Bonneville
Burnaby, BC, V3N 4J7
Tel: (604) 421-0988
Fax: (604) 421-0987
Contact: Bob Williams
E-Mail: tideland@istar.ca
Website: www.tidelandsignal.com
No of Employees: 4; Years in Operation: 20
Manufacture and supply DC solar powered equipment, marine navigation lights, buoys, lighthouses, fog signals and marine entry beacons

TN Conseil
5227, Notre-Dame Est Suite 200
Montréal, QC, H1N 3P2
Tel: (514) 254-3541
Fax: (514) 257-7729
Contact: Pierre Hosatte
E-Mail: tn.conseil@sympatico.ca
No of Employees: 3; Years in Operation: 10
TN conseil provides the energy industry with technical and management support, conducting surveys on both technical and market approaches and offers also an autonomous system to remove snow from PV modules.

Total North Communications Ltd.
311 Black Street
Whitehorse, YT, Y1A 2N1
Tel: (867) 6685175
Fax: (867) 6684710
Contact: Vernon Marshall
E-Mail: totalnor@yknet.yk.ca
Website: www.totalnorth.ca
No of Employees: 15; Years in Operation: 26
Whether in the Arctic's Beaufort Sea, or on Canada's highest massif—Mount Logan, or on the rugged Queen Charlotte Islands, Total North holds and unrivalled legacy for developing and installing worldclass systems in some of the most demanding environments found anywhere.

Trans-Canada Energie
228, rue Jean-Talon
St-Luc, QC, J2W 1J2
Tel: (450) 348-2370
Fax: (450) 348-4405
Contact: Claude Charbonneauy
E-Mail: ronbatterie@yahoo.com
No of Employees: 23; Years in Operation: 34

Trigen Energy Canada Inc.
246 Waterloo Street
London, ON, N6B 2N4
Tel: (519) 434-9194
Fax: (519) 661-0178
Contact: Derek Macartney
No of Employees: 13; Years in Operation: 60+
Supplier of District Heating and Cooling to core area of London

Trillium Windmills Inc.
R.R. 2
Orillia, ON, L3V 6H2
Tel: (705) 326-6513
Fax: (705) 326-2778
Contact: Peter R. Sanguinetti
E-Mail: gowind@bconnex.net
No of Employees: 3; Years in Operation: 12
Trillium Windmills distributes the rutland windchargers throughout North America. We sell Koenders windmills for pond aeration and water pumping in Ontario and market sonic bird control systems to airports.

V

Vestas-American Wind Technology
RR #5
Kincardine, ON, N2Z 2X6
Tel: (519) 396-6922
Fax: (519) 396-6158
Contact: Philipp Andres
E-Mail: pandres@swbi.net

Vision Quest Windelectric Inc.
3160-246 Stewart Green SW
Calgary, AB, T3H 3C8
Tel: (403) 686-1485
Fax: (403) 686-0087
Contact: Frederick Gallagher
E-Mail: ceedfmg@greenenergy.com

W

Ward & Associates
400 - 1444 Alberni St.
Vancouver, BC, V6G 2Z4
Tel: (604) 689-0991
Fax: (604) 688-1286
Contact: Peter Ward
No of Employees: 2; Years in Operation: 15
Ward & Associates specialises in the design of micro & mini hydro projects with emphasis on water availability, energy output, intake design and layout. Extensive experience abroad.

Wenvor Technologies Inc.
120 Regal Road
Box 31015
Guelph, ON, N1H 8K1
Tel: (519) 767-2114
Fax: (519) 767-0529
Contact: Albert Paulissen
E-Mail: sunenergy-wenvor@sympatico.ca
No of Employees: 4; Years in Operation: 6
Wenvor develops opportunities in the rapidly growing renewable energy market. Wenvor's solar projects include domestice markets as well as installations in Haiti, Africa, Central and South America. Wenvor recently began manufacturing pitch regulated wind turbines in partnership with Vergnet SA of France.

West Kootenay Power Ltd.
PO Box 130
1290 Esplanade
Trail, BC, V1R 4L4
Tel: (250) 368-3321
Fax: (250) 364-2314
Contact: Keith A Veerman
E-Mail: kveerman@wkpower.com
Website: www.wkpower.com
No of Employees: 400+; Years in Operation: 101
WKP is an integrated electric utility that offers a full range of PowerSense programs, and "green" and "Time-of-Use" rate options.

Westrock Battery Ltd. (Solar Division)
601 Logan Ave
Winnipeg, MB, R3A 0S6
Tel: (204) 784-2310
Fax: (204) 775-9698
Contact: Ron Osadac
No of Employees: 17; Years in Operation: 48
Design, sales, service and installation of solar electrical systems since 1985, over 6700 systems sold.Specialising in 1 to 2.5 kilowatt AC photovoltaic/inverter systems and high efficiency lighting products

Wind Power Inc.
Box 609
Pincher Creek, AB, T0K 1W0
Tel: (403) 627-2923
Fax: (403) 627-3239
Contact: Dale Johnson
No of Employees: 10; Years in Operation: 16

Y

Yalakom Appropriate Technologies
Box 1509
Lillooet, BC, V0K 1V0
Tel: (250) 2560258
Fax: (250) 2560268
Contact: Scott Davis

York WindPower Corporation
3410-1000 De La Gauchetiere O.
Montreal, QC, H3B 4W5
Tel: (514) 875-9057
Fax: (514) 879-3365
Contact: Dave Ward

Z

Zephyr North
4034 Mainway
Burlington, ON, L7M 4B9
Tel: (905) 335-9670
Fax: (905) 335-0119
Contact: Jim Salmon
E-Mail: zephyr.north@sympatico.ca
No of Employees: 3; Years in Operation: 18
Zephyr North is an independent consulting company providing a full start-to-finish wind prospecting and assessment service for the wind energy industry.

Zerowatt Energy
5224 Radcliffe Rd.
RR #1, Snodgrass Site, C7
Sechelt, BC, V0N 3A0
Tel: (604) 740-0857
Contact: Robin Krause
E-Mail: robin_krause@sunshine.net
Website: www.greenhabitat.com
No of Employees: 1; Years in Operation: 2
Design and Consulting Services for sustainable architecture and renewable energy systems in residential and commercial buildings. Also sales and installation of solar thermal and photovoltaic systems.

ZZZAP Power
PO Box 397
Montreal, QC, H3Z 2T5
Tel: (514) 3694919
Fax: (514) 3694817
Contact: Pierre Bussieres
E-Mail: info@zzzap.com